NOISE CONTROL

HANDBOOK OF PRINCIPLES AND PRACTICES

NOISE CONTROL
HANDBOOK OF PRINCIPLES AND PRACTICES

edited by

David M. Lipscomb, Ph. D
University of Tennessee

Arthur C. Taylor, Jr., Ph. D.
Virginia Military Institute

Van Nostrand Reinhold Environmental Engineering Series

VNR VAN NOSTRAND REINHOLD COMPANY
NEW YORK CINCINNATI ATLANTA DALLAS SAN FRANCISCO
LONDON TORONTO MELBOURNE

Van Nostrand Reinhold Company Regional Offices:
New York Cincinnati Atlanta Dallas San Francisco

Van Nostrand Reinhold Company International Offices:
London Toronto Melbourne

Library of Congress Catalog Card Number: 77-20200
ISBN: 0-442-24811-3

Manufactured in the United States of America

Published by Van Nostrand Reinhold Company
135 West 50th Street, New York, N.Y. 10020

Published simultaneously in Canada by Van Nostrand Reinhold Ltd.

15 14 13 12 11 10 9 8 7 6 5 4 3 2 1

Library of Congress Cataloging in Publication Data
Main entry under title:

Noise control: handbook of principles and practices.

(Van Nostrand Reinhold environmental engineering series)
Includes index.
1. Noise control—Handbooks, manuals, etc.
I. Lipscomb, David M. II. Taylor, Arthur C.
TD892.H36 620.2'3 77-20200
ISBN 0-442-24811-3

Van Nostrand Reinhold Environmental Engineering Series

THE Van Nostrand Reinhold Environmental Engineering Series is dedicated to the presentation of current and vital information relative to the engineering aspects of controlling man's physical environment. Systems and subsystems available to exercise control of both the indoor and outdoor environment continue to become more sophisticated and to involve a number of engineering disciplines. The aim of the series is to provide books which, though often concerned with the life cycle–design, installation, and operation and maintenance–of a specific system or subsystem, are complementary when viewed in their relationship to the total environment.

Books in the Van Nostrand Reinhold Environmental Engineering Series include titles concerned with the engineering of mechanical systems designed (1) to control the environment within structures, including those in which manufacturing processes are carried out, (2) to control the exterior environment through control of waste products expelled by inhabitants of structures and from manufacturing processes. The series will include books on heating, air conditioning and ventilation, control of air and water pollution, control of the acoustic environment, sanitary engineering and waste disposal, illumination, and piping systems for transporting media of all kinds.

Van Nostrand Reinhold Environmental Engineering Series

ADVANCED WASTEWATER TREATMENT, by Russell L. Culp and Gordon L. Culp

ARCHITECTURAL INTERIOR SYSTEMS—Lighting, Air Conditioning, Acoustics, by John E. Flynn and Arthur W. Segil

SOLID WASTE MANAGEMENT, by D. Joseph Hagerty, Joseph L. Pavoni and John E. Heer, Jr.

THERMAL INSULATION, by John F. Malloy

AIR POLLUTION AND INDUSTRY, edited by Richard D. Ross

INDUSTRIAL WASTE DISPOSAL, edited by Richard D. Ross

MICROBIAL CONTAMINATION CONTROL FACILITIES, by Robert S. Rurkle and G. Briggs Phillips

SOUND, NOISE, AND VIBRATION CONTROL (Second Edition), by Lyle F. Yerges

NEW CONCEPTS IN WATER PURIFICATION, by Gordon L. Culp and Russell L. Culp

HANDBOOK OF SOLID WASTE DISPOSAL: MATERIALS AND ENERGY RECOVERY, by Joseph L. Pavoni, John E. Heer, Jr., and D. Joseph Hagerty

ENVIRONMENTAL ASSESSMENTS AND STATEMENTS, by John E. Heer, Jr. and D. Joseph Hagerty

ENVIRONMENTAL IMPACT ANALYSIS: A New Dimension in Decision Making, by R. K. Jain, L. V. Urban and G. S. Stacey

CONTROL SYSTEMS FOR HEATING, VENTILATING, AND AIR CONDITIONING (Second Edition), by Roger W. Haines

WATER QUALITY MANAGEMENT PLANNING, edited by Joseph L. Pavoni

HANDBOOK OF ADVANCED WASTEWATER TREATMENT (Second Edition), by Russell L. Culp, George Mack Wesner and Gordon L. Culp

HANDBOOK OF NOISE ASSESSMENT, edited by Daryl N. May

NOISE CONTROL: HANDBOOK OF PRINCIPLES AND PRACTICES, edited by David M. Lipscomb and Arthur C. Taylor, Jr.

AIR POLLUTION CONTROL TECHNOLOGY, by Robert M. Bethea

POWER PLANT SITING, by John V. Winter and David A. Conner

Foreword

With the passage of the Noise Control Act of 1972, the problems of noise affecting human health and welfare changed from a mainly local concern to one of national interest and importance. The act itself reflected a congressional compromise among the widely varying and often intense differences of opinion on both the magnitude of noise effects as an environmental issue and how noise was to be controlled. Among the key provisions of the enacted legislation are those relating to establishing noise emission control regulations based on analysis of technology, costs, and health and welfare benefits; also, the reserving to state and local governments of responsibility and authority for ambient noise. Underlying those key elements are provisions for public participation in rulemaking by the federal government.

As part of the U.S. Environmental Protection Agency's actions to promote broad understanding of the act and to employ talents of those in the academic community on its implementation, Dr. David Lipscomb of the University of Tennessee and Col. Arthur Taylor of the Virginia Military Institute served for several summers with the Office of Noise Abatement and Control, under the Intergovernmental Personnel Act. They contributed in a major degree to the work on the document identifying levels of noise not to be exceeded in order to protect public health and welfare.

A need for a general understanding of the fundamentals of noise and means of control exists for members of various professions not necessarily specializing in environmental noise. This handbook has been designed to meet that need. The various authorities who have contributed to it have likewise been associated with the implementation of the Noise Control Act, and are aware of the importance of informed application of fundamental knowledge of noise as an environmental problem.

In viewing this work, it is important to keep in mind that noise is different from other environmental pollutants in many ways. In noise (unlike in air) pollution, there is no well-defined "normal atmosphere"; not all sounds are noise, and thus human subjective response is involved. There are no residuals left for long-term buildup. As with other pollutants, the marginal costs of control can be quite expensive, and the judgmental processes to balance cost and social benefit are difficult because of the wide range of exposures, which can produce varying degrees of effects.

This handbook will be of great value to the engineers, scientists, lawyers, physicians, and others concerned with noise. Likewise, it will be of value to students, and to all those who wish to have a better understanding of the acoustical environment.

ALVIN F. MEYER, JR.
President, A. F. Meyer and Associates, Inc.
McLean, Virginia
(*Formerly Deputy Assistant Administrator for Noise Control*
U.S. Environmental Protection Agency)

Preface

The multidimensional demands on planners, managers, engineers, architects, environmentalists, and members of government regulatory agencies make it necessary that they have available reliable and authoritative resource materials. Noise in the environment is but one of the problems with which they must deal. It is the purpose of this book to provide such professional personnel with a problem-solving guide that will lead them through the steps necessary to prepare, enforce, and/or meet noise-control measures.

Noise abatement and control is a topic that is far too complex to be completely covered in a single volume. Presently, numerous introductory books that discuss the noise problem in lay terms are available, as are very sophisticated handbooks intended for acoustic engineers and acousticians. This text deals with the middle ground. We have prepared each chapter with the assumption of very little prior knowledge on the part of the reader. The topics are discussed at increasingly sophisticated levels so that the readers will be elevated in their understanding of the subject. This will allow those who are uninitiated in the study of noise to develop a more adequate understanding so that, if necessary, they can move on to the high-level reference and application resource books.

However, this handbook is not only tutorial. We have made a concerted effort to provide workable guidelines for the development of environmental impact statements and the preparation of comprehensive noise-control legislation. To many readers, these sections will prove to be of great value. To others, the discussions of practical applications of noise-control techniques will be most important. It should be noted that these treatments are introductory and that further study will be necessary for one who wishes to become fully adept in noise-control activities.

After establishing a background in the concepts and terminology involved in noise control, which comprises the first section, the book then presents useful material to foster an appreciation of means whereby noise exposure can be lessened through engineering, acoustical treatment, and personnel management. There are several major noise sources in the environment such as industrial and construction noise, aircraft noise, and vehicular and railway noise. We have chapters devoted to each of these in order to describe past efforts in controlling these sources and to set forth principles for future control. In the final two sections detailed information is given on the problems and solutions for noise in the community, the home, and recreational activities.

Each chapter has been prepared by professionals whose daily work involves the application of concepts and principles that are described in their writing. The editors have judiciously selected the contributors, giving every consideration to the demonstrated skill and quality of work that characterizes the authors' professional status. We are very proud of the team assembled for the preparation of this reference work. We hope that application of the material found herein will lead to a quieter environment for all to enjoy.

David M. Lipscomb
Arthur C. Taylor, Jr.

Contents

NOISE CONTROL

HANDBOOK OF PRINCIPLES AND PRACTICES

I

The nature of sound

Richard B. Minnix, Ph.D.

Physics Department, Virginia Military Institute
Lexington, Virginia

INTRODUCTION

A *wave* is defined as any disturbance from an equilibrium condition that travels or propagates with time from one point in space to another. By itself, it is not a material thing, but a name for an oscillatory state of motion of something else. The fundamental aspect of wave motion is that it provides a mechanism for the transfer of energy and momentum from one point to another without the transport of matter between the points.

A wide variety of wave phenomena can be observed in nature. Yet, there are a number of features common to all kinds of waves. Some of these general properties, which will be considered initially, are: the mechanism of propagation of mechanical waves; the manner in which the paths of waves are altered due to the phenomena of reflection, refraction, and diffraction; the superposition of two or more waves; and transmission and absorption of waves by obstacles located in their path.

Most of the remainder of the chapter is devoted to the specific nature of sound waves: how they propagate; their velocity and its dependence on the medium and its temperature; the characteristics of sound waves; the resultant effect of two or more sound waves in the same region of space, which leads to the phenomenon of interference and the production of standing waves; and the resonance condition. Particular attention is devoted to sound intensity, sound power, sound pressure, and the decibel scale because of their importance in sound monitoring and measurement. The concept of *spectrum analysis* and types of sound fields are also discussed. Sonic booms are considered as an example of impact or impulse sound, and, finally, there is a brief discussion of ultrasound and infrasound.

PROPAGATION OF MECHANICAL WAVES

Waves that propagate through deformable or elastic media are termed *mechanical waves*. Sound waves are one example of this type of wave. Whenever some portion of an elastic medium is displaced from its normal position, that portion oscillates back and forth about its equilibrium position. Because of the elastic properties of the medium, the disturbance is transmitted from one portion of the medium to the next.

For the case of a single disturbance or pulse traveling through a medium, each particle remains at rest until the disturbance reaches it, then oscillates for a short time, and then returns to the rest (or equilibrium) position. If a series of regularly timed disturbances is introduced, a periodic wave train is produced. The most interesting example of a periodic wave train is one in which the particles of the medium vibrate in *simple harmonic motion* where the displacement, y, of each particle is given as a function of time, t, by

$$y = A \sin (2\pi ft + \delta) \tag{1-1}$$

where A is the amplitude (or maximum displacement) of the motion, f is the frequency of the periodic disturbance, and δ is the phase angle. The value of δ is determined by the initial conditions on the motion; e.g., if the displacement is a maximum ($y = A$) at $t = 0$, then it is found that δ must be $\pi/2$ by substitution into the equation. When the vibrations of the particles are at right angles to the direction of travel of the wave, the wave is called *transverse*. If the particles oscillate parallel to the direction of propagation, the wave is called *longitudinal*.

Consider a pulse propagating in three dimensions. The surface formed by connecting all points undergoing the same disturbance at the same instant is called the *wave front*. Of course, these surfaces move at the same rate at which the pulse propagates. The direction of propagation is always perpendicular to the wave front for homogeneous, isotropic media. For a wave train, there will be a whole series of wave fronts. Disturbances that propagate in a single direction produce plane wave fronts and are therefore termed *plane waves*. A pebble dropped on the surface of a pond will produce circular wave fronts. Similarly, *spherical waves* result from a point source producing a disturbance that propagates in three dimensions. At large distances from the point source, spherical wave fronts have small enough curvature that they may be regarded as plane.

Successive positions of an advancing wave front may be found by employing *Huygen's principle:*

> Every point on an advancing wave front can be considered as point sources of secondary spherical wavelets. After a time t those secondary wavelets can be represented by an arc of radius $r = vt$ where v is the velocity of the wave front in the medium. The new position of the wave front (at time t) will be the surface of tangency to these secondary wavelets.

ALTERATION OF WAVE PATHS

Reflection

Whenever an advancing wave front encounters an obstructing barrier or surface, the wave front's path is altered in a way that is termed *reflection*. The application of Huygen's principle shows that wave fronts of various shapes are reflected from plane barriers without alteration of shape; that is, spherical waves remain spherical, plane waves remain plane, and so on.

Consider a plane wave front impinging on a perfectly reflecting (no transmission) smooth barrier (see Fig. 1-1). A *ray* is defined as a line drawn normal to the

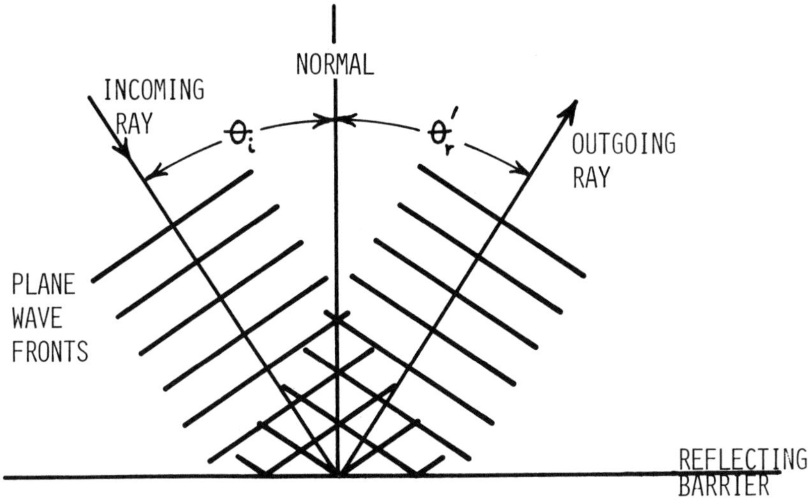

Fig. 1-1 Reflection of plane waves from a smooth barrier.

advancing wave fronts and in the direction of propagation. The application of Huygen's principle to this situation leads to the laws of reflection:

1. The incoming ray, the reflected ray, and the normal to the barrier all lie in the same plane.
2. The angle between the incoming ray and the normal to the barrier, termed the *angle of incidence* (θ_i), equals the angle between the outgoing ray and the normal to the barrier, termed the *angle of reflection* (θ_r').

Refraction

In general, when an advancing wave train traveling in one medium impinges upon the smooth boundary surface of a second medium, part of the wave is reflected and the remainder is *refracted* (or bent) into the second medium. This change in direction occurs whenever the velocity of the wave differs in the two media and can be understood in terms of the following analogy. Consider a column of people who march from a paved area into a muddy field. If the column does not approach the boundary perpendicularly, those persons in each row who enter the mire first will be slowed while their compatriots in the same

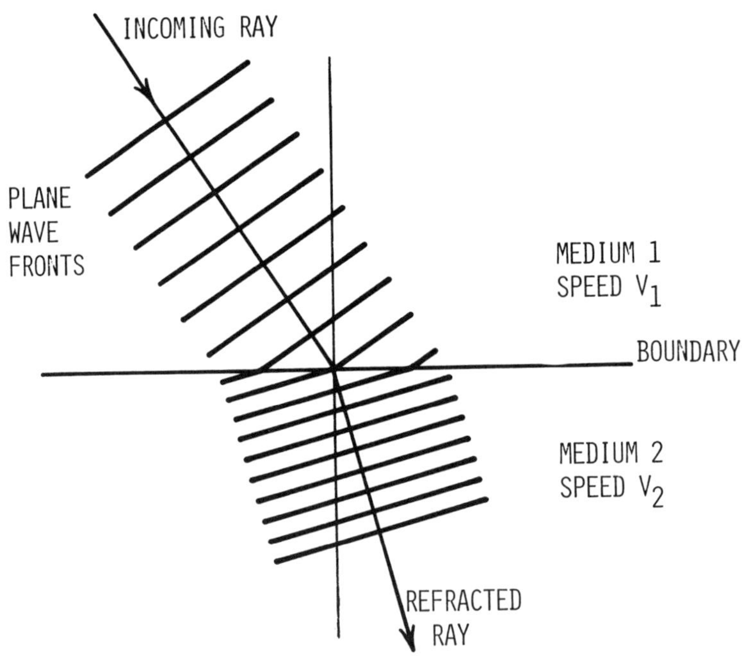

Fig. 1-2 Refraction of plane waves (or marching column) as they move from a fast to a slow medium ($v_1 > v_2$).

row will continue to move at a faster rate until they too enter the mud. As a result, the direction of the column will be altered, the change of direction becoming more pronounced as the difference in velocities in the two media increases (see Fig. 1-2).

Huygen's principle can also be applied to the analysis of refraction. Define the refracted ray as a line drawn perpendicular to the refracted wave fronts in the second medium. Then, the angle of refraction is defined to be the angle between the refracted ray and the normal to the boundary. Refer to Fig. 1-3. The laws of refraction can then be stated as:

1. The incident ray, the refracted ray, and the normal to the boundary all lie in the same plane.
2. The angle of incidence (θ_i) and the angle of refraction (θ_r) are related by

$$\frac{\sin \theta_i}{v_1} = \frac{\sin \theta_r}{v_2} \tag{1-2}$$

   where v_1 and v_2 are the wave velocities in medium 1 and medium 2, respectively.

Refraction of waves can also occur in a single medium if the speed of the wave varies between different regions. Refraction of sound waves in this manner is often observed while boating on a lake or river, due to the greater speed of sound in warm air than in cold (see Equation 1-7). At night, the temperature increases with elevation, with the coldest air near the water. Consequently, the portion of a wave front at the higher elevation travels faster than that portion

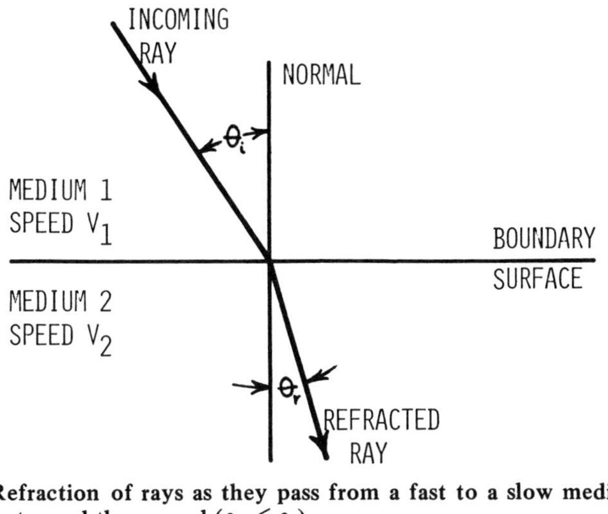

Fig. 1-3 Refraction of rays as they pass from a fast to a slow medium results in a bending toward the normal ($\theta_r < \theta_i$).

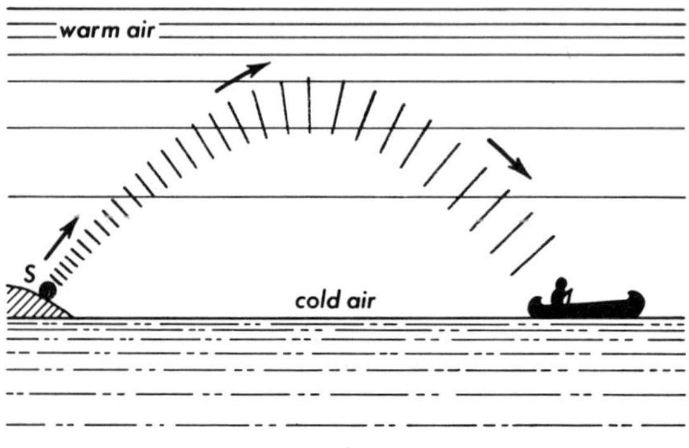

Fig. 1-4 The higher velocity of sound waves in warm air causes them to be refracted (bent) downward at night. (From *Modern College Physics*, Sixth Edition, 1972, by Harvey E. White. Used with permission of Van Nostrand Reinhold, New York.)

nearer the water, bending (refracting) the wave down as shown in Fig. 1-4. During the day, the temperature decreases with elevation, the air nearer the water is warmer, and the wave bends (refracts) away from the water as shown in Fig. 1-5.

Diffraction

Diffraction can be defined as any deviation of wave motion away from straight-line propagation. It becomes most evident when waves pass through apertures or around obstacles, for it is in such cases that waves are observed in regions where they would not be expected if strictly straight-line propagation occurred.

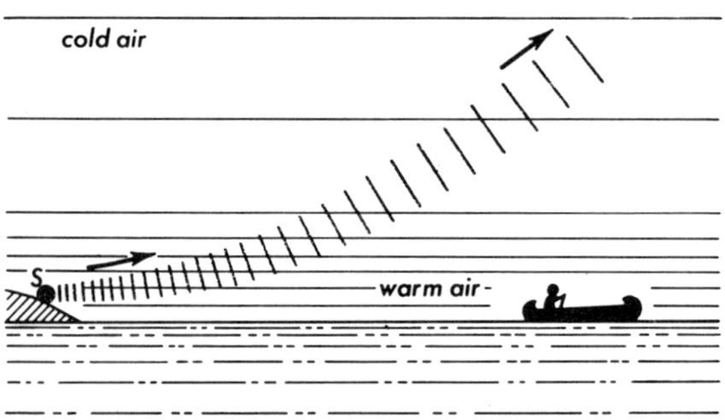

Fig. 1-5 The higher velocity of sound waves in warm air causes them to be refracted (bent) upward in the daytime. (From *Modern College Physics*, Sixth Edition, 1972, by Harvey E. White. Used with permission of Van Nostrand Reinhold, New York.)

The phenomenon of diffraction can be most easily understood employing Huygen's principle. Figure 1-6 shows a plane wave impinging on a barrier whose size is large compared to the wavelength of the wave. Very little of the wave disturbance enters the region behind the barrier, so a "shadow" is cast and little diffraction occurs. Contrast this case with those shown in Figs. 1-7 and 1-8. Here, the size of the aperture and obstacle is comparable to the wavelength of the wave. In the case of the aperture, the slit is small compared with the wavelength so it can be assumed that a single secondary wavelet of low energy is propagated forward from the slit. In the case of the obstacle, its small size compared to the wavelength results in the waves bending around it and continuing essentially unobstructed.

A typical wavelength of visible light is about 5×10^{-7} m; for audible sound it is about 1 m. Since the size of the obstacles (corners, doorways, etc.) in every-

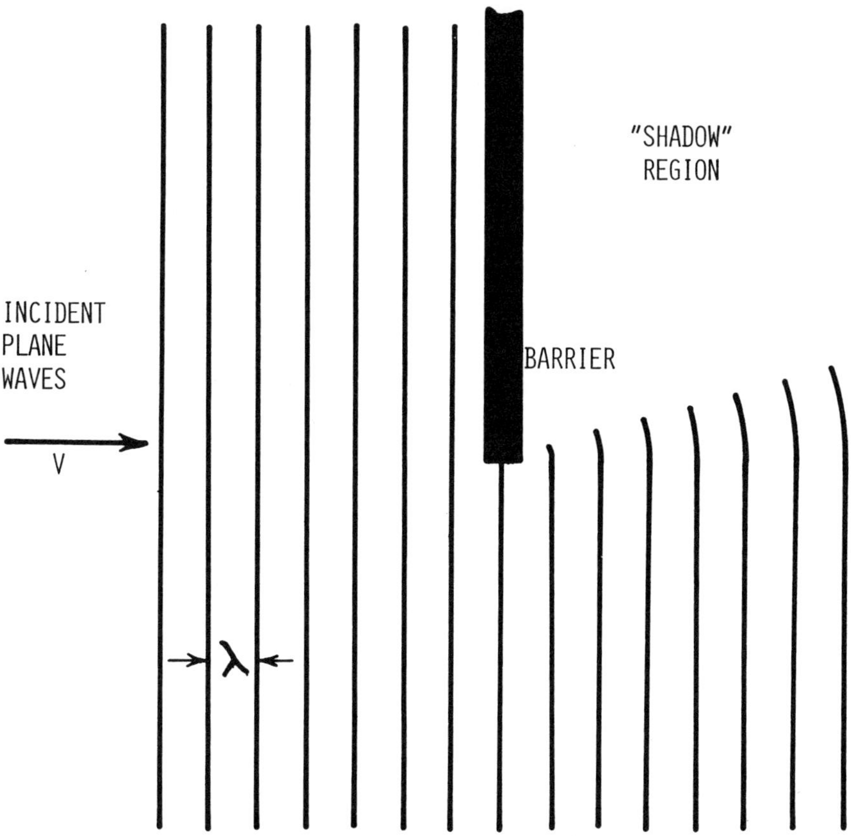

Fig. 1-6 Plane waves encountering a barrier whose width is large compared with the wavelength, λ.

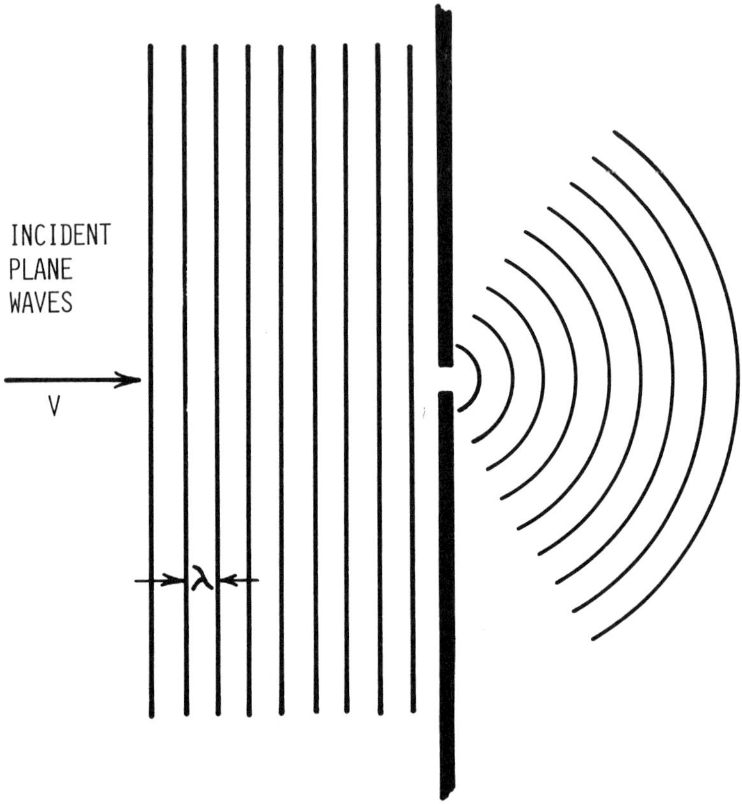

Fig. 1-7 Plane waves incident on a barrier with an opening comparable to the wavelength, λ. Waves to the right of the barrier are circular waves concentric about the opening as if there were a point source at the opening.

day life is close to the latter figure, sound is diffracted around corners but light is not. This explains why we can hear another person located around the corner when we cannot see him.

TRANSMISSION AND ABSORPTION

When an advancing wave train impinges upon an obstructing barrier or surface, part of the wave is reflected, and the remainder is *transmitted* into the barrier. If the barrier consists of hard, rigid, and flat surfaces, such as concrete or glass, then almost all of the incident energy striking the barrier is reflected. If, however, the barrier is nonrigid and consists of soft, porous materials and fabrics, then a considerable portion of the incident energy will be transmitted into the

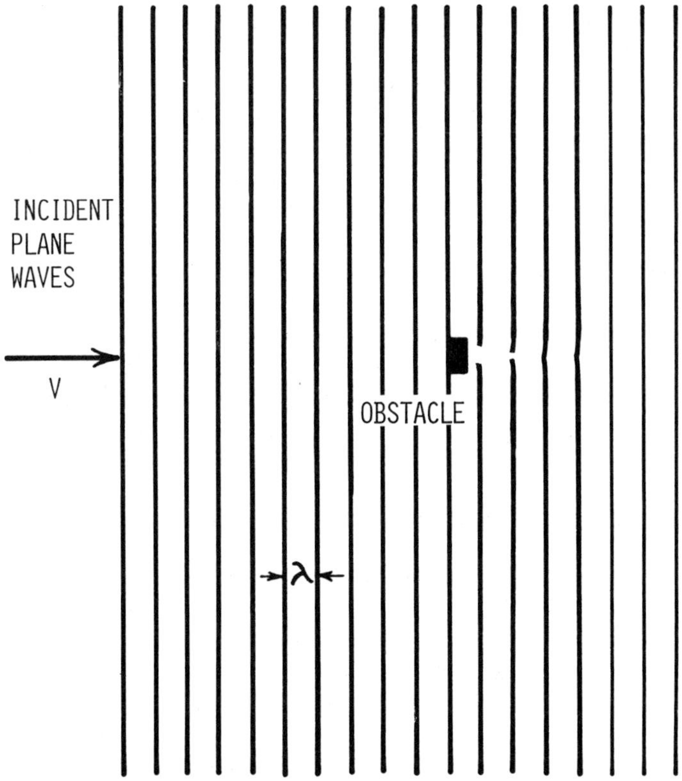

Fig. 1-8 Plane waves incident on an obstacle with a size comparable to the wavelength, λ.

barrier. The transmitted wave will emerge from the other side of the barrier with reduced energy because of *absorption*. Here, absorption refers to the conversion of wave energy into some other form, usually heat, in passing through the material of the barrier or on striking its surface. The amount of heat produced by the conversion of wave energy is extremely small.

THE SUPERPOSITION PRINCIPLE

What happens when two or more wave trains move simultaneously through the same region of space? Experiments show that each wave proceeds independently, as if the others were not present. Because the waves act independently, the resulting displacement at any point and time is simply the vector sum of the displacements that the individual waves alone would produce. The phrase *vector sum* is used to emphasize that the phase of each wave train must be taken into account. This process of vector addition of the displacements is called *superposition.*

The superposition principle applies to mechanical waves in deformable media as long as the displacement amplitude is not too great. Superposition fails for large amplitudes because the mathematical relation between the deformation and the restoring force becomes nonlinear. Thus, for example, the superposition principle cannot be applied to extremely loud sound waves or shock waves.

In 1807, the French mathematician Jean Fourier showed that any periodic function can be expressed as the sum of sine and cosine terms. Since a *harmonic wave* is defined as a wave that can be expressed in terms of sines and cosines, Fourier's result means that any periodic wave can be expressed as a superposition of harmonic waves. Hence, in principle, the response of a physical system to any periodic wave can be determined if its response to a harmonic wave is understood. Herein lies the importance of the superposition principle.

LONGITUDINAL WAVES IN AIR (SOUND)

Longitudinal waves can be generated in an air column enclosed by a long narrow tube fitted with a vibrating piston or plunger at one end as shown in Fig. 1-9. Suppose the piston is forced to undergo a simple harmonic vibration of frequency f parallel to the axis of the tube. Motion of the piston causes a displacement of the air mass near it, which is transmitted to the other air masses in the tube by molecular collisions. As the piston moves further into the tube, a region called a *condensation* is formed where the air molecules are pushed closer together, thus increasing the local density and pressure of the gas above equilibrium values. Following the production of a condensation, a region called a *rarefaction* is formed when the piston moves in the opposite direction (out of the tube). Air molecules in this region are drawn further apart, thereby reducing the local density and pressure of the gas below equilibrium values.

Condensations and rarefactions move down the tube with a constant velocity v, which is governed by the nature of the vibrating medium–in this instance, air. The distance between two successive condensations or two successive rarefactions is called the *wavelength*, λ. Since a condensation, traveling with constant velocity v, advances a distance of one wavelength in the time T for the piston to complete one oscillation, it follows that

$$v = \lambda/T = f\lambda \tag{1-3}$$

where $f = 1/T$. This result is valid for all types of periodic waves.

The pressure wave propagated down the tube can be described mathematically by the expression

$$p = P \sin(kx - \omega t) \tag{1-4}$$

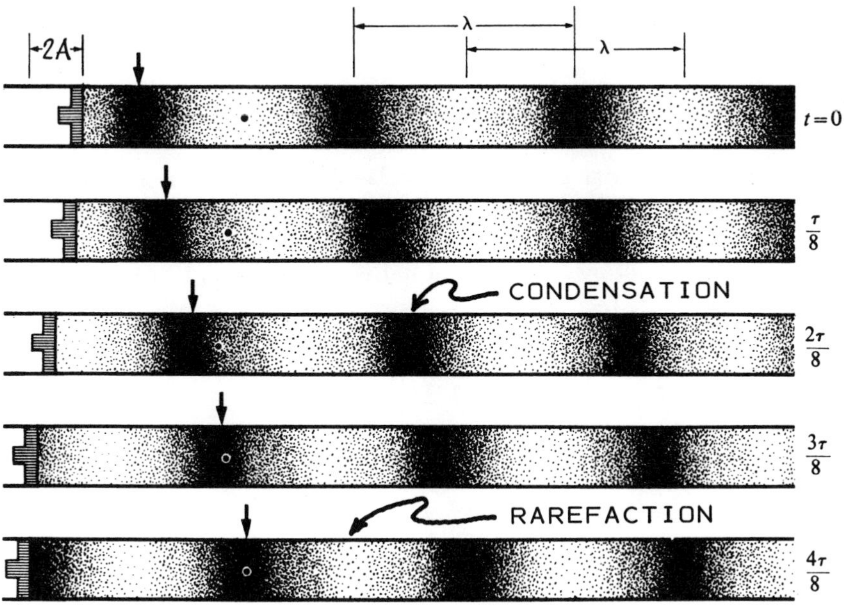

Fig. 1-9 A sinusoidal longitudinal wave traveling toward the right, shown at intervals of one-eighth of a period. (From *College Physics*. Fourth Edition, 1974, by Sears, Zemansky, and Young. Used with permission of Addison-Wesley Publishing Co., Reading, Massachusetts.)

where

$k = 2\pi/\lambda$, $\omega = 2\pi f$, p is the change in pressure above or below the equilibrium value when there is no wave, and P is the maximum value of this change. The constant k is called the *wave number.*

VELOCITY OF SOUND WAVES

The speed of compressional waves through a solid or liquid is given by

$$v = \sqrt{\frac{E}{\rho}} \tag{1-5}$$

where v is the velocity of propagation in m/sec, E is the bulk modulus of the medium in N/m^2, and ρ is the equilibrium density of the medium in kg/m^3. Since the bulk modulus is a measure of how hard it is to compress a material, this expression shows that the speed of sound is greatest in "hard" materials such as steel.

By treating air as an ideal gas and, moreover, assuming that the compressional vibrations are rapid enough to be regarded as adiabatic, it can be shown that the speed of sound in air is

$$v = \sqrt{\frac{1.41\,RT}{M_o}} \tag{1-6}$$

where R is the universal gas constant (=8,314 J/kg-mole °K), T is the absolute temperature in °K, and M is the molecular weight (about 29 kg/mole for air). Note that this result predicts the speed of sound to be independent of pressure.

A convenient result may be obtained if T is replaced by $273 + t$ where t is the Celsius temperature. Then

$$v = \sqrt{\frac{1.41\,R}{M_o}}\ \sqrt{273\left(1 + \frac{t}{273}\right)} = \sqrt{\frac{1.41\,R(273)}{M_o}}\ \sqrt{1 + \frac{t}{273}}$$

The quantity under the first radical is v_0, the velocity at 0°C. If $t \ll 273$ or $t/273 \ll 1$, then the second radical can be expanded by the binomial theorem to give (approximately)

$$v = v_o + \frac{v_o}{546}\,t$$

For air, $v_o = 331$ m/sec, so

$$v = 331 + 0.61\,t;\ (t \ll 273) \tag{1-7}$$

gives the speed of sound in air at different Celsius temperatures. At room temperature ($t = 20$°C), this gives the velocity of sound in air as 343 m/sec.

CHARACTERISTICS OF SOUND WAVES

Audible sound waves are produced by vibrating bodies—either strings, air columns, or plates and membranes. The human ear is sensitive to sound waves in the frequency range from about 20 to 20,000 Hz although the upper limit of the range decreases with advancing age. The frequency range of several sound sources is shown in Fig. 1-10.

Treating sound as an auditory sensation produced by these vibrating bodies, it is characterized by its *pitch*, *loudness*, and *tone quality*. However, these characteristics cannot be objectively measured because they are psychological sensations and depend upon the ear and judgment of the individual observer. In physical terms, sound is defined as a fluctuation in pressure (or particle displacement) in an elastic medium. This definition leads to the objective characteristics of sound: *frequency*, *intensity*, and *waveform*.

Even though one set of characteristics is psychological and the other is physical, they are related. In general, the pitch of a sound wave increases as its frequency increases but there is no one-to-one correspondence. Similarly, loudness

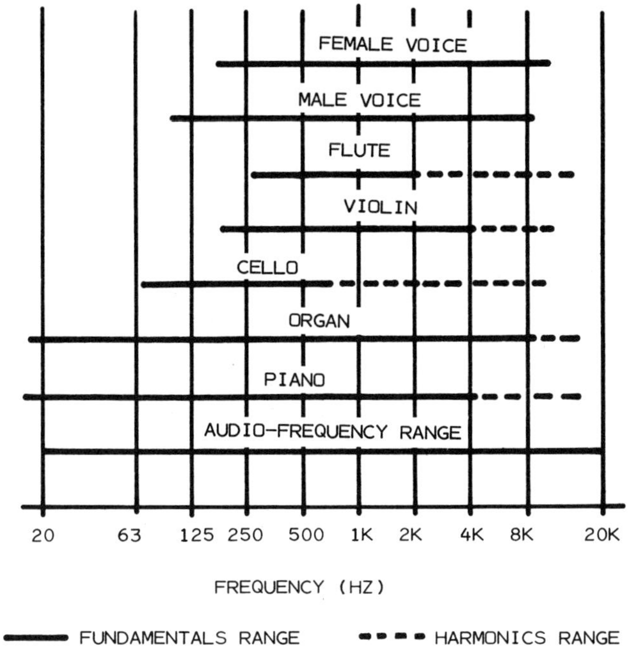

Fig. 1-10 Frequency range of several sources of sound. (From *Environmental Acoustics* by Doelle, 1972. Used with permission of McGraw-Hill Book Co., New York.)

varies with intensity; however, here the relationship is more nearly logarithmic. And tone quality changes as the complexity of the waveform increases.

A *pure tone* is the result of a harmonic disturbance of a single frequency as shown in Fig. 1-11 where A represents the amplitude of the wave. Reference to Equation 1-3 shows that, if the conditions of the medium are maintained so that the velocity remains constant, the wavelength increases as the frequency decreases and vice versa. Also, it can be shown that the intensity of such a harmonic wave is proportional to the square of the frequency and to the square of the amplitude of the wave. Thus, the loudness of a pure tone depends upon both its amplitude and frequency.

Figure 1-11 also depicts a more complex periodic wave, one that is a combination of three pure tones of different amplitude and frequency. Although the complex wave has the same period as the dominant pure tone shown, it would exhibit a different tone quality. The tone quality of a sound wave is determined in part by the number of high frequency components or overtones present.

Noise waveforms differ from those shown in Figure 1-11 because they are aperiodic; that is, they are irregular so that the disturbance is not repeated in

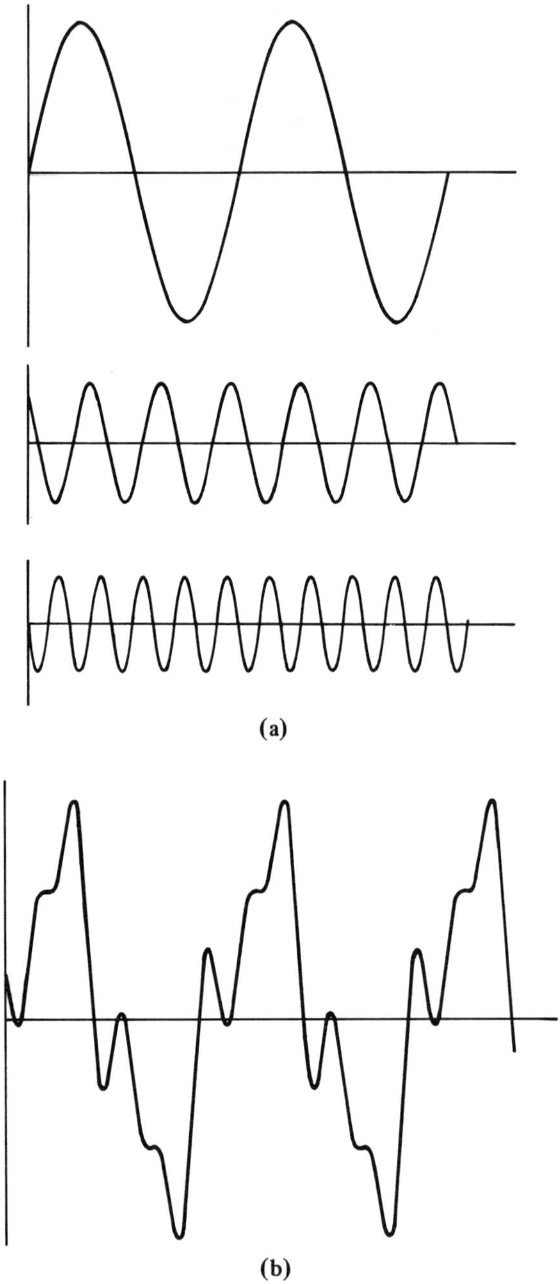

Fig. 1-11 Combination of pure tones: *a*, three pure tones of different amplitude and frequency; *b*, superposition of the pure tones in (*a*).

equal intervals of time. There is something in humans' psychological makeup that causes them to regard periodic waves as pleasant or musical whereas aperiodic waves are regarded as unpleasant.

WAVE INTENSITY

Traveling waves transmit energy in the direction of propagation of the wave. The *intensity* I of a traveling wave is the time average rate at which energy is transmitted through a unit area normal to the direction of propagation; that is, it is the energy flow, per unit time, per unit area, perpendicular to the direction of flow. Since the rate at which work is done (or energy emitted) is defined as *power*, the intensity I is also the power W transmitted per unit area normal to the direction of propagation.

$$I = \frac{W}{A} \tag{1-8}$$

Units of intensity are J/sec-m^2 or W/m^2.

INVERSE SQUARE LAW

Consider a point source of power W emitting periodic waves in three dimensions. Energy from the source is assumed to be transmitted equally in all directions. At a distance r from the source, the energy is uniformly distributed over a sphere of area $4\pi r^2$. The intensity I at this distance from the source will be

$$I = \frac{W}{4\pi r^2} \tag{1-9}$$

The intensity thus decreases as $1/r^2$—this result is called the *inverse square law* for the wave intensity.

SOUND INTENSITY

A sound wave may be described either in terms of the displacement of air particles produced at a point in space or in terms of the variation of air pressure at the same point. The air pressure rises above and falls below atmospheric pressure with the same frequency as the displacement. The maximum amount by which the pressure differs from atmospheric pressure is called the pressure amplitude, P. In general, P is directly proportional to the displacement amplitude and inversely proportional to the wavelength of the wave.

The power per unit area in a sound wave equals the product of the excess pressure and the particle velocity. Averaging this product over one cycle, the result for the *sound intensity* is

$$I = \frac{P^2}{2\rho v} \tag{1-10}$$

where P is the maximum pressure amplitude in N/m^2, ρ is the average density of the air (~1.20 kg/m^3 for air at 20°C), and v is the propagation velocity of the wave.

The human ear can tolerate a rather large range of sound wave intensities. From Equation 1-10, the pressure amplitude for a sound wave of intensity 10^{-12} W/m^2, which is usually taken to be the threshold of hearing, is about 3×10^{-5} N/m^2. An intensity of about 1 W/m^2, which produces a sensation of pain in most people, gives a pressure amplitude of about 30 N/m^2. These very small pressure variations are superimposed on the normal constant atmospheric pressure (~1.01×10^5 N/m^2). These figures show the ear to be an extremely sensitive organ.

MEASUREMENT OF SOUND WAVES

The Decibel Scale

A common method used to compare one quantity Q to another quantity Q_0 (possibly a reference) is to measure their ratio in *bels* where

$$\text{Bel} = \log_{10} \frac{Q}{Q_0}$$

This practice is especially useful when the quantity Q assumes values over a very wide range. Because the bel represents large differences in Q, a smaller unit, called the *decibel* (dB), is more commonly employed.

$$\text{Decibel} = 10 \log_{10} \frac{Q}{Q_0} \tag{1-11}$$

Since $\log_{10} 10 = 1$, $\log_{10} 100 = 2$, $\log_{10} 1{,}000 = 3$, and so on, a quantity Q 10 times larger than Q_0 has a dB level of 10; one 100 times larger than Q_0 has a dB level of 20; and one 1000 times larger has a dB level of 30. So, a 10-dB increase corresponds to an increase in Q by a factor of 10.

Sound Power Level

Employing the decibel scale, *sound power level*, L_W, is defined by

$$L_W = 10 \log_{10} \left(\frac{W}{W_0}\right) \tag{1-12}$$

where W is the power in watts of the sound source, and W_0 is the reference power level, usually 10^{-12} W.

Sound Intensity Level

Sound intensity level, L_I, is defined in similar fashion by

$$L_I = 10 \log_{10}\left(\frac{I}{I_0}\right) \tag{1-13}$$

where I is the sound intensity in W/m^2 and I_0 is the reference intensity, usually taken as 10^{-12} W/m^2.

On this scale, the *threshold of hearing* ($I = I_0 = 10^{-12}$ W/m^2) occurs at

$$L_I = 10 \log_{10}\left(\frac{I}{I_0}\right) = 10 \log 1 = 0 \text{ dB}$$

Similarly, the *threshold of pain* ($I = 1$ W/m^2) occurs at

$$L_I = 10 \log_{10}\left(\frac{1}{10^{-12}}\right) = 10 \log_{10}(10^{12}) = 10(12) = 120 \text{ dB}$$

Sound Pressure Level

According to Equation 1-10, the acoustic intensity, I, is directly proportional to the maximum acoustic pressure squared. Based on this relationship, *sound pressure level*, L_P, is defined by

$$L_P = 10 \log_{10}\left(\frac{P^2}{P_0^2}\right) \tag{1-14}$$

where P is the maximum acoustic pressure in N/m^2, and P_0 is the reference acoustic pressure. Note in this case that an increase in acoustic pressure by a factor of 10 results in a change in sound pressure level of 20 dB.

Sound pressure level is measured by a sound level meter and is thus the most commonly used decibel scale. However, these meters read a time-averaged pressure called the root-mean-square pressure, P_{rms}. Since the root-mean-square (rms) pressure and the maximum pressure amplitude are related by the equation $p_{rms} = P/\sqrt{2}$ for a harmonic wave, the equation applies equally well for rms pressure, that is,

$$L_P = 10 \log_{10}\left(\frac{p_{rms}^2}{p_{0_{rms}}^2}\right) \tag{1-15}$$

where p_{rms} is the rms acoustic pressure in N/m^2, and $p_{0_{rms}}$ is the rms reference acoustic pressure, usually taken as 2×10^{-5} N/m^2. Typical values of sound pressure levels of some common sounds are shown in Table 1-1.

TABLE 1-1. Sound pressure levels of representative sounds and noises.*

Source of Sound	dB	Description
Large rocket engine (nearby)	180	
Jet takeoff (nearby)	150	
Pneumatic riveter; machine gun	130	
Rock concert with amplifiers (2 m); jet takeoff (60 m)	120	Pain threshold
Construction noise (3 m)	110	
Subway train	100	
Heavy truck (15 m); Niagara Falls	90	Constant exposure endangers hearing
Noisy office with machines; average factory	80	
Busy traffic	70	
Normal conversation (1 m)	60	
Quiet office	50	Quiet
Library	40	
Soft whisper (5 m)	30	Very quiet
Rustling leaves	20	
Normal breathing	10	Barely audible
	0	Hearing threshold

*(From *Physics*, by Tipler, 1976. Used with permission of Worth Publishers, New York.)

Relationship Among Sound Power, Sound Intensity, and Sound Power Level

Consider a uniform spherical wave propagating from a point source. (Most sources can be considered as point sources at distances that are large compared to the size of the source itself.) According to Equation 1-9,

$$W = I(4\pi r^2) = IS \tag{1-16}$$

where S is the surface area of a sphere in m^2. Substitution of this expression into Equation 1-12 leads to

$$L_W = 10 \log_{10} \left(\frac{W}{W_0}\right) = 10 \log_{10} \left(\frac{IS}{I_0 S_0}\right) = 10 \log_{10} \frac{I}{I_0} + 10 \log_{10} \left(\frac{S}{S_0}\right) \tag{1-17}$$

where S_0 is the area of a reference sphere, taken as $S_0 = 1 m^2$. Thus,

$$L_W = L_I + 10 \log_{10} S \tag{1-18}$$

From this relationship, the sound intensity level, L_I, can be determined at a distance r ($S = 4\pi r^2$) from a source of sound power level, L_W.

Substituting Equation 1-10 for I in Equation 1-13,

$$L_I = 10 \log_{10} \frac{P^2}{2\rho v I_0} = 10 \log_{10} \frac{p_{\text{rms}}^2}{\rho v I_0}$$

$$= 10 \log_{10} \left(\frac{p_{\text{rms}}^2}{p_{0_{\text{rms}}}^2}\right)\left(\frac{p_{0_{\text{rms}}}^2}{\rho v I_0}\right)$$

$$= 10 \log_{10} \left(\frac{p_{\text{rms}}^2}{p_{0_{\text{rms}}}^2}\right) + 10 \log_{10} \left(\frac{p_{0_{\text{rms}}}^2}{\rho v I_0}\right)$$

Using $p_{0_{\text{rms}}} = 2 \times 10^{-5}\ \text{N/m}^2$,

$$\rho v = 412 \frac{\text{kg}}{\text{m}^2\text{-sec}} \text{ (for air at 1 atmosphere and 20°C), and}$$

$$I_0 = 10^{-12}\ \text{W/m}^2,$$

results in

$$L_I = L_P - 0.13 \tag{1-19}$$

Thus, for all practical purposes, the sound intensity level and sound pressure level are equal in magnitude under these conditions.

Addition of Harmonic Waves

A mathematical expression for the pressure variation in a harmonic wave at a given location can be written as

$$p_1 = P_1 \sin(\omega_1 t + \delta_1) \tag{1-20}$$

where P_1 is the maximum pressure amplitude, ω_1 is the angular frequency, t is the time, and δ_1 is the phase angle. According to the superposition principle, the result of combining two harmonic waves of different amplitude, frequency, and phase angle would be

$$p = P_1 \sin(\omega_1 t + \delta_1) + P_2 \sin(\omega_2 t + \delta_2) \tag{1-21}$$

Figure 1-12 graphically illustrates this process of addition. Squaring both sides of Equation 1-21 and taking the time average leads to the following expression:

$$(p^2)_{\text{avg}} = \left(\frac{P_1}{\sqrt{2}}\right)^2 + \left(\frac{P_2}{\sqrt{2}}\right)^2 \tag{1-22}$$

But $(P_1/\sqrt{2})^2 = (p_{1_{\text{rms}}})^2$, and $(P_2/\sqrt{2})^2 = (p_{2_{\text{rms}}})^2$. So,

$$(p^2)_{\text{avg}} = (p_{1_{\text{rms}}})^2 + (p_{2_{\text{rms}}})^2 \tag{1-23}$$

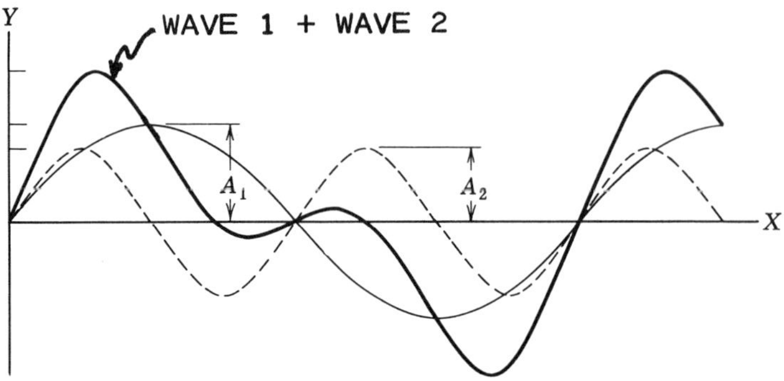

Fig. 1-12 Superposition of two waves moving in the +X-direction. In this example, wave 2 has 3/4 the amplitude and 1/2 the wavelength of wave 1. (From *Elements of Physics*, Fourth Edition, 1965, by Shortley and Williams. Used with permission of Prentice Hall, Inc., Englewood Cliffs, New Jersey.)

Taking the square root of both sides gives

$$p_{\mathrm{rms}} = \sqrt{(p_{1\,\mathrm{rms}})^2 + (p_{2\,\mathrm{rms}})^2} \tag{1-24}$$

where p_{rms} is the rms pressure of the resultant wave. This equation plays an important role in sound level measurements since sound level meters measure rms pressure amplitudes.

Consider two sources that, when operating alone, produce harmonic waves of rms sound pressure level 75 dB and 80 dB, respectively. With both sources operating simultaneously, the resultant sound pressure level will not be 155 dB. Sound pressure levels on the dB scale cannot be added when combining two or more waves because the dB scale is nonlinear. The resultant sound pressure can be determined as follows. Use Equation 1-15 to write the sound pressure level of each source when operating alone. Then,

$$75 = 10 \log_{10}\left(\frac{p_{1\,\mathrm{rms}}^2}{p_{0\,\mathrm{rms}}^2}\right) \text{(Source 1)}$$

$$80 = 10 \log_{10}\left(\frac{p_{2\,\mathrm{rms}}^2}{p_{0\,\mathrm{rms}}^2}\right) \text{(Source 2)}$$

Taking the antilog of each equation gives

$$\frac{p_{1\,\mathrm{rms}}^2}{p_{0\,\mathrm{rms}}^2} = \text{antilog}\left(\frac{75}{10}\right) = 3.16 \times 10^7 \text{ (Source 1)}$$

$$\frac{p_{2\,\mathrm{rms}}^2}{p_{0\,\mathrm{rms}}^2} = \text{antilog}\left(\frac{80}{10}\right) = 1.00 \times 10^8 \text{ (Source 2)}$$

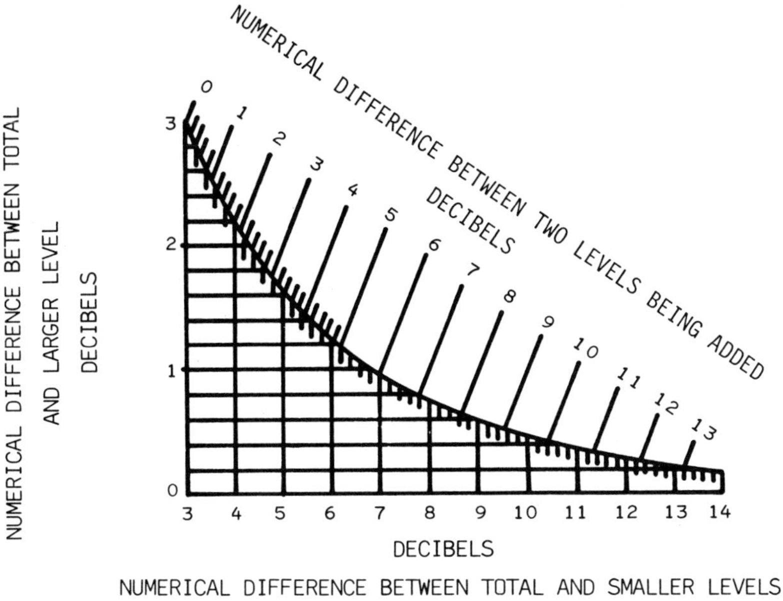

TO ADD LEVELS

Enter the chart with the NUMERICAL DIFFERENCE BETWEEN TWO LEVELS BEING ADDED. Follow the line corresponding to this value to its intersection with the curved line, then left to read the NUMERICAL DIFFERENCE BETWEEN TOTAL AND LARGER LEVEL. Add this value to the larger level to determine the total.

Example: Add 75 dB and 80 dB. The difference is 5 dB. The 5-dB line intersects the curved line at 1.2 dB on the vertical scale. Thus the total value is 80 + 1.2 or 81.2 dB.

TO SUBTRACT LEVELS

Enter the chart with the NUMERICAL DIFFERENCE BETWEEN TOTAL AND LARGER LEVELS if this value is less than 3 dB. Enter the chart with the NUMERICAL DIFFERENCE BETWEEN TOTAL AND SMALLER LEVELS if this value is between 3 and 14 dB. Follow the line corresponding to this value to its intersection with the curved line, then either left or down to read the NUMERICAL DIFFERENCE BETWEEN TOTAL AND LARGER (SMALLER) LEVELS. Subtract this value from the total level to determine the unknown level.

Example: Subtract 81 dB from 90 dB. The difference is 9 dB. The 9-dB vertical line intersects the curved line at 0.6 dB on the vertical scale. Thus the unknown level is 90-0.6 or 89.4 dB.

Fig. 1-13 Graphical addition and subtraction of dB sound levels. (From *Handbook of Noise Measurement*, Sixth Edition, 1967, by Peterson and Gruss. Used with permission of General Radio Company, West Concord, Massachusetts.)

Dividing both sides of Equation 1-23 by p_0^2 and substituting for each source gives

$$\frac{(p^2)_{avg}}{p_0^2} = \frac{p_{1\,rms}^2}{p_0^2} + \frac{p_{2\,rms}^2}{p_0^2}$$

$$\frac{(p^2)_{avg}}{p_0^2} = 3.16 \times 10^7 + 1.00 \times 10^8 = 1.32 \times 10^8.$$

Hence, the resultant rms sound pressure level is

$$L_P = 10 \log_{10} \frac{(p^2)_{avg}}{p_0^2} = 10 \log_{10} (1.32 \times 10^8)$$

$$= 10\,(8.12) = 81.2 \text{ dB}$$

A more convenient way to add or subtract two sound levels is provided by Fig. 1-13. This graph and associated instructions makes it apparent that, when combining two sound levels that differ by 10 dB or more, the presence of the lower-level source may be neglected.

INTERFERENCE

An interesting special case of Equation 1-21 occurs when ω_1 equals ω_2 and δ_1 and δ_2 are both constant. Then, the phase relation between the two waves remains constant in time, and the two waves are said to be *coherent*. Under these conditions, a phenomenon known as *interference in space* can occur. That is, at certain points in the radiation field of the two sources, the disturbances produced by each source will act in the same direction at all times to produce a net disturbance greater than either source alone would produce; *constructive interference* occurs at these points. However, there are other points where the disturbances produced by each source always act in opposite directions at all times to produce a net disturbance smaller than either source alone would produce; *destructive interference* occurs at these points. In fact, if the amplitudes of the two waves are the same, complete cancellation of the two waves occurs at the latter points, and the interference is said to be *total destructive interference*. The location of points of constructive and destructive interference depends only upon the frequency and separation of the sources, and the relative location of each point with respect to the sources.

Most sound sources act independently and thus emit incoherent waves that do not produce interference in space. This circumstance is fortunate in many cases; for example, it means that there are no seats in a concert hall where the sound waves produced by two instruments (such as two violins) interfere destructively over appreciable time intervals during a concert.

The superposition principle also leads to *interference in time.* If two waves of slightly different frequency traverse the same space, the phase relation between the two waves at a given location varies with time. That is, the two waves are constantly changing from in-phase to out-of-phase much as man and woman walking together continually shift from in-step to out-of-step. This change of phase produces variations in the resultant amplitude of disturbance, which, in turn, produces variations in loudness called *beats.*

Standing Waves

Consider a source of sound waves located near a smooth reflecting barrier. Waves reflected from the barrier will have the same frequency and virtually the same amplitude as those emitted by the source. Thus, there will be two waves of the same frequency, speed, and amplitude traveling in opposite directions in the same medium. These are the necessary conditions to produce *standing waves.*

Longitudinal waves are emitted from a source and reflected from a barrier as shown in Fig. 1-14. Note that displacement amplitude is plotted on the ordinate. At certain locations, called *nodes*, the two waves interfere destructively; at other locations, called *antinodes*, they interfere constructively. The separation between these points depends only upon the wavelength of the sound wave; it can be shown that the distance between a node and an adjacent antinode is one-quarter wavelength.

A standing wave pattern similar to that shown in Fig. 1-14 is produced only for certain distances of separation between the source and reflecting barrier. A similar pattern, but with one less "loop," would result if the barrier were moved one-half wavelength closer to the source. However, if the barrier is moved some distance less than one-half wavelength closer, the standing wave pattern would not be produced.

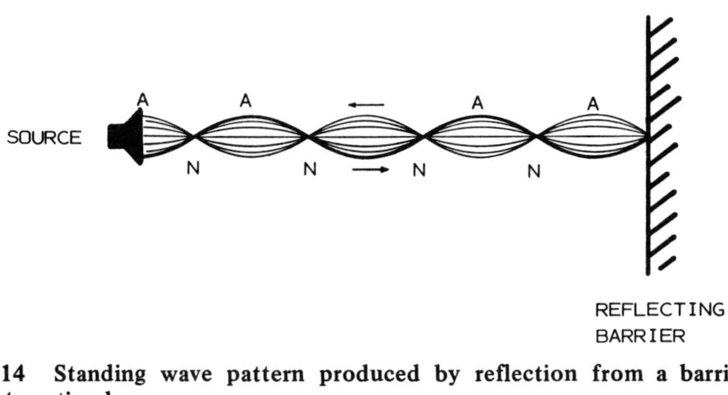

Fig. 1-14 Standing wave pattern produced by reflection from a barrier. *N*: node; *A*: antinode.

Energy is not transported to the right or to the left in a standing wave, for energy cannot flow past the nodal points, which are permanently at rest. Hence, the energy remains "standing," which leads to the name, *standing waves.*

It is clear that the conditions to produce standing waves can be met whenever reflecting barriers impede the flow of sound waves. The resulting pressure amplitude at the location of antinodes can become quite uncomfortable. Alteration of the standing wave pattern can be accomplished by either changing the wavelength of the sound waves emitted by the source or by altering the distance between the source and reflecting barrier.

RESONANCE

Whenever a system capable of oscillating is acted on by a series of impulses having a frequency equal or nearly equal to one of the natural frequencies of oscillation of the system, the system is set into oscillation with a relatively large amplitude. This phenomenon is called *resonance*, and the system is said to resonate with the applied impulses.

A common example of mechanical resonance is provided by a child on a rope swing. The swing is considered to be a simple pendulum whose single natural frequency is determined by the length of the rope. The amplitude of the swing can be made quite large if a series of regularly spaced pushes is given with a frequency equal to that of the swing. If the frequency of the pushes differs from the natural frequency of the swing, however, the pushes will hinder the motion as often as not, and the amplitude of oscillation will be quite small.

In like fashion, mechanical structures will resonate with nearby sound sources if the source is oscillating at or near one of the structure's many natural frequencies. The resonance condition can be eliminated either by changing the frequency of the source or by altering the natural frequencies of the mechanical structure. The latter alternative can usually be accomplished by changing the "stiffening" of the structure.

SOUND SPECTRA

Figure 1-15 shows the result of superposition of two coherent harmonic waves (see Equation 1-21). Another convenient way to display this information is by means of a *spectrum plot* as shown in Fig. 1-16. Here, relative sound intensity (or P^2) is plotted along the ordinate, and the frequency is plotted along the abscissa. This particular spectrum plot indicates the presence of two harmonic waves with frequencies ω_1 and ω_2; ω_1 is the dominant wave.

A source of sound waves will vibrate with many frequencies at the same time—it is a rare occurrence for a body to vibrate with only one frequency. Musical

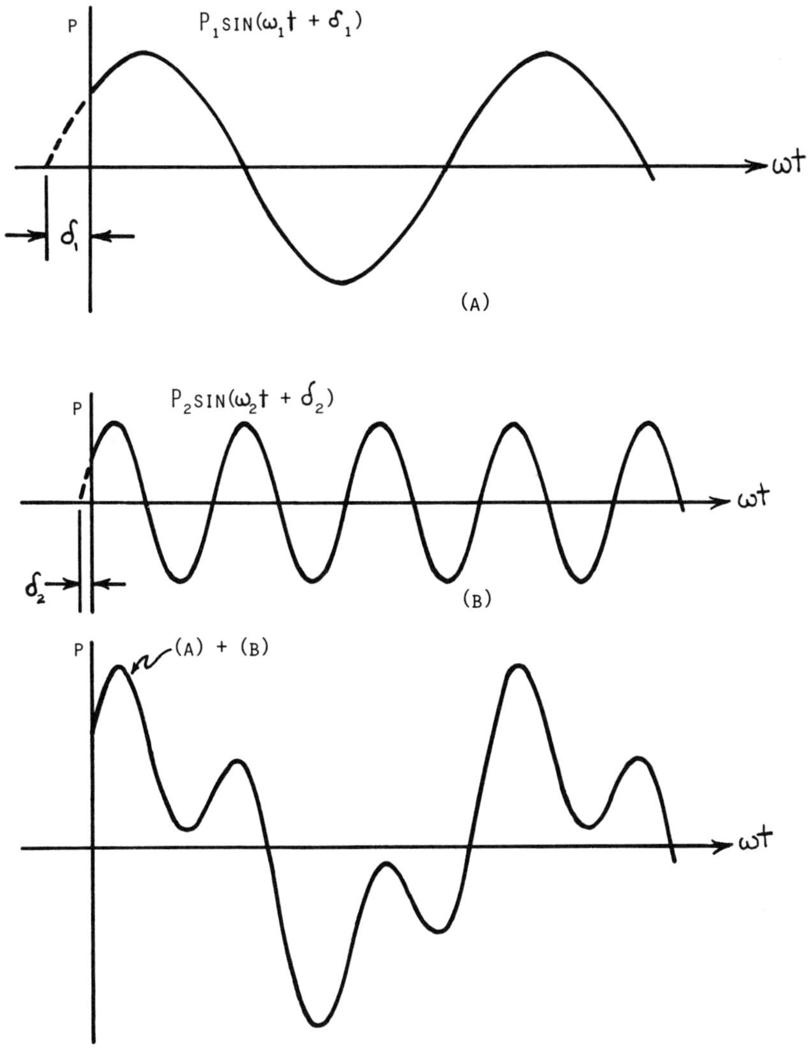

Fig. 1-15 Superposition of two coherent harmonic pressure waves.

instruments are no exception, producing several or more pure tones when a note is played, the number of pure tones produced depending upon the instrument. The lowest frequency component is called the *fundamental* frequency, whereas higher frequency components are called *overtones.* If the frequency of an overtone is an integer multiple (for example, n) of the fundamental frequency, it is also called the nth *harmonic.* Figure 1-17 shows the waveforms for a tuning fork, a clarinet, and a cornet, each at a frequency of 440 Hz and at approxi-

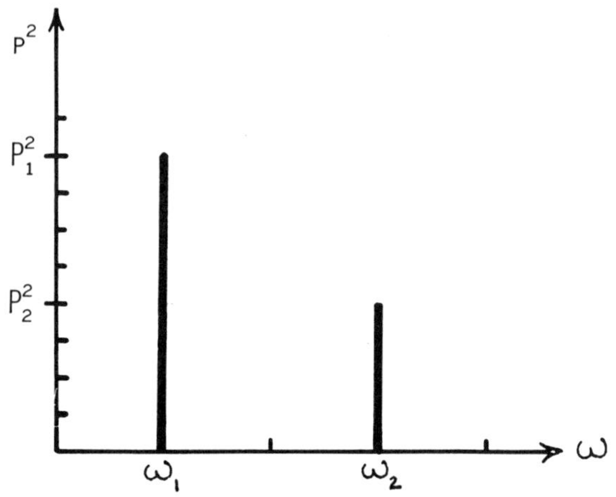

Fig. 1-16 Spectrum plot of superposition shown in Fig. 1-15.

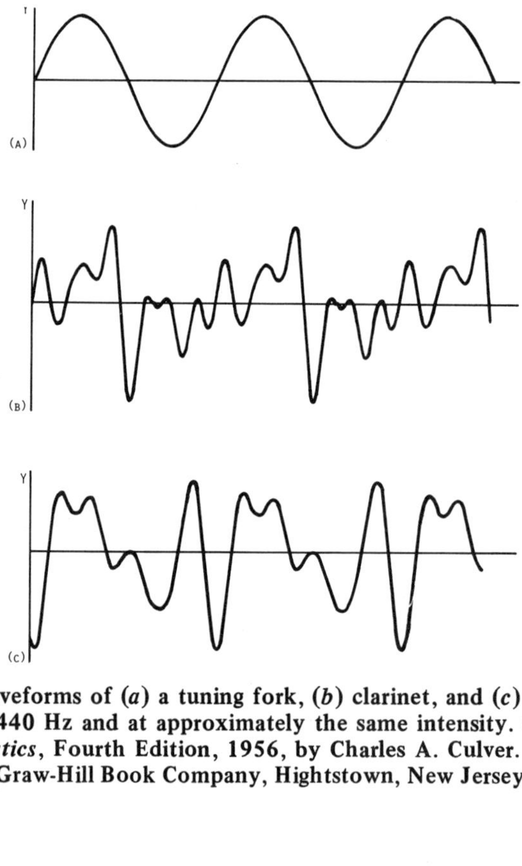

Fig. 1-17 Waveforms of (*a*) a tuning fork, (*b*) clarinet, and (*c*) cornet, each at frequency of 440 Hz and at approximately the same intensity. (Redrawn from *Musical Acoustics*, Fourth Edition, 1956, by Charles A. Culver. Used with permission of McGraw-Hill Book Company, Hightstown, New Jersey.)

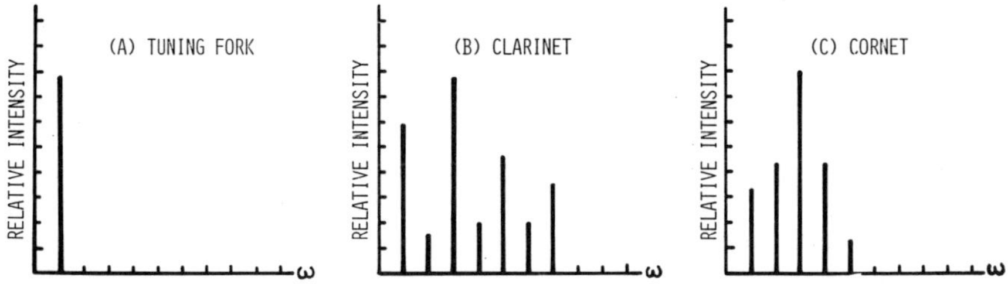

Fig. 1-18 Relative intensities of the harmonics in the waveforms shown in Fig. 1-17 for the tuning fork, clarinet, and cornet. (Redrawn from *Musical Acoustics*, Fourth Edition, 1956, by Charles A. Culver. Used with permission of McGraw-Hill Book Company, Hightstown, New Jersey.)

mately the same intensity. Figure 1-18 shows the relative intensities of the harmonics in the waveforms shown in Fig. 1-17. Such spectrum plots clearly show the distribution of acoustical energy as a function of frequency.

Everyday, nonmusical sounds experienced in the home, outdoors, or in industrial environments contain a multitude of frequencies. A spectrum plot for many sounds, such as running water and industrial background noise, would show a continuous spectrum of frequencies.

TYPES OF SOUND FIELDS

Three types of sound fields are distinguishable: the *near field*, *far field*, and *diffuse field*.

The *near field* is defined as that region where the source cannot be treated as a point source of acoustic energy and is a region in which the inverse square law does not apply. Usually, this region is located within a few wavelengths of the source or within a few diameters of the source, whichever is larger.

The *far field* is defined as that region beyond the near field where the inverse square law is obeyed. Multiple reflections from obstacles or barriers may produce a *diffuse sound field* in which the acoustic energy per unit volume is essentially constant throughout. Diffuse fields are most likely to occur in enclosed situations, whereas far fields are more likely to occur in outdoor situations.

IMPULSE OR IMPACT SOUND

The terms *impulse sound* or *impact sound* refer to transient acoustical events of short duration (usually less than 0.5 sec) and involve a change in sound pressure level above some minimum reference value (usually 40 dB). Examples of impulse sound are sonic booms produced by aircraft and gunshots.

Two types of impulse sound are distinguished. The first, *type A impulse*, usually involves a rapid rise to a peak sound pressure level followed by a decay to a negligible magnitude. Figure 1-19, which depicts such an event, also shows a subsequent negative pressure wave of smaller amplitude that is produced in some events. The duration of type A events is usually taken as the duration of the initial peak, disregarding the negative pressure wave. The other type of impulse sound, *type B*, is oscillatory. A type B event is also shown in Fig. 1-19. The duration of a single type B event is usually taken as the time required for the envelope of the oscillation curve to decay to a value 20 dB below the peak value of the pressure level. Special techniques must be employed to study impulse sound because of its short duration (see Chapter 3).

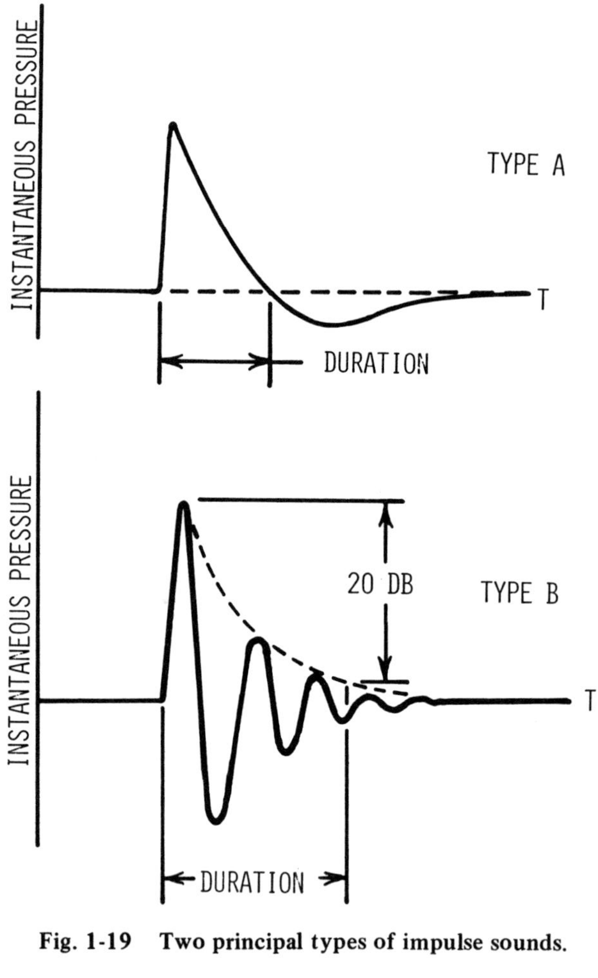

Fig. 1-19 Two principal types of impulse sounds.

Sonic Booms

The flow of air around an aircraft or other object whose speed exceeds the speed of sound (supersonic) is characterized by the existence of discontinuities in the air known as *shock waves.* These discontinuities result from the sudden encounter of an impenetrable body with air. At subsonic speeds, the air seems to be forewarned; thus, it begins its outward flow before the arrival of the leading edge. At supersonic speeds, however, the air in front of the aircraft is undisturbed, and the sudden impulse at the leading edge creates a region of overpressure (see Fig. 1-20) where the pressure is higher than atmospheric pressure. This overpressure region travels outward with the speed of sound, creating a conical-shaped shock wave called the *bow wave* that changes the direction of airflow. A second shock wave, the *tail wave*, is produced by the tail of the aircraft and is associated with a region where the pressure is lower than normal. This underpressure discontinuity causes the air behind the aircraft to move sideways.

Major pressure changes are experienced at the ear as the bow and tail shock waves reach an observer. Each of these pressure deviations produce the sensa-

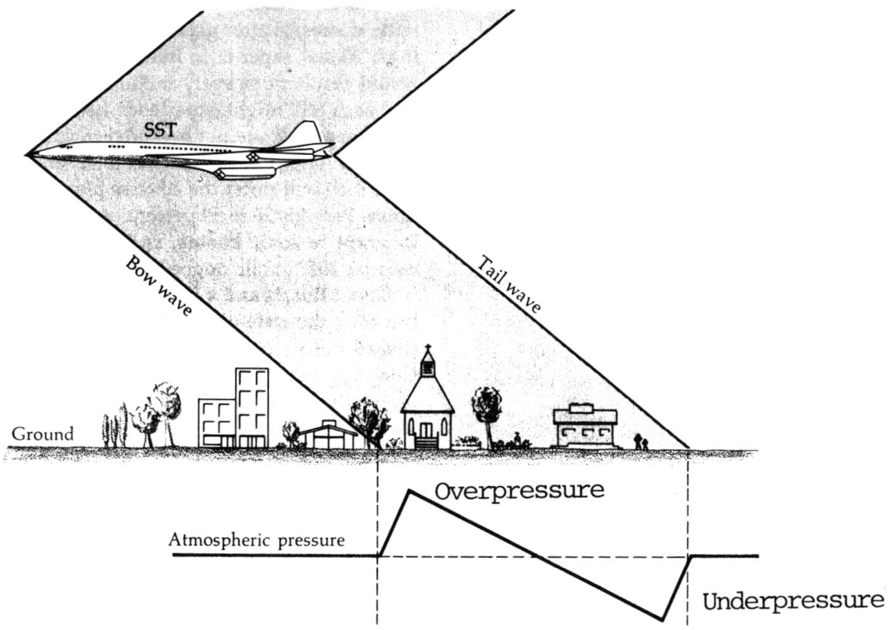

Fig. 1-20 Cross-sectional representation of a supersonic transport in flight, showing bow and tail shock waves, and a plot of the pressure at ground level. (From *Physics*, by Tipler, 1976. Used with permission of Worth Publishers, New York.)

tion of an explosive sound–the double crack of a *sonic boom.* Small objects moving at supersonic speeds also produce both bow and tail shock waves; however, they occur so close together in time that only one crack is heard.

A common misconception is that a sonic boom is produced when the aircraft first accelerates past the speed of sound. Actually, the shock waves exist during the entire period the aircraft is traveling supersonically, and a sonic boom is heard each time the shock waves sweep over an observer.

ULTRASOUND AND INFRASOUND

The term *sound* is usually understood to apply to those longitudinal mechanical waves in the frequency range from about 20 Hz to 20,000 Hz, which can stimulate the human ear and brain to the sensation of hearing.

Ultrasound is concerned with very high frequency (above about 20,000 Hz) longitudinal mechanical waves whose corresponding wavelength is a few centimeters or less. With their short wavelengths, these waves can provide information about the medium through which they travel; indeed, the smaller the wavelength of the ultrasound, the greater the detail that can be "seen." Penetration through the medium decreases as the wavelength is reduced; thus, ultrasound provides detailed information only at short range.

Ultrasonic waves are assuming increasing practical importance for such diverse uses as the detection and study of flaws in the interior of metals, the production of stable emulsions of normally immiscible liquids, the coagulation of aerosols and dust precipitation, and the cleaning of surgical instruments, watches, and the like.

Some of the most interesting applications of ultrasound lie in the field of medicine, where ultrasonic waves are now routinely used for both diagnosis and therapy. For example, reflection of such waves from interior regions of the body can be used to detect tumors, to study heart-valve action and measure bloodflow, and to detect and monitor the development of the human fetus. Ultrasonic waves have proven to be more sensitive than X-rays in distinguishing among various kinds of tissue as well as less hazardous to the human body. Therapeutical use of ultrasound is based on its ability to destroy pathological tissue at high power levels.

Probably the most exciting recent application is in *acoustical holography*, where coherent ultrasonic waves are used to obtain three-dimensional pictures analogous to optical holography. Such holographic techniques have interesting possibilities over a wide range, including underwater exploration, medical diagnosis, nondestructive testing, and biological and physical ultrasonic microscopy.

Infrasound is concerned with very low frequency (below about 20 Hz) longitudinal mechanical waves where sound is felt rather than heard. Infrasound waves easily penetrate large buildings or other barriers because, with their very long wavelength (about 15 m or more), they are not absorbed by normal building materials.

Powerful aircraft engines, volcanic eruptions, earthquakes, tornadoes, intense storms, and other weather phenomena are sources of infrasonic waves. There is some evidence that infrasonic waves cause great discomfort in human beings by affecting the inner ear and also by producing painful resonances inside the body where internal organs rub against one another producing intense friction. Perhaps infrasonic waves are the source of the unease that some people and animals are reported to experience prior to an earthquake or intense storm.

ENVIRONMENTAL NOISE

Music and speech are two of humanity's contributions to the world of sound. Unfortunately, they are not the only ones. Even from earliest days, our advancing technology has been accompanied by an increase in environmental noise. It is only within recent years, however, that the problem has been recognized as a serious danger to our well-being. There are three ways of attacking the problem of noise: (1) reduce its production through design, (2) confine it to an area or exclude it from others, or (3) set a limit, by legislation, to the amount that will be tolerated. Only through an increased understanding of the nature of sound and its interaction with our environment can the first two ways be fully employed to control the mushrooming problem of environmental noise.

REFERENCES

General:

1. Feather, Norman. *The Physics of Vibrations and Waves.* Edinburgh: Edinburgh University Press, 1961 (also Penguin Books, London, 1964).
2. Resnick, Robert, and Halliday, David. *Physics, Part I.* New York: John Wiley & Sons, 1966.
3. Sears, Francis W., Zemansky, Mark W., and, Young, Hugh D. *College Physics*, Fourth Ed. Reading, Massachusetts: Addision-Wesley Publishing Co., 1974.
4. Shortley, George, and Williams, Dudley. *Elements of Physics*, Fourth Ed. Englewood Cliffs, New Jersey: Prentice-Hall, 1965.
5. Tipler, Paul A. *Physics.* New York: Worth Publishers, 1976.
6. White, Harvey E. *Modern College Physics*, Sixth Ed. New York: Van Nostrand Reinhold, 1972.

Specialized:

7. Chedd, Graham. *Sound*, Doubleday Science Series. Garden City, New York: Doubleday & Co., 1970.

8. Culver, Charles A. *Musical Acoustics*. New York: McGraw-Hill Book Co., 1956.

9. Doelle, Leslie L. *Environmental Acoustics*. New York: McGraw-Hill Book Co., 1972.

10. Josephs, Jess J. *The Physics of Musical Sound*, Momentum Books. New York: Van Nostrand Reinhold, 1967.

11. Peterson, Arnold, P. G. and Gross, Erving E., Jr., *Handbook of Noise Measurement*, Sixth Ed. West Concord, Massachusetts: General Radio Co., 1963.

12. Rayleigh 3rd, Baron (J. W. Strutt). *The Theory of Sound*, Vols. I and II, Second Ed. London: Macmillan and Co., 1894 and 1896 (also Dover, New York, 1945).

13. Roederer, Juan G. *Introduction to the Physics and Psychophysics of Music*, Heidelberg Science Library, Vol. 16. New York: Springer-Verlag, 1973.

2

An introduction to the assessment and measurement of sound

Kent Williams, Ph.D.

Noise Program Coordinator, Region IV
U.S. Environmental Protection Agency
Atlanta, Georgia

INTRODUCTION

Sound is a highly variable component of the total human environment. As such, the quantification of sound levels for the purpose of determining whether they are hazardous to health and welfare, or exceed local legally adopted limits, can be a complex task. Depending upon both the purpose of the measurement and the desired accuracy of the result, a wide range of measurement techniques and sound level descriptors should be understood. In addition, the instrumentation available to perform the assessments should be known to potential users so that their selection of same is compatible with the required task. It is the purpose of this chapter to categorize in basic terms some of the instrumentation available and the measurement fundamentals employed in the assessment of both indoor and outdoor acoustic environments.

The measurement of the acoustic characteristics of an environment consisting of a single sound source (rooftop fan on an industrial concern) or of a multitude of sources (looms in the weaving room of a textile factory) becomes increasingly difficult as the sophistication in the measurement increases. A simple indicating meter may provide the observer with the maximum sound level emitted by a source. The central tendency of the indicating meter needle may yield an "average" sound level. In many situations, neither of these two measures of the characteristics of the source is sufficient to allow an appropriate assessment of the acoustic environment. As a result, a consideration of the instrumentation necessary to obtain more meaningful data about both the source and the environment is required.

INSTRUMENTATION

The simplest instrumentation available to determine sound levels is referred as a *sound level meter*. The meter consists of the following elements: (1) a transducer (microphone) to sense the sound-wave pressure and convert pressure fluctuations into an electrical voltage; (2) an input amplifier to raise the electrical signal to a usable level; (3) a calibrated attenuator to adjust the amplification to a value appropriate to the sound level being measured; (4) an indicating instrument (meter) to display the measured level; (5) weighting networks to modify the frequency characteristics of the instrument's response; (6) an output amplifier and calibrated output attenuator; and (7) an output jack to carry the signal to additional instrumentation for further analysis. In some cases, portable external filters may be added to the basic system. A block diagram of a simple sound level meter is given in Fig. 2-1.

More complex circuitry is involved for more sophisticated sound level meters, e.g., impulse and/or precision meters. Four different classifications are utilized in the description of sound level meters:[1]

Type 1: Precision
Type 2: General purpose
Type 3: Survey
Type 4: Special purpose

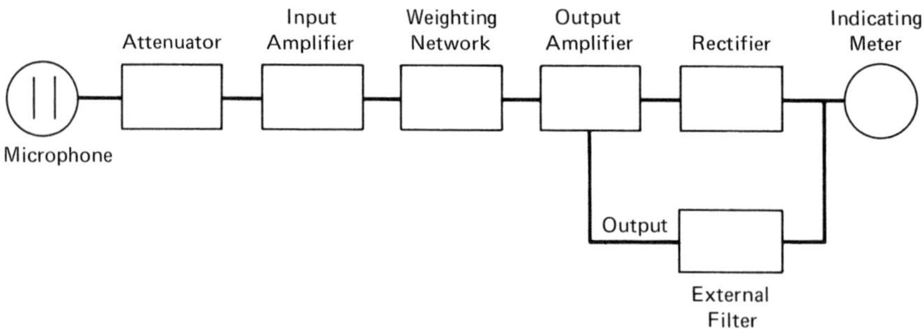

Fig. 2-1 Schematic of sound level meter.

The response of the meter, the characteristics of the indicating instrument, and the tolerances of the weighting networks vary significantly depending upon whether the instrument is of type 1, 2, or 3. "Slow" and "Fast" dynamic meter characteristics are provided for in all type 1, 2, and 3 meters. Impulse characteristics, and peak-hold and impulse-hold features are special additions obtainable on some type 1 impulse precision sound level meters. The indicating meter, where the sound level is observed, should be of the square-law type.

For meaningful data to be obtained, the sound level meter should be kept in constant calibration. All meter manufacturers provide calibration procedures, which usually involve two steps. The first step is a battery check and amplifier calibration; the second involves the calibration of the sound level meter including the microphone. In the field, this calibration is accomplished using a pistonphone and/or oscillator-type calibrator. These instruments supply a pure tone of fixed intensity and, in some cases, variable frequency, which may be used to standardize or calibrate the system. When using the pistonphone, care should be taken to make the required correction for atmospheric pressure. When using either the pistonphone or stabilized oscillator with a precision sound level meter, the system may be calibrated to within ±0.25 dB. (Note that this does not imply that measurements of sound levels may be performed as accurately.)

The microphone is a crucial part of any measuring system, and several different types are currently available. Three types used most commonly are condenser, electret, and piezoelectric.

In the condenser microphone, the incident pressure wave causes the microphone diaphragm to deflect slightly, thus changing the capacitance between the diaphragm and a flat electrode positioned parallel to and behind the diaphragm. There are several methods through which the change in capacitance may be converted to an electrical signal. The most common one is to maintain a constant charge on the capacitor by a polarization voltage.

Rather than operating with an air gap between the diaphragm and a parallel electrode, an electret microphone employs a solid dielectric stretched over a backplate. Because the dielectric material (electret foil) is prepolarized, there is no need to supply the constant polarization voltage required in a condenser microphone. An electrical voltage is produced when the electret foil moves relative to the backplate as the result of an incident sound wave.

Piezoelectric microphones are so named because they use piezoelectric materials such as natural crystals as part of the construction. The force resulting from the incident pressure wave on the diaphragm strains the piezoelectric material and generates an electric charge. This type of construction is more rugged than either an electret or condenser microphone because the piezoelectric material is behind the diaphragm and tends to support it.

While microphone construction is of interest, performance in terms of stability, precision, and frequency response are characteristics to be studied when a particular application is required. Specifically discussing condenser and electret microphones, the following information is available.

The widest frequency range for the pressure response of a condenser microphone is obtained when the diaphragm is critically damped. The high-frequency range can be extended by simply decreasing the size of the diaphragm, although the sensitivity is also decreased. Condenser microphones have excellent long-term stability and a very uniform sensitivity as a function of temperature, i.e., typically 0.008 dB/°C between -50°C and 150°C.[2] The change with respect to humidity (usually less than 0.5 dB between 0 and 90 percent relative humidity) and the effect of ambient pressure change -0.0003 dB/mm Hg are indicative of its suitability for extended long-term monitoring situations. It is, however, sensitive to moisture; at times, condensation can appear as a result of rapidly moving the microphone from a cold to a warm environment. The condensation manifests itself in "popping" or "noise."

Electret microphones vary considerably in their long-term stability and sensitivity as a function of temperature. Typically, their temperature coefficient would be 0.03 dB/°C from 0° to 55°C, and sensitivity may vary by ±1 dB over an extended time period.[2] As a result, they are not as well suited as condenser microphones to performing measurements in an environment that experiences moderately large temperature variations. Electret microphones also differ somewhat from condenser microphones in their frequency response; i.e., their linear response at high frequency is considerably less than a condenser microphone with the same diameter diaphragm. However, the electret microphone does not suffer as significant a loss in sensitivity as the diaphragm size is reduced.

The microphone performance just considered did not address the critical factor of how the diaphragm is oriented with respect to the sound field being measured. The response of a microphone depends upon its orientation to the incident sound energy. Depending upon the mechanical design of the microphone cartridge, a condenser microphone can be either a pressure microphone or a free-field microphone.

When the microphone is placed in a sound field, the sound wave is partly reflected from the surface of the diaphragm, which causes an increase in the effective sound pressure. The magnitude of the pressure increase depends on the microphone dimensions, the wavelength of the incident sound energy, and the angle between the incident sound and the diaphragm. The free-field response is equal to the pressure response plus any pressure increase caused by reflections at a certain angle of incidence.[3] The two angles selected are 90° (grazing incidence) and 0° (perpendicular incidence), and corrections to the pressure response may be made. Prior knowledge of the microphone type being utilized can remove the need for this correction by proper orientation of the diaphragm so that the most uniform frequency response is obtained.

Five measurement conditions may be identified as occurring most commonly in the field. They are: perpendicular incidence (0°), omnidirectional incidence, pressure, grazing incidence (90°), and random incidence.[4] To determine the sound field directly from a single source with no reflections, a perpendicular incidence free-field measurement should be made with a free-field microphone. If the total sound from a large number of sources separated at various angles is to be measured, an omnidirectional microphone (one that responds to sound equally, regardless of the angle of incidence) should be used. This also applies to indoor measurements such as would be found in a factory environment. Under certain conditions, a random incidence corrector may be required as well.

When earphones or audiometers require calibration, a pressure microphone should be used. The pressure microphone provides the most uniform frequency response when closely coupled to the source—an ideal application for this microphone.

Pressure microphones should also be used for the grazing incidence measurement condition such as identified with a moving sound source. At grazing incidence (90°), the response of the microphone is the same for all source locations (omnidirectional response). This grazing free-field configuration also tends to attenuate high frequency ground reflections very well.[4]

When measurements are to be made in a reverberation room (measurement of acoustic power, for example), a pressure microphone should be used. Because of the highly reverberant character of the sound field, the sound direction can be ignored. The pressure microphone, with its relatively uniform random incidence response, is acceptable for such a measurement condition.

Manufacturers generally provide sufficient information about the microphones that are part of a sound level meter kit, and give guidance on how to properly orient the diaphragm. Some judgment must be exercised, however, in assessing what further safeguards are required to insure the integrity of the measurement. Of particular concern for outdoor measurements is the use of a windscreen. The pressure fluctuations characterizing the sound wave are modified by the turbulence of the air passing over the microphone diaphragm and suffering a velocity change, and by the interference between the variations in the air pressure caused by the wind-velocity changes and the turbulence.[5] The influence of these factors may be reduced by covering the microphone diaphragm with a windscreen (cloth or polyurethane ball). The windscreen presents a large resistance to the wind itself but has a very small acoustic impedance. As a result, accurate acoustic measurements may be made under less than calm atmospheric conditions where wind direction is not of great concern. For internal directed flows, such as ventilation ducts and pipeflows, a nose cone may be employed. It is recommended that a windscreen be used for all outdoor measurements, and it is not bad practice to use them for indoor measurements since they protect the microphone cartridge from contaminants such as dust and dirt. Under certain wind conditions (often specified by a national or international test methodology or procedure), the measurements of environmental noise or source noise levels are not valid, regardless of whether a windscreen has been used. Again, it is not uncommon for information on windscreen performance to be provided by instrumentation manufacturers.

Earlier, it was suggested that to determine the sound level at a particular location due to a noise source, all that was required was the electronic instrumentation just described. There is in fact much more to it than that. First, whenever a measurement of "level" is being obtained, the level is a relative quantity fixed only when a reference condition is established. This is normally done electronically and in general corresponds to the acoustic pressure found to be analogous to the threshold of hearing at 1,000 Hz. Second, different sound levels are obtainable from a measurement of the same source if different electronic weighting filters are employed (see Fig. 2-2).[6] If, for example, the acoustic signal was unaltered over a frequency range of, for instance, 10 Hz to 20,000 Hz, the resulting meter indication would be representative of the sound pressure level. It has been found, however, that this characterization of the acoustic field does not correlate well with subjective effects such as annoyance. As a result, some sound level meters have as many as four separate weighting networks that electronically alter the frequency characteristics of the source prior to its display at the meter output. What is actually being measured under these conditions is a weighted sound pressure level–most commonly called the *sound level.* It is readily apparent (as one views the relative response characteristics of these networks) that the specification of a sound level must also be accompanied by the weighting network employed. Further amplification of how the measure-

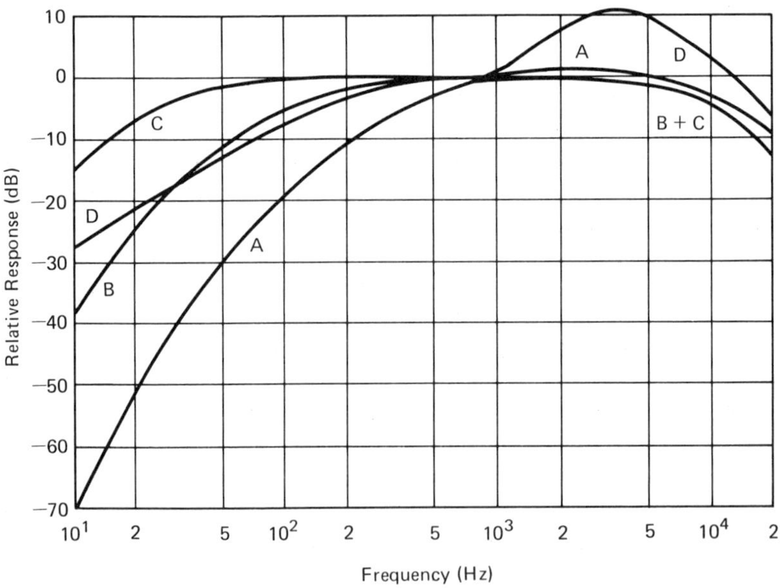

Fig. 2-2 A-, B-, C-, and D-weighting networks[6].

ment of the physical acoustic environment may be transformed into a subjective analysis will be presented in some detail later in this chapter. It is, however, worthwhile to note that all of the weighting networks do not share equal utility. The B-weighting network (originally presented as how individuals might respond to sounds of moderate intensity) is rarely if ever used. The C-weighting network, generally assumed to be a good approximation to a linear response or the sound-pressure level, is used occasionally when there is reason to believe that the low-frequency energy content of a source is being overly discriminated against by the low frequency roll-off of the A-weighted filter. Thus, certain annoying noise sources with a large percentage of their energy below 200–250 Hz may have a misleadingly low A-weighted sound level (e.g., diesel locomotive). For this type of source, a C-weighted sound level is very informative.

The D-weighted networks (note that there exists more than one D-weighting network) have yet to gain complete universal acceptance. They were designed primarily to provide a single-number subjective measure of aircraft noise. Although it is available on some sound level meters, the D-weighting network is currently used only rarely and for very specific measurement applications.

The most commonly used weighting network for both indoor (occupational) and outdoor (environmental) noise measurements is the A-weighting network. Federal standards for daily occupational noise limits are specified in terms of this measure. These federal criteria, developed by the Occupational Safety and Health Administration of the U.S. Department of Labor, are designed to protect against hearing loss, and hearing-loss data for individuals obtained over many years of study correlate well with this measure. In addition, environmental noise intrusions are evaluated in light of the A-weighting network. In a recent publication,[7] the U.S. Environmental Protection Agency (EPA) expressed its reasons for selecting the A-weighted sound level as the appropriate measure of environmental noise:

> With respect to both simplicity and adequacy for characterizing human response, a frequency-weighted sound level should be used for evaluation of environmental noise. Several frequency weightings have been proposed for general use in the assessment of response to noise, differing primarily in the way sounds at frequencies between 1000 and 4000 Hz are evaluated. The A-weighting, standardized in current sound level meter specifications, has been widely used for transportation and community noise description. For many noises the A-weighted sound level has been found to correlate as well with human response as more complex measures such as calculated perceived noise level or the loudness level derived from spectral analyses. It is concluded that a frequency weighted sound pressure level is the most reasonable choice for describing the magnitude of environmental noise. In order to use available standardized instrumentation for direct measurement, the A-frequency weighting is the only suitable choice at this time.

Even though a particular measurement situation may dictate a different procedure, generally it can be assumed that the A-weighted response will give the required information. Because of its universal acceptance, all sound level meters have A-weighting filters, and, when used appropriately, they satisfy most commonly encountered applications.

The measurement of sound levels from a noise source can give a crude approximation of the frequency distribution of the sound. For example, if the A-weighted sound level is considerably less than the C-weighted sound level (other physical conditions being the same), it may be concluded that the source contains a large amount of lower frequency energy, i.e., below 500 Hz. Likewise, if the A-weighted and C-weighted levels are nearly the same, most of the acoustic energy of the source is likely to be in the frequency range above 500 Hz. Other than being informative in a crude sense, this type of analysis is unacceptable when more detailed information is required concerning the spectral characteristics of the source. Consequently, electronic filters have been developed for use in conjunction with sound level meters to provide a more accurate assessment of the noise source in terms of its frequency content or distribution. This type of analysis, called *spectrum analysis*, yields a plot of amplitude as a function of frequency and is particularly useful in obtaining engineering solutions to noise problems.

Spectrum analyzers fall into two general categories: those with a constant bandwidth (fixed number of Hz) over the entire tuning range, and those analyzers having constant percentage bandwidth filters, where the upper-band-edge frequency bears a constant ratio to the lower-band-edge frequency over the tuning range. In the case of an octave-band analyzer (an octave describing a frequency ratio of 2:1), the upper-band-edge frequency is twice the lower-band-edge frequency. Since this alone is not a definitive description for a standard analysis, the preferred center frequencies of an octave-band analyzer, as well as those for $\frac{1}{2}$ and $\frac{1}{3}$ octave-band analyzers, are given in Table 2-1.[8] For those situations where high amplitudes representing possible pure tone components are separated only slightly in frequency, the type analysis just described may not be narrow enough. Narrow-band filters (1 percent or possibly 1 Hz) are frequently required to portray an accurate picture of the tonal characteristics of a source. This may eventually lead to digital-computer–aided real-time analysis if the expense of an even narrower analysis is justified.

Certain parameters are almost always present to affect the type of discrete measurements just considered. Background noise is one such parameter. In any measurement of the sound field of a particular source, cognizance should be taken of all other sources that together make up the ambient or background noise level. If it is possible to specifically identify secondary sources, this should be done; otherwise, the total of all contributing sources is assumed to be the background noise level. What effect does the background noise level have on the measurement of the acoustic field from a stationary source? To determine

TABLE 2-1 Preferred Frequencies in Hz, at Various Intervals and Band Numbers for Acoustical Measurements and for Center Frequencies of Filter Pass Bands.[8]

Preferred Frequencies	Band No.	1	½ Octave	⅓	Preferred Frequencies	Band No.	1	½ Octave	⅓	Preferred Frequencies	Band No.	1	½ Octave	⅓
16	12	X	X	X	160	22			X	1,600	32			X
18	12.5				180	22.5		X		1,800	32.5			
20	13			X	200	23			X	2,000	33	X	X	X
22.4	13.5		X		224	23.5				2,240	33.5			
25	14			X	250	24	X	X	X	2,500	34			X
28	14.5				280	24.5				2,800	34.5		X	
31.5	15	X	X	X	315	25			X	3,150	35			X
35.5	15.5				355	25.5		X		3,550	35.5			
40	16			X	400	26			X	4,000	36	X	X	X
45	16.5		X		450	26.5				4,500	36.5			
50	17			X	500	27	X	X	X	5,000	37			X
56	17.5				560	27.5				5,600	37.5		X	
63	18	X	X	X	630	28			X	6,300	38			X
71	18.5				710	28.5		X		7,100	38.5			
80	19			X	800	29			X	8,000	39	X	X	X
90	19.5		X		900	29.5				9,000	39.5			
100	20			X	1,000	30	X	X	X	10,000	40			X
112	20.5				1,120	30.5				11,200	40.5		X	
125	21	X	X	X	1,250	31			X	12,500	41			X
140	21.5				1,400	31.5		X		14,000	41.5			
160	22			X	1,600	32			X	16,000	42	X	X	X

Note: Higher and lower preferred frequencies are obtained by successive multiplication or division by 1000.

this effect accurately, the correct procedure is to eliminate the noise source of interest (turn it off if possible, or remove it from the vicinity) and then measure the background level. This should be done employing the same measuring system to be used for the actual measurement–if octave-band data are to be obtained, the background noise level should be measured in octave bands as well. (In fact, if an octave-band analyzer is available, it is good practice to obtain the octave-band sound pressure levels for the background noise, even though only A-weighted data are required from the source.) If the difference between the measured level for the source and background together is greater by 10 dB than the background alone in all frequency bands, then the background noise will not significantly affect the measurements. If it is found that in some frequency bands the difference between background noise and the source under test is less than 10 dB, a correction must be made to obtain the correct source levels. The corrections are given in Fig. 2-3 (this figure is essentially the same as Fig. 1-14).[9]

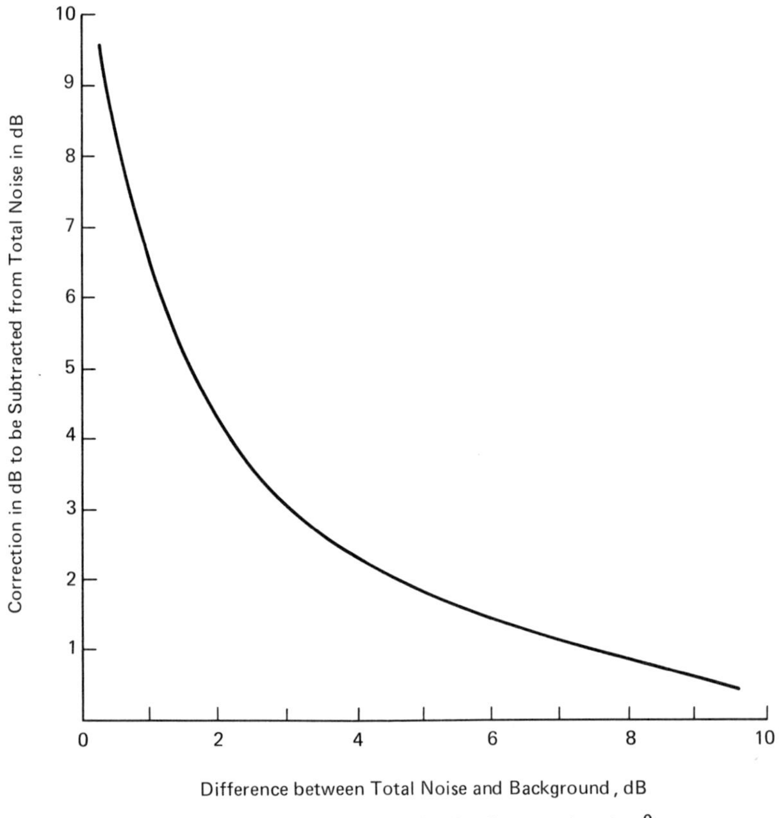

Fig. 2-3 Correction factors for background noise.[9]

If it is physically impossible to either turn off the source or remove it from the area in order to increase the difference in noise levels between the source and background, an alternate procedure may prove helpful. The microphone itself may be moved closer to the source, but kept far enough away to remain in the far field. (See Chapters 1 and 5 for discussion of near and far fields.) This may effectively increase the difference in sound level between the source and the ambient, thereby eliminating the need for a large correction.

In addition to background noise, reflections may occur that can affect the measurements. If there are objects in the vicinity of the microphone that have physical dimensions that are about equal to the wavelength of the sound or larger, it is possible for reflections to occur, thus disturbing the sound field. The magnitude of the reflection depends upon the shape and orientation of the reflecting object, as well as its absorptive properties. If the measurement is being made in a diffuse field or if the harmonic content of the noise shows it to be broadband (i.e., having no discrete frequencies), there is generally no serious problem with reflections. Under these conditions, the primary constraining factor is the quality of the instrumentation. Experimentally, it has been found that the maximum sound reflections from the human body occur at around 400 Hz.[10] When measurements are being made in that frequency range, errors of up to 6 dB may result if the person operating the instrument stands too close to the microphone.

SUBJECTIVE MEASURES AND RATING SCHEMES

Weighted Sound Pressure Levels

Although the *sound pressure level* of a source depends upon both the distance from the source to the microphone and the particular physical environment in which the measurement is made, it is nevertheless an *objective* measurement. The same may be said for octave-band and $^1/_3$-octave-band measurements. These measures of sound-source characteristics do not, however, give much indication of how we would perceive the noise source, i.e., our perception of the noise-source characteristics is a subjective phenomenon. Consequently, what might be considered raw, objective noise data are modified through subjective rating methodologies or measures to more adequately assess the reaction the noise intrusion is likely to evoke in a statistically significant number of individuals. The first attempt at such a subjective analysis resulted in the electronic weighting networks (A, B, C, and later D) described earlier and led to the now familiar concept of sound level (or weighted sound pressure level). The most commonly used sound level utilizes the A-weighting network—a response approximating the 40-phon equal loudness contour. (Loudness contours will be discussed more fully later.) The apparent loudness we attribute to a sound varies not only with sound pressure (intensity) but also with frequency. In addition, the way loudness varies with frequency also depends on the sound pressure. In general, people are less sensitive to sounds of low frequency, such as 100 Hz, than to sounds of higher frequency, such as 1000 Hz. In addition, at high frequencies, such as 8000 Hz, sensitivity decreases.[7] An acceptable approach to compensate for this difference in response to different frequencies is to apply a weighting to the overall spectrum in such a way that the sounds of various frequencies are weighted in much the same way as the human ear hears them. The A-weighted network weighs the contributions of sounds of different frequencies so that the response of the average human ear is simulated. Each frequency of the sound then contributes to the total sound level an amount approximately proportional to the subjective response associated with that frequency. While this single-number rating concept—the A-weighted sound level—is useful, it fails to adequately express such subjective feelings as how loud a sound really is. Thus, methods have been developed for rating loudness.

Equal Loudness Contours (Phons and Sones)

Initial attempts by Fletcher and Munson[11] produced equal loudness contours where loudness level was defined as the intensity level of a tone sounding equally loud when compared with a reference tone having the single frequency of 1000 Hz. (See Fig. 2-4).[12] Loudness level has been described in units of phons, and from the figure it may be seen that a tone at 200 Hz and 60-dB sound pressure level would sound equally as "loud" as a 51-dB tone at 1000 Hz. Thus, we would say that the 200 Hz tone had a loudness level of 51 phons. These contours are rarely used now since more precise and refined methods have been developed. Methodology developed by S. S. Stevens[13] introduced the idea of loudness as contrasted to the just considered concept of loudness level. The unit of loudness is the *sone*. To establish a reference, the loudness of a pure tone at 1000 Hz with a sound pressure level of 40 dB re. 20 μ Pascals is taken to be 1 sone. It should be noted that this corresponds to a loudness level of 40 phons. Subjects were asked to make adjustments to a comparison sound until it was twice as loud as the reference or four times as loud, etc. Thus, conceptually, a loudness scale is established with a 4-sone sound that seems to a listener to be four times as loud as a 1-sone sound. The relationship between loudness in sones and loudness level in phons for sounds of loudness level equal to or greater than 40 phons is:

$$S = 2^{\left(\frac{L_S - 40}{10}\right)} \text{ sones} \tag{2-1}$$

Because most sounds are more complex than the pure tone examples considered above, more sophisticated methods are actually employed for real noise sources.

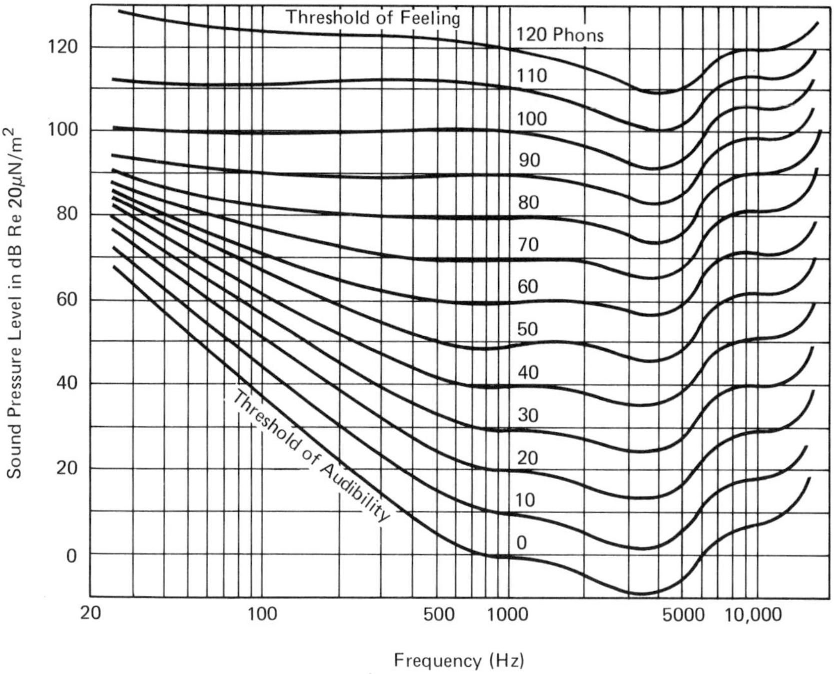

Fig. 2-4 Equal loudness level contours (Fletcher-Munson curves).

These utilize octave-band and ⅓-octave-band data. For example, Stevens's method adds to the sone value of the loudest frequency band a fractional portion of the sum to the sone values of the remainder of the bands:

$$S = S_m + f\left(\sum S - S_m\right) \tag{2-2}$$

where

ΣS = summation of the sone values in all frequency bands
S_m = maximum sone value in any one band
f = fractional multiplier that depends on bandwidth ($f = 0.3$ for octave-band data and $f = 0.15$ for ⅓-octave-band data).

The appropriate sone values corresponding to each band level are given in Table 2-2.[14] (An example illustrating the use of Equation 2-2 and Table 2-2 will appear later in the chapter.)

Perceived Noisiness

A method similar to that previously discussed for loudness was developed by Kryter.[15] A series of subjects were asked to compare noises in terms of "noisi-

TABLE 2-2

Band Level (dB)	Band Loudness Index									Loudness (Sones)	Loudness Level (Phons)
	31.5	63	125	250	500	1000	2000	4000	8000		
20						.18	.30	.45	.61	.25	20
21						.22	.35	.50	.67	.27	21
22					.07	.26	.40	.55	.73	.29	22
23					.12	.30	.45	.61	.80	.31	23
24					.16	.35	.50	.67	.87	.33	24
25					.21	.40	.55	.73	.94	.35	25
26					.26	.45	.61	.80	1.02	.38	26
27					.31	.50	.67	.87	1.10	.41	27
28				.07	.37	.55	.73	.94	1.18	.44	28
29				.12	.43	.61	.80	1.02	1.27	.47	29
30				.16	.49	.67	.87	1.10	1.35	.50	30
31				.21	.55	.73	.94	1.18	1.44	.54	31
32				.26	.61	.80	1.02	1.27	1.54	.57	32
33				.31	.67	.87	1.10	1.35	1.64	.62	33
34			.07	.37	.73	.94	1.18	1.44	1.75	.66	34
35			.12	.43	.80	1.02	1.27	1.54	1.87	.71	35
36			.16	.49	.87	1.10	1.35	1.64	1.99	.76	36
37			.21	.55	.94	1.18	1.44	1.75	2.11	.81	37
38			.26	.62	1.02	1.27	1.54	1.87	2.24	.87	38
39			.31	.69	1.10	1.35	1.64	1.99	2.38	.93	39
40		.07	.37	.77	1.18	1.44	1.75	2.11	2.53	1.00	40
41		.12	.43	.85	1.27	1.54	1.87	2.24	2.68	1.07	41
42		.16	.49	.94	1.35	1.64	1.99	2.38	2.84	1.15	42
43		.21	.55	1.04	1.44	1.75	2.11	2.53	3.0	1.23	43
44		.26	.62	1.13	1.54	1.87	2.24	2.68	3.2	1.32	44
45		.31	.69	1.23	1.64	1.99	2.38	2.84	3.4	1.41	45
46	.07	.37	.77	1.33	1.75	2.11	2.53	3.0	3.6	1.52	46
47	.12	.43	.85	1.44	1.87	2.24	2.68	3.2	3.8	1.62	47
48	.16	.49	.94	1.56	1.99	2.38	2.84	3.4	4.1	1.74	48
49	.21	.55	1.04	1.69	2.11	2.53	3.0	3.6	4.3	1.87	49
50	.26	.62	1.13	1.82	2.24	2.68	3.2	3.8	4.6	2.00	50
51	.31	.69	1.23	1.96	2.38	2.84	3.4	4.1	4.9	2.14	51
52	.37	.77	1.33	2.11	2.53	3.0	3.6	4.3	5.2	2.30	52
53	.43	.85	1.44	2.24	2.68	3.2	3.8	4.6	5.5	2.46	53
54	.49	.94	1.56	2.38	2.84	3.4	4.1	4.9	5.8	2.64	54
55	.55	1.04	1.69	2.53	3.0	3.6	4.3	5.2	6.2	2.83	55
56	.62	1.13	1.82	2.68	3.2	3.8	4.6	5.5	6.6	3.03	56
57	.69	1.23	1.96	2.84	3.4	4.1	4.9	5.8	7.0	3.25	57
58	.77	1.33	2.11	3.0	3.6	4.3	5.2	6.2	7.4	3.48	58
59	.85	1.44	2.27	3.2	3.8	4.6	5.5	6.6	7.8	3.73	59
60	.94	1.56	2.44	3.4	4.1	4.9	5.8	7.0	8.3	4.00	60
61	1.04	1.69	2.62	3.6	4.3	5.2	6.2	7.4	8.8	4.29	61
62	1.13	1.82	2.81	3.8	4.6	5.5	6.6	7.8	9.3	4.59	62
63	1.23	1.96	3.0	4.1	4.9	5.8	7.0	8.3	9.9	4.92	63
64	1.33	2.11	3.2	4.3	5.2	6.2	7.4	8.8	10.5	5.28	64

TABLE 2-2 *(Continued)*

Band Level (dB)	Band Loudness Index									Loudness (Sones)	Loudness Level (Phons)
	31.5	63	125	250	500	1000	2000	4000	8000		
65	1.44	2.27	3.5	4.6	5.5	6.6	7.8	9.3	11.1	5.66	65
66	1.56	2.44	3.7	4.9	5.8	7.0	8.3	9.9	11.8	6.06	66
67	1.69	2.62	4.0	5.2	6.2	7.4	8.8	10.5	12.6	6.50	67
68	1.82	2.81	4.3	5.5	6.6	7.8	9.3	11.1	13.5	6.96	68
69	1.96	3.0	4.7	5.8	7.0	8.3	9.9	11.8	14.4	7.46	69
70	2.11	3.2	5.0	6.2	7.4	8.8	10.5	12.6	15.3	8.00	70
71	2.27	3.5	5.4	6.6	7.8	9.3	11.1	13.5	16.4	8.6	71
72	2.44	3.7	5.8	7.0	8.3	9.9	11.8	14.4	17.5	9.2	72
73	2.62	4.0	6.2	7.4	8.8	10.5	12.6	15.3	18.7	9.8	73
74	2.81	4.3	6.6	7.8	9.3	11.1	13.5	16.4	20.0	10.6	74
75	3.0	4.7	7.0	8.3	9.9	11.8	14.4	17.5	21.4	11.3	75
76	3.2	5.0	7.4	8.8	10.5	12.6	15.3	18.7	23.0	12.1	76
77	3.5	5.4	7.8	9.3	11.1	13.5	16.4	20.0	24.7	13.0	77
78	3.7	5.8	8.3	9.9	11.8	14.4	17.5	21.4	26.5	13.9	78
79	4.0	6.2	8.8	10.5	12.6	15.3	18.7	23.0	28.5	14.9	79
80	4.3	6.7	9.3	11.1	13.5	16.4	20.0	24.7	30.5	16.0	80
81	4.7	7.2	9.9	11.8	14.4	17.5	21.4	26.5	32.9	17.1	81
82	5.0	7.7	10.5	12.6	15.3	18.7	23.0	28.5	35.3	18.4	82
83	5.4	8.2	11.1	13.5	16.4	20.0	24.7	30.5	38	19.7	83
84	5.8	8.8	11.8	14.4	17.5	21.4	26.5	32.9	41	21.1	84
85	6.2	9.4	12.6	15.3	18.7	23.0	28.5	35.3	44	22.6	85
86	6.7	10.1	13.5	16.4	20.0	24.7	30.5	38	48	24.3	86
87	7.2	10.9	14.4	17.5	21.4	26.5	32.9	41	52	26.0	87
88	7.7	11.7	15.3	18.7	23.0	28.5	35.3	44	56	27.9	88
89	8.2	12.6	16.4	20.0	24.7	30.5	38	48	61	29.9	89
90	8.8	13.6	17.5	21.4	26.5	32.9	41	52	66	32.0	90
91	9.4	14.8	18.7	23.0	28.5	35.3	44	56	71	34.3	91
92	10.1	16.0	20.0	24.7	30.5	38	48	61	77	36.8	92
93	10.9	17.3	21.4	26.5	32.9	41	52	66	83	39.4	93
94	11.7	18.7	23.0	28.5	35.3	44	56	71	90	42.2	94
95	12.6	20.0	24.7	30.5	38	48	61	77	97	45.3	95
96	13.6	21.4	26.5	32.9	41	52	66	83	105	48.5	96
97	14.8	23.0	28.5	35.3	44	56	71	90	113	52.0	97
98	16.0	24.7	30.5	38	48	61	77	97	121	55.7	98
99	17.3	26.5	32.9	41	52	66	83	105	130	59.7	99
100	18.7	28.5	35.3	44	56	71	90	113	139	64.0	100
101	20.3	30.5	38	48	61	77	97	121	149	68.6	101
102	22.1	32.9	41	52	66	83	105	130	160	73.5	102
103	24.0	35.3	44	56	71	90	113	139	171	78.8	103
104	26.1	38	48	61	77	97	121	149	184	84.4	104
105	28.5	41	52	66	83	105	130	160	197	90.5	105
106	31.0	44	56	71	90	113	139	171	211	97	106
107	33.9	48	61	77	97	121	149	184	226	104	107
108	36.9	52	66	83	105	130	160	197	242	111	108
109	40.3	56	71	90	113	139	171	211	260	119	109
110	44	61	77	97	121	149	184	226	278	128	110

TABLE 2-2 *(Continued)*

Band Level (dB)	Band Loudness Index									Loudness (Sones)	Loudness Level (Phons)
	31.5	63	125	250	500	1000	2000	4000	8000		
111	49	66	83	105	130	160	197	242	298	137	111
112	54	71	90	113	139	171	211	260	320	147	112
113	59	77	97	121	149	184	226	278	343	158	113
114	65	83	105	130	160	197	242	298	367	169	114
115	71	90	113	139	171	211	260	320		181	115
116	77	97	121	149	184	226	278	343		194	116
117	83	105	130	160	197	242	298	367		208	117
118	90	113	139	171	211	260	320			223	118
119	97	121	149	184	226	278	343			239	119
120	105	130	160	197	242	298	357			256	120
121	113	139	171	211	260	320				274	121
122	121	149	184	226	278	343				294	122
123	130	160	197	242	298	367				315	123
124	139	171	211	260	320					338	124
125	149	184	226	278	343					362	125

ness" or to provide a measure of the average perceived noisiness of sounds. A scale was developed expressing the concept of perceived noisiness (PN), which was based on occurrences of sounds of equal duration. Analogous to the sone, the unit of perceived noisiness is the *noy*, and a sound judged to be subjectively equal in noisiness to an octave band of random noise centered about 1000 Hz and a sound pressure level of 40 dB is given a value of 1 noy. (Contrast this to the reference level of 40 dB with a *pure tone* of 1000 Hz to establish the level of 1 sone.) A noise judged four times as noisy is 4 noys, and so on. The logarithmic scale analogous to the phon scale for loudness level is the perceived noise level (PNL).

> The judged PNL of a given sound is equal numerically to the maximum overall sound pressure level of a reference sound that is judged by listeners at any given point in time to have the same perceived noisiness as the given sound, the reference sound being random "pink" noise precisely one octave in width centered at 1000 Hz and of comparable temporal characteristics as the given sound, i.e., rise and decay times and total duration.[12]

The determination of PNL has been particularly fruitful in the assessment of aircraft noise although PNL was found to be deficient as well when jet noise was considered. (This application will be addressed later.) The computational procedure for determining PNL is vertually the same as Stevens' procedure for loudness with a few modifications. The first difference is that the octave-band levels used for evaluating aircraft data are to be the maximum values obtained in each band during the event regardless of when the peak in the spectrum actually occurred. The second variation is that the index assigned to each measured band level is in units of noys. Thus, for an octave-band analysis of a noise event:

$$\text{Noys} = 0.3 \sum N + 0.7\, N_{max}$$

where

ΣN = sum of noys for each band level
N_{max} = largest noy value of all bands.

The total summed noy value may be converted to perceived noise level PNL in dB by utilizing the equivalent 1000 Hz octave-band-center frequency and its resultant band level (see Table 2-3).[14] The PNL in PNdB is similar conceptually to the loudness level in phons, for an increase of 10 PNdB is roughly equivalent to a doubling of its noy value.

Example Calculations of Loudness, Loudness Level, Perceived Noise, and Perceived Noise Level

For the purpose of illustration, three distinct noise sources with widely differing frequency characteristics will be considered. This should allow the reader an opportunity to assess the effect of the energy distribution on these commonly used subjective measures.

Center Frequency (Hz)	OBSPL* (dB) I	dB re 20 μ Pascals II	III
31.5	76	97	113
63.0	80	102	110
125	85	104	90
250	89	97	86
500	92	90	79
1000	98	85	86
2000	100	80	90
4000	95	75	95
8000	85	70	80

*Octave-band sound-pressure level.

Since the A-weighted sound level may be determined from the octave-band analysis (the reverse of this procedure is, of course, not possible), an illustration of this procedure employing Source II may prove beneficial as well. Conceptually, the steps are:

1. Correct each octave-band sound pressure level relative to the A-weighted filter characteristics.

TABLE 2-3

Perceived Noise (noys)									
Band Level (dB)	Octave Band								
	31.5	63	125	250	500	1000	2000	4000	8000
20							.3	.4	
21							.4	.5	
22							.4	.6	
23							.5	.6	
24							.5	.7	
25					.1	.1	.6	.7	.1
26					.2	.2	.7	.8	.2
27					.2	.2	.7	.9	.2
28					.3	.3	.8	.9	.3
29				.1	.3	.3	.8	1.0	.3
30				.2	.4	.4	.9	1.1	.4
31				.2	.4	.4	.9	1.2	.4
32				.3	.5	.5	1.0	1.2	.5
33				.3	.6	.6	1.1	1.3	.6
34				.4	.6	.6	1.2	1.4	.7
35			.1	.4	.7	.7	1.3	1.5	.8
36			.1	.5	.7	.7	1.3	1.6	.9
37			.2	.5	.8	.8	1.4	1.8	1.0
38			.2	.6	.9	.9	1.5	1.9	1.2
39			.3	.7	.9	.9	1.6	2.0	1.3
40			.3	.8	1.0	1.0	1.8	2.2	1.4
41			.4	.8	1.1	1.1	1.9	2.3	1.6
42			.4	.9	1.1	1.1	2.0	2.5	1.7
43			.5	1.0	1.2	1.2	2.2	2.7	1.8
44			.6	1.0	1.3	1.3	2.3	2.9	2
45			.6	1.1	1.4	1.4	2.5	3.1	2.2
46			.7	1.2	1.5	1.5	2.7	3.3	2.3
47			.7	1.3	1.6	1.6	2.9	3.5	2.5
48		.1	.8	1.4	1.7	1.7	3.1	3.8	2.7
49		.2	.9	1.5	1.9	1.9	3.3	4.1	2.9
50		.2	1.0	1.6	2.0	2.0	3.5	4.3	3.1
51		.3	1.0	1.7	2.1	2.1	3.8	4.7	3.3
52		.4	1.1	1.8	2.3	2.3	4.1	5.0	3.5
53		.4	1.2	2.0	2.5	2.5	4.3	5.3	3.8
54		.5	1.3	2.1	2.6	2.6	4.7	5.7	4.1
55		.6	1.5	2.3	2.8	2.8	5.0	6.1	4.3
56		.6	1.6	2.4	3.0	3.0	5.3	6.6	4.7
57		.7	1.7	2.6	3.2	3.2	5.7	7.0	5.0
58		.8	1.8	2.8	3.5	3.5	6.1	7.5	5.3
59		.9	2.0	3.0	3.7	3.7	6.6	8.1	5.7

TABLE 2-3 *(Continued)*

Band Level (dB)	Perceived Noise (noys) Octave Band 31.5	63	125	250	500	1000	2000	4000	8000
60		1.0	2.2	3.2	4.0	4.0	7.0	8.7	6.1
61		1.1	2.3	3.5	4.3	4.3	7.5	9.3	6.6
62		1.2	2.5	3.7	4.6	4.6	8.1	10.0	7.0
63		1.3	2.7	4.0	4.9	4.9	8.7	10.7	7.5
64		1.5	2.9	4.3	5.3	5.3	9.3	11.4	8.1
65		1.6	3.2	4.7	5.7	5.7	10.0	12.2	8.7
66		1.8	3.4	5.0	6.1	6.1	10.7	13.1	9.3
67		2.0	3.7	5.4	6.5	6.5	11.4	14.1	10.0
68	.1	2.2	4.0	5.9	7.0	7.0	12.2	15.1	10.7
69	.2	2.4	4.3	6.4	7.5	7.5	13.1	16.1	11.4
70	.3	2.6	4.6	6.9	8.0	8.0	14.1	17.3	12.2
71	.4	2.8	5.0	7.4	8.6	8.6	15.1	18.5	13.1
72	.5	3.1	5.4	8.0	9.2	9.2	16.1	19.3	14.1
73	.6	3.4	5.9	8.6	9.8	9.8	17.3	21.2	15.1
74	.7	3.7	6.4	9.2	10.6	10.6	18.5	22.7	16.1
75	.8	4.1	7.0	9.8	11.3	11.3	19.8	24.3	17.3
76	1.0	4.5	7.6	10.6	12.1	12.1	21.2	26.1	18.5
77	1.1	4.9	8.2	11.3	13.0	13.0	22.7	27.9	19.8
78	1.3	5.4	9.0	12.1	13.9	13.9	24.3	29.8	21.2
79	1.5	5.9	9.7	13.0	14.9	14.9	26.1	31.9	22.7
80	1.7	6.4	10.5	13.9	16.0	16.0	27.9	34.2	24.3
81	1.9	7.0	11.3	14.9	17.1	17.1	29.9	36.6	26.1
82	2.2	7.7	12.1	16.0	18.4	18.4	32.0	39.1	27.9
83	2.5	8.4	13.0	17.1	19.7	19.7	34.2	42	29.8
84	2.8	9.1	13.9	18.4	21.1	21.1	36.6	45	31.9
85	3.1	9.9	14.9	19.7	22.6	22.6	39.1	48	34.2
86	3.5	10.8	16.0	21.1	24.3	24.3	42	51	36.6
87	3.9	11.7	17.2	22.6	26.0	26.0	45	55	39.1
88	4.4	12.7	18.4	24.3	27.9	27.9	48	58	42
89	4.9	13.7	19.7	26.0	29.9	29.9	51	62	45
90	5.5	14.7	21.1	27.9	32.0	32.0	55	67	48
91	6.2	15.9	22.6	29.9	34.3	34.3	58	71	51
92	6.9	17.1	24.3	32.0	36.8	36.8	62	76	55
93	7.7	18.3	26.0	34.3	39.4	39.4	67	82	58
94	8.6	19.7	27.9	36.8	42.2	42.2	71	88	62
95	9.5	21.1	29.9	39.4	45.3	45.3	76	94	67
96	10.6	22.6	32.0	42	48.5	48.5	82	100	71
97	11.7	24.3	34.3	45	52	52	88	107	76
98	12.9	26.0	36.8	43	56	56	94	115	82
99	14.1	27.9	39.4	52	60	60	100	123	88

TABLE 2-3 *(Continued)*

Band Level (dB)	Perceived Noise (noys) Octave Band								
	31.5	63	125	250	500	1000	2000	4000	8000
100	15.5	29.9	42	56	64	64	107	132	94
101	16.9	32.0	45	60	69	69	115	142	100
102	18.4	34.3	48	64	74	74	123	152	107
103	19.9	36.8	52	69	79	79	132	163	115
104	21.5	39.4	56	74	84	84	142	174	123
105	23.2	42	60	79	91	91	152	187	132
106	25.1	45	64	84	97	97	163	200	142
107	27.0	48	69	91	104	104	174	215	152
108	29.0	52	74	97	111	111	187	230	163
109	31.2	56	79	104	119	119	200	247	174
110	33.6	60	84	111	128	128	215	264	187
111	36.1	64	91	119	137	137	230	283	200
112	39	69	97	128	147	147	247	304	215
113	42	74	104	137	158	158	264	325	230
114	45	79	111	147	169	169	283	349	247
115	49	84	119	158	181	181	304	374	264
116	52	91	128	169	194	194	325	401	283
117	57	97	137	181	208	208	349	429	304
118	61	104	147	194	223	223	374	460	325
119	66	111	158	208	239	239	401	493	349
120	71	119	169	223	256	256	429	529	374
121	76	128	181	239	274	274	460	567	401
122	82	137	194	256	294	294	493	607	429
123	88	147	208	274	315	315	529	651	460
124	94	158	223	294	338	338	567	698	403
125	101	169	239	315	362	362	607	748	529
126	108	181	256	338	388	388	651	801	567
127	117	194	274	362	416	416	698	859	607
128	125	208	294	388	446	446	748	921	651
129	135	223	315	416	478	478	801	987	698
130	145	239	338	446	512	512	859	1057	748
131	156	256	362	478	549	549	921	1133	801
132	168	274	388	512	588	588	987	1215	859
133	180	294	416	549	630	630	1057	1302	921
134	193	315	446	588	676	676	1133	1395	987
135	207	338	478	630	724	724	1215	1495	1057
136	223	362	512	676	776	776	1302	1603	1133
137	239	383	549	724	832	832	1395	1720	1215
138	256	416	588	776	891	891	1495	1840	1302
139	274	446	630	832	955	955	1603	1970	1395

TABLE 2-3 (Continued)

Band Level (dB)	Perceived Noise (noys) — Octave Band								
	31.5	63	125	250	500	1000	2000	4000	8000
140	294	478	676	891	1024	1024	1720	2110	1495
141	315	512	724	955	1098	1098	1840		1600
142	338	549	776	1024	1176	1176	1970		1720
143	362	588	832	1098	1261	1261	2110		1840
144	388	630	891	1176	1351	1351			1970
145	416	676	955	1261	1448	1448			2110
146	446	724	1024	1351	1552	1552			
147	478	776	1098	1448	1663	1663			
148	512	832	1176	1552	1783	1783			
149	549	891	1261	1663	1911	1911			
150	588	955	1351	1783	2048	2048			

2. Add the resulting corrected levels for each center frequency by energy addition, using Fig. 1-14.

Center Frequency (Hz)	A-Weighting Relative Response (dB)*	OBSPL (dB)	Corrected Level (dB)
31.5	-39.4	97	57.6
63	-26.2	102	75.8
125	-16.1	104	87.9
250	- 8.6	97	88.4
500	- 3.2	90	86.8
1000	0	85	85.0
2000	+ 1.2	80	81.2
4000	+ 1.0	75	76.0
8000	- 1.1	70	68.9

*From Fig. 2-2.

Rank levels to be added in descending order, and add adjacent pairs, as follows: 88.4, 87.9, 86.8, 85.0, 81.2, 76.0, 75.8, 68.9, 57.6.

Consider the first pair: Difference = 88.4 - 87.9 = 0.5 dB.
From Fig. 2-3, the amount to be added to the larger is 2.8 dB.
In terms of *energy addition*, 88.4 + 87.9 ⇒ 88.4 + 2.8 = 91.2 dB.
Taking the second pair: Difference = 86.8 - 85.0 = 1.8 dB.
From Fig. 2-3, the amount to be added to the larger is 2.2 dB.
Again, in terms of *energy addition*, 86.8 + 85 ⇒ 86.8 + 2.2 = 89 dB.

Continuing:
81.2 + 76 ⇒ 81.2 + 1.3 = 82.5 dB and 75.8 + 68.9 ⇒ 75.8 + 0.7 = 76.5 dB.

Note that the corrected OBSPL from the 31-Hz center frequency will not contribute to the total energy in any significant amount (i.e., it is 76.5 - 57.6 = 18.9 dB below the next smallest value, and from Fig. 2-3 the contribution to the total level is too small to be concerned with).

Combining the remaining two pairs,
91.2 + 89.0 ⇒ 91.2 + 2 = 93.2 dB and 82.5 + 76.5 ⇒ 82.5 + 1 = 83.5 dB.

Finally, then 93.2 + 83.5 ⇒ 93.2 + .40 = 93.6 dB.
The octave-band data characterizing source II are equivalent to an A-weighted sound level of 93.6 dBA.

In a similar manner, the results for sources I and III are
I 103.9 dBA
III 97.8 dBA.
See Chapter 3 for another method of combining decibel values.

Determination of Loudness and Loudness Level

Octave Band Center Frequency (Hz)	Band Level (dB) I	II	III	Band Loudness Index (from Table 2-2) I	II	III
31.5	76	97	113	3.2	14.8	59.0
63.0	80	102	110	6.7	32.9	<u>61.0</u>
125	85	104	90	12.6	<u>48.0</u>	17.5
250	89	97	86	20.0	35.3	16.4
500	92	90	79	30.5	26.5	12.6
1000	98	85	86	61.0	23.0	24.7
2000	100	80	90	<u>90.0</u>	20.0	41.0
4000	95	75	95	77.0	17.5	77.0
8000	85	70	80	44.0	15.3	30.5
			ΣS =	345.0	233.3	339.7

Recalling Equation 2-2 and noting that $f = 0.30$,

$$S = 0.3 \sum S + 0.7\, S_{max} \tag{2-3}$$

(S_{max} = the maximum band loudness index for each source as underlined to facilitate understanding the calculation procedure).

I $S = (0.3)(345) + (0.7)(90) = 166.5$ sones
Loudness level = 113.8 phons

II $S = (0.3)(233.3) + (0.7)(48) = 103.6$ sones
Loudness level = 106.8 phons

III $S = (0.3)(339.7) + (0.7)(61) = 144.6$ sones
Loudness level = 111.7 phons

Determination of Perceived Noise and Perceived Noise Level

Octave Band Center Frequency (Hz)	Band level (dB) I	Band level (dB) II	Band level (dB) III	Perceived Noise (noys) (from Table 2-3) I	II	III
63	80	102	110	–	–	60.0
	86.2	106.1	110.1			
125	85	104	90	16.1	64.0	–
250	89	97	86	26.0	45.0	21.1
500	92	90	79	36.8	32.0	14.9
1000	98	85	86	56.0	22.6	24.3
2000	100	80	90	107.0	27.9	55.0
4000	95	75	95	94.0	24.3	94.0
8000	85	70	80	34.2	12.2	24.3
			$\Sigma N =$	370.1	228.0	293.6

Following Kryter's method, the levels in the 63-Hz and 125-Hz bands are combined, and the 125-Hz level is replaced by the combined level with the 63-Hz level ignored. If, however, the original 63-Hz level was greater than the 125-Hz level (see case II), replace the 63-Hz level by the combined level, and ignore the 125-Hz level.

$$N = 0.3 \sum N + 0.7\, N_{\max} \tag{2-4}$$

I $N = (0.3)(370.1) + (0.7)(107) = 185.9$ noys
Perceived noise level = 115.4 dB

II $N = (0.3)(228) + (0.7)(64) = 113.2$ noys
Perceived noise level = 108.3 dB

III $N = (0.3)(293.6) + (0.7)(94) = 153.9$ noys
Perceived noise level = 112.6 dB

In summary,

Source	A-Level (dBA)	Loudness (sones)	Loudness Level (phons)	Noisiness (noys)	Perceived Noise Level (PNdB)
I	103.9	166.5	113.8	185.9	115.4
II	93.6	103.6	106.8	113.2	108.3
III	97.8	144.6	111.7	153.9	112.6

Subjective Rating Methodologies Employing Perceived Noise

The advent of jet aircraft noise, with a complex spectrum in which pure tone components were possible, required a refinement of the PNdB concept. In addition, duration of the flyover was an additional parameter of importance. Thus, the effective perceived noise level (EPNL) in EPNdB was conceived. EPNL = PNL + C (tone correction) + D (duration correction) in EPNdB.

Adequate assessment of community response to aircraft noise is a problem that has occupied researchers for many years, and has resulted in planning tools being developed for use by airport personnel as new airports, or the expansion of existing facilities, are contemplated. The first such method is called the *composite noise rating* or *CNR*.[16] This methodology employs perceived noise level (PNdB) contours for the particular aircraft utilizing an airport in order to identify zones where the subjective response of individuals may be predicted. The basic PNdB contours (unique for each aircraft type) are modified or corrected to account for such factors as whether the flights are during the day (0700–2200 hours) or at night (2200–0700 hours) (10-dB penalty for night flights), percent runway utilization, and number of flights. The result is a series of CNR contours in PNdB superimposed upon the airport and surrounding land uses to identify those areas where the subjective response may be predicted. The most widely displayed set of contours for takeoffs and landings identify CNR Zone 1 (PNdB $\leqslant$ 100), CNR Zone 2 (100 $<$ PNdB $<$ 115), and CNR Zone 3 (PNdB $\geqslant$ 115). The contours may also be developed for ground run-up operations as seen in Fig. 2-5, which expresses an estimate of community response.[16]

In equation form:

$$\mathrm{CNR} = \overline{\mathrm{PNL}}_{\max} + 10 \log_{10} N_f - 12 \tag{2-5}$$

where $\overline{\mathrm{PNL}}_{\max}$ = arithmetic mean maximum perceived noise level (PNL) at a point where it is assumed that the range of $\mathrm{PNL}_{\max}$ is relatively small, and

$$N_f = (N_d' + 16.7\, N_n) = (15\, \overline{n}_d + 150\, \overline{n}_n) \tag{2-6}$$

Composite Noise Rating: Takeoffs and Landings	Composite Noise Rating: Run-ups	Zone	Description of Expected Response
<100	<80	1	Essentially no complaints would be expected. The noise may, however, interfere occasionally with certain activities of the residents.
100 to 115	80 to 95	2	Individuals may complain, perhaps vigorously. Concerted group action is possible.
>115	>95	3	Individual reactions would likely include repeated, vigorous complaints. Concerted group action might be expected.

Fig. 2-5 Chart for estimating response of residential communities from composite noise rating.

where N_d' and $\overline{n}_d$ = total number and average number per hour, respectively, of flights during the time period (0700–2200 hours), and N_n and $\overline{n}_n$ = total number and average number per hour, respectively, of flights during the time period (2200–0700 hours).

Recall that the complex spectrum of jet aircraft noise included pure tone components that were not adequately treated by an analysis involving only PNL, and hence EPNL was conceived. In the same spirit, a second method of forecasting community response to aircraft noise was developed utilizing the EPNdB descriptor. The *noise exposure forecast* (*NEF*)[17] is defined for a given runway and one or two dominant aircraft types as:

$$\text{NEF} = \overline{\text{EPNL}} + \log N_f - 88.0 \tag{2-7}$$

where $\overline{\text{EPNL}}$ = arithmetic mean of EPNL for each single event at the point in question. Here, it is assumed that the range of EPNL divided by N_f is less than about 1.5 EPNdB. When this is not the case, the NEF value is computed for each aircraft type separately and the total value is the result of a logarithmic sum.

The constant $-$ 88.0 dB includes an arbitrary -75 scale changing constant (to prevent confusion with CNR) and a reference number of daytime flights of 20. The 16.7 factor in N_f accounts for a 10-to-1 weighting factor for flights during the 9-hour night period.

The single event noise level defined by the preceding EPNL is specified approximately by the following relationship:

$$\text{EPNL} = \text{PNL}_{\text{max}} + \log \frac{t_{10}}{20} + F \text{ EPNdB} \tag{2-8}$$

where

PNL_{max} = maximum PNL during flyover in PNdB
t_{10} = 10 dB down duration of the PNL time history, in seconds
F = pure tone correction, typically + 3 dB.

Equivalent Sound Level

In the assessment of environmental noise exposure, it is necessary to develop measures of more general utility than those whose value is limited to aircraft noise. Toward this end, the EPA chose a form of A-weighted descriptor designed to satisfy the following constraints presented in the EPA "levels document."[7]

1. The measure should be applicable to the evaluation of pervasive long-term noise in various defined areas and under various conditions over long periods of time.
2. The measure should correlate well with known effects of the noise environment on the individual and the public.

3. The measure should be simple, practical and accurate. In principle, it should be useful for planning as well as for enforcement or monitoring purposes.
4. The required measurement equipment, with standardized characteristics, should be commercially available.
5. The measure should be closely related to existing methods currently in use.
6. The single measure of noise at a given location should be predictable, within an acceptable tolerance, from knowledge of the physical events producing the noise.
7. The measure should lend itself to small, simple monitors which can be left unattended in public areas for long periods of time.

These considerations, when coupled with the physical attributes of sound that influence human response, lead EPA to the conclusion that the magnitude of sound is of most importance insofar as cumulative noise effects are concerned. Long-term average sound level, henceforth referred to as equivalent sound level (L_{eq}), is considered the best measure for the magnitude of environmental noise to fulfill the above seven requirements.

The *equivalent sound level* (L_{eq}) is the *constant* level that, over a given time period, transmits to the receiver the same amount of acoustic energy as the actual time-varying sound. (Note that equivalent level is a concept that may be applied to all weighted levels or sound pressure level—however, as used by the EPA, the A-weighted level is employed exclusively.) The unit of L_{eq} is the decibel, and inherent in its definition is the "equal energy rule." For example, two sounds, one of which contains twice as much energy but lasts only one-half as long would have the same L_{eq}. While NEF and CNR cannot be obtained *directly* by measurement, it is possible to determine L_{eq} with hand-held instrumentation (see Chapter 3).

An extension of this equivalent-noise-level concept has been utilized to more fully express the concept of annoyance since nighttime noise intrusions are inherently more disruptive than those in the daytime. When a 10-dB penalty is applied to the 9 nighttime hours (2200–0700 hours), the 24-hour equivalent sound level L_{eq} (24) becomes the average day-night sound level L_{dn}. Because of a logarithmic relationship, the 10-dB penalty applied to nighttime noise intrusions, generally significantly affects L_{dn}.

Because L_{dn} is at times used as a measure of exposure to aircraft noise, a comparison of L_{dn} with both CNR and NEF may prove useful. The following *approximate* relationships are acceptable for aircraft flyover noise with a tolerance of ±3 dB:

$$L_{dn} \doteq \text{NEF} + 35 = \text{CNR} - 35 \tag{2-9}$$

Care should be exercised in such comparisons, however, because of the inherent differences in A-weighted sound level, PNL, and EPNL. In general, the PNL in

PNdB exceeds the A-weighted sound level by about 13 dB with a range of typically 11-17 dB,[18] the amount depending primarily upon the amount of correction for pure tones. For aircraft, the 13-dB correction is considered acceptable.

SUMMARY

The purpose of this chapter has been to acquaint the reader with some of the basic measures of sound, how these measures are determined (instrumentation and measurement), and how the measurements may be used to compute other measures with more significant subjective response. In addition, analytical predictions based upon flight data were identified insofar as the average community response to composite aircraft noise levels is known. The treatment has been necessarily cursory in all of these areas, and the interested reader is encouraged to delve into the existing literature for greater depth of understanding.

REFERENCES

1. American National Standards Institute, S1.4-1971, Specifications for Sound Level Meters.
2. Hruska, G. R. et al., Environmental Effects on Various Microphone Constructions, NBSIR 76-1090 (EPA 550/9-76-009), July 1976.
3. Bruel, P. V. and Rasmussen, G., Free field response of condenser microphones, *Bruel and Kjaer Technical Review*, Nos. 1 and 2, 1959.
4. Schneider, A., Microphone orientation in the sound field, *Sound and Vibration*, February 1970.
5. Skøde, F., Windscreening of outdoor microphones, *Bruel and Kjaer Technical Review*, No. 1, 1966.
6. *Instructions and Applications–Model 2209 Sound Level Meter*, Bruel and Kjaer Instruments, June 1972.
7. U.S. Environmental Protection Agency, *Information on Levels of Environmental Noise Requisite to Protect Public Health and Welfare with an Adequate Margin of Safety*, EPA 550/9-74-004, March 1974.
8. American National Standards Institute, S1.6-1967, Preferred Frequencies and Band Numbers for Acoustical Measurements.
9. Harris, C. M., *Handbook of Noise Control*, Chapter 17, McGraw-Hill Book Company, New York, 1957.
10. Brock, J. T., *Acoustic Noise Measurements*, Second Edition, Bruel and Kjaer Instruments, January 1971.

11. Fletcher, H. and Munson, W. A., Loudness, its definition, measurement and calculation, *Journal of Acoustical Society of America*, **5** (2): 82–108 (October 1933).

12. U.S. Dept. of Commerce, *Fundamentals of Noise: Measurement, Rating Schemes, and Standards*, NTID 300.15, December 31, 1974.

13. Stevens, S. S., Procedure for calculating loudness: Mark VI, *Journal of Acoustical Society of America* **33** (11): 1577–1585 (November 1961).

14. Peterson, A. P. G. and Gross, E. E., *Handbook of Noise Measurement*, 6th Edition, General Radio Company, West Concord, Massachusetts, 1967.

15. Kryter, K. D., *The Effects of Noise on Man*, Academic Press, Inc., New York, 1970.

16. Galloway, W. J. and Pietrasanta, A. C., Land Use Planning Relating to Aircraft Noise, Technical Report No. 821, Bolt, Beranek, and Newman, Inc., published by FAA, October 1964.

17. Galloway, W. J. et al., *Noise Exposure Forecasts: Evolution, Extensions, and Land Use Interpretations*, FAA-NO-70-9, August 1970.

18. U.S. Environmental Protection Agency, *Public Health and Welfare Criteria for Noise*, EPA 550/9-73-002, July 27, 1973.

3

The use and measurement of equivalent sound level

Arthur C. Taylor, Jr., Ph.D.
Head, Department of Mechanical Engineering
Virginia Military Institute
Lexington, Virginia

and

David M. Lipscomb, Ph.D.
Professor, Department of Audiology and Speech Pathology
University of Tennessee
Knoxville, Tennessee

INTRODUCTION

Noise is a word used to describe sound conditions in certain instances. The most common operational definition of noise is that it is *unwanted sound*. Although numerous attempts have been made to quantify and delineate those sounds that constitute noise, there is an element of subjectivity that precludes an all-encompassing description of noise other than in the preceding operational definition.

Sound itself is a definable and measurable physical parameter (see Chapters 1 and 2). Basically, it is comprised of pressure waves radiating from a vibrating body. These waves move through the air or other propagating media as oscillatory (back and forth) motion with relative high- and low-pressure phases. The rapidity of the pressure oscillations is called the *frequency* of the sound, which corresponds to the psychological perception of pitch. Thus, sound pressure waves that oscillate slowly are perceived to be lower in pitch than those that oscillate more rapidly. The second major component of sound is its *amplitude*, which is perceived as loudness. A quiet sound source radiates pressure waves that are small in magnitude because the pressure the sound exerts on the ear is not great. Intense sound, however, causes considerably more pressure against the ear tissues since the energy of the pressure wave is relatively large.

A third dimension of sound is its *duration*. The interaction of frequency, amplitude, and duration of several sounds results in sound complexity. The complex interaction of numerous sound waves containing several frequencies at widely ranging amplitudes imparts a rising and falling nature to sound levels. At certain moments, many sound sources combine to cause relatively high amplitude. At other moments, the sound sources are sporadic in their generation of sound waves; thus, their transience is imparted to the total sound in the environment. Therefore, sound is both variable and transient. Due to this complex series of generally unpredictable interactive events, sound levels in the environment are constantly changing. Only in rare locations is environmental sound at a constant level for long periods of time. Thus, the physical nature of sound is quite complicated and is summarized here only briefly to foster an appreciation of the difficulty that is inherent in considering sound quality for the purpose of generating and enforcing noise control regulations.

COMPARISON OF NOISE AND OTHER POLLUTANTS

Noise in the environment is a unique situation that makes it difficult to adequately compare noise with other environmental contaminants. Although it is tempting to consider analogies to water, air, or solid waste problems, noise should be considered a totally separate entity. In Table 3-1, some differences between noise and air pollution are noted.

It is possible to purify air and water, and all solid waste can be removed from the landscape. Sound, however, can only be eliminated by creating a vacuum—a condition in which humans cannot exist. Sound must be a part of the environment. In fact, too little sound is considered by many to be as undesirable as too much sound. Therefore, in the consideration of noise regulations, one must necessarily incorporate the uniqueness of sound itself as attempts are made to control excessive sound in the environment. Since noise must be treated as a separate and distinct environmental problem, it can only be controlled by regulations specific to noise problems with special enforcement approaches geared to the unique conditions caused by noise in the environment.

TABLE 3-1 Comparison of Noise and Air Pollution.

	Air Pollution	Noise
Characteristics	Composed of potentially toxic chemicals which are both cumulative and residual. Subject to meteorological dispersion/ diffusion laws.	Made up of physical energy with wide variation in brief periods of time. The sound energy is additive. Sound decays over distance and ceases as soon as the sound source is stilled.
Measurement in the Environment	Time-concentration dose levels are easily assessed since variation does not occur rapidly.	Rapid changes in sound level preclude time-concentration assessment except with the use of cumulative measures, which are then related to an equivalent steady-state sound energy.
Measurement in the Human Body	Presence of toxicants in the body can be directly related to measurable reference concentrations and is often immediately identified by overt physiological reaction (e.g., respiratory attacks during high ambient chemical concentrations).	Observed in some instances as hearing impairment. No other readily identifiable physiological changes directly attributable to noise exposure. Annoyance cannot be correlated with noise levels alone since human reaction to noise varies greatly within and between persons.
Geographical Control	Air quality regions have been established to monitor and control ambient air pollutants traveling great distances and crossing geographical and political boundaries.	Noise travels only a relatively short distance from its source and, in most cases, within existing political and geographical entities.

Sound can be reliably measured and quantified in the physical sense by electronic equipment that is presently available. However, the variability among individuals in reaction to sound creates an immense additional problem. Also, individual variation on a day-to-day basis is an equally imposing factor. Because of attitude, emotional conditions, hearing ability, and other mental and physical health factors, a particular sound that is ignored some days will be very bothersome on other days. This, of course, is not due to the physics of sound; rather, it is a factor of human response to sound.

A general understanding of the physical parameters of sound, such as that given in Chapter 1, is requisite to setting noise control regulations. Although human reaction to sound in the environment is highly variable, an appreciation of the human factor will also contribute importantly to the preparation of noise control regulations.

FOUNDATION DOCUMENTS

The U.S. Environmental Protection Agency (EPA) has sought to collect information on the effect of noise on people for the use of federal, state, and local officials. To this end, and at the behest of Congress as specified by the Noise Control Act of 1972,[1] the EPA published two important background documents. In July 1973, a document entitled *Public Health and Welfare Criteria for Noise* (*550/9-73-002*)[2] was published. The purpose of this document (herein called the "criteria document") was to summarize present knowledge of the various effects of noise. Later, in March 1974, the EPA published *Information on Levels of Environmental Noise Requisite to Protect Public Health and Welfare with an Adequate Margin of Safety* (*550/9-74-004*).[3] This latter document, referred to as the "levels document," sets forth those levels of sound in the environment that, according to knowledge at that time, are necessary if all the population is to be protected from any deleterious effects of noise exposure. A detailed rationale based on health and welfare was developed without any consideration of cost and technical feasibility.

Both documents, particularly the latter one, contain a sophisticated treatment of the subject of the effects of sound on health and welfare. Although this technical backup is vital to the establishment of requisite levels of sound in the environment for the protection of public health and welfare, it contains numerous concepts that are highly technical and sometimes difficult to understand by persons who are in the position of generating laws and regulations for the control of noise.

One purpose of this chapter is to interpret in understandable terms those features of the levels document that provide an important foundation for the formulation of noise regulations. Every attempt has been made to simplify the concepts in the levels document while maintaining the accuracy of the information and guidelines. It is anticipated that this effort will make the levels document usable and comprehensible to interested agencies as they undertake to generate meaningful and enforceable rules, regulations, and ordinances for the control of unnecessary and disturbing sounds in the environment.

In addition to interpreting the previous information promulgated by the EPA, this chapter will set forth basic techniques in sound measurement specific to certain sound descriptors in order to assist regulating agencies in their quest to select and enforce the proper levels in their noise ordinances. In this sense, some of the material of this chapter will be an expansion of the *equivalent sound level* introduced in Chapter 2.

In providing information on the levels of noise requisite to protect public health and welfare with an adequate margin of safety, the levels document was prepared without consideration of two very important ingredients–technological feasibility and economic impact. Therefore, the levels in the document should not be simply enacted into law by regulating agencies; first, technical feasibility and economics should be considered. Thus, the distinction must be made between the all-inclusive levels found in the levels document and the realistic (enforceable) levels to be used for the purpose of noise control regulations.

In the levels document, the equivalent sound level descriptors, L_{eq} and L_{dn}, were selected as the measure for cumulative noise analysis. The ultimate range of applications for these descriptors are yet to be determined. They appear to have promise for projections in land use planning and certain community noise assessments, to mention two major areas. These descriptors were formally proposed for use by all federal agencies in a position statement in the summer of 1975 by Russell Train, Administrator of EPA. For discrete, or short-term, measurements not subject to cumulative use, A-weighted sound measures were recommended.

In recent years, cumulative sound measures have increased dramatically (see Chapters 2 and 9). An underlying purpose of this chapter is the recognized need for uniformity of methodology for measuring and assessing cumulative community environmental noise. To this end, and in agreement with the rationale set forth in the levels document, it is suggested that the equivalent sound level, L_{eq}, be used for all cumulative measures of sound in the environment. For discrete event measures, it is suggested that regulations utilizing maximum D- or A-weighted sound levels generated by sound sources be used.

By narrowing the choices of sound-evaluating methodology, it is anticipated that the confusion surrounding the myriad measurement and analytic schemata will be abated to a significant degree. Thus, there are two basic sets of sound measurement protocol advanced in this chapter: (1) cumulative measures based on the L_{eq} concept and (2) discrete event measures based on the use of maximum D- or A-weighted sound levels.

THE EQUIVALENT SOUND LEVEL (L_{eq})

The magnitude of environmental sound almost always varies with time. Over the years many single-number measures for time-varying sound levels have been developed. Some of these measures include: noise exposure forecast (NEF), composite noise rating (CNR), community noise equivalent level (CENL), the level exceeded 90, 50, or 10 percent of the time (L_{90}, L_{50}, L_{10}), and equivalent sound level (L_{eq}).

The Noise Control Act of 1972 (Section 4[c]) names the EPA as the coordinating agency for all federal agencies involved in noise research, control, and

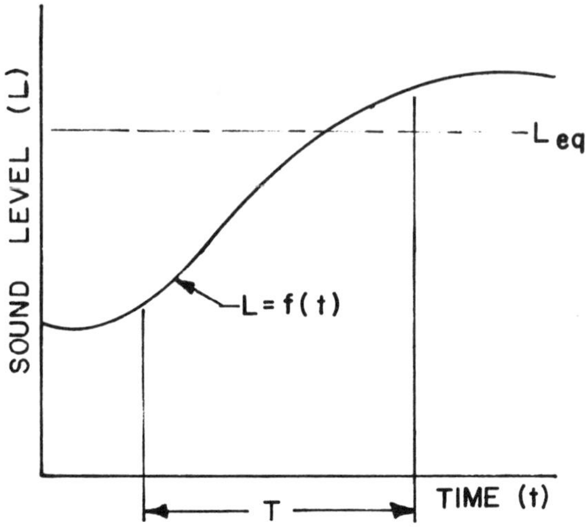

Fig. 3-1 A continuous variation of sound level with time.

regulation. To foster uniformity and simplicity of measurement and monitoring, the equivalent sound level concept has been advanced to supersede other cumulative measures.

Suppose the A-weighted sound level (L) varies with time (t) as shown in Fig. 3-1. The equivalent sound level for the interval of time (T) is defined as:

$$L_{eq} = 10 \log_{10}\left(\frac{1}{T}\int_0^T 10^{L/10}dt\right). \tag{3-1}$$

Although this equation seems formidable, it is a necessary way to describe L_{eq} when the level varies as a known function of time. The integral sign (or the summation symbol) and the division by T indicate that it is some kind of an average. However, it is not an arithmetical average because logarithms are involved in the calculation. The equation does, in fact, produce a level that has the same energy as the time-varying level (L). If L is constant, then $10^{L/10}$ comes outside the integral and the equation becomes:

$$L_{eq} = 10 \log_{10} 10^{L/10} \frac{T}{T} = L$$

This is the expected result because when the sound level is constant, the level that produces the same energy is the identical constant level.

Often the variation of sound level is not given as a function of the time but instead is presented as shown in the histogram in Fig. 3-2. In this figure, the level is plotted as the abscissa (horizontal axis) instead of the ordinate (vertical

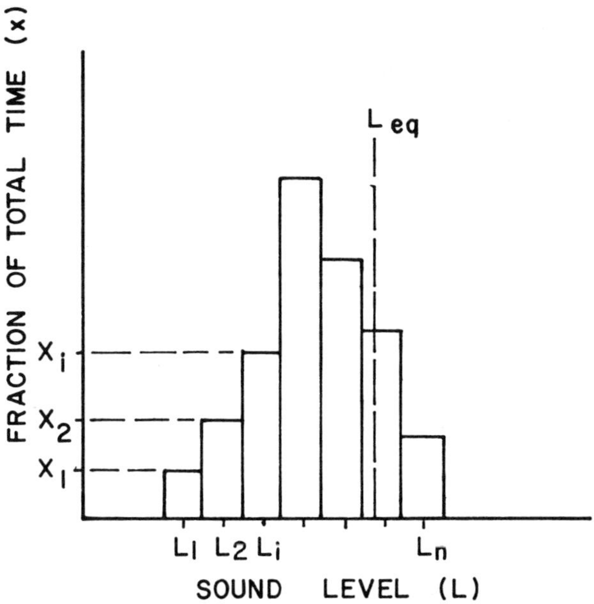

Fig. 3-2 A variation of sound level with discrete intervals.

axis) as in Fig. 3-1. Discrete levels, L_1, L_2, L_3, up to L_n, are given with L_i the symbol for any of the levels. In this figure, x is plotted as the ordinate and is the fraction of time that a particular sound level occurs. This fraction is the duration sound stays at that level divided by the total time. The symbol x_i is the fraction of time at the ith level. It should be clear that the sum of all x's should equal unity (one) or:

$$\sum_{i=1}^{i=n} x_i = 1.$$

Because of the discrete levels in place of a continuous function, Equation 3-1 becomes:

$$L_{eq} = 10 \log_{10}\left(\sum_{i=1}^{i=n} x_i \; 10^{L_i/10} \right). \tag{3-2}$$

In the case of a single constant sound level, L_i, which occurs for x_i fraction of the total time and for which the rest of the time the sound level is negligible (zero background level), Equation 3-2 becomes:

$$L_{eq} = L_i + 10 \log_{10} x_i \tag{3-3}$$

If, on the other hand, the background level is not zero, or if other sounds occur for other fractions of the time, then this must be considered a partial L_{eq}

to be added to the other partial L_{eq} values by a method designed for adding decibel levels. (This is discussed later as Fig. 3-4.)

As a hypothetical example, if a single sound of 120 dBA lasts for a 1-second interval in a 24-hour day with no significant sound for the rest of the day, then x becomes 1/24(3600) since there are 3600 seconds in an hour. Substitution in Equation 3-3 gives:

$$L_{eq} = 120 + 10 \log_{10} \frac{1}{24(3600)}$$

$$L_{eq} = 120 - 10 \log_{10} (24[3600])$$

$$L_{eq} = 120 - 10\,(4.941) = 70.6 \text{ dB.}$$

This is the L_{eq} only if the 1-second sound is the only sound within 20 dB of 70 dB in the 24-hour period.

THE IDENTIFIED LEVELS

The yearly average levels given in the levels document were identified to protect the public health and welfare with an adequate margin of safety. Economic considerations and technical feasibility were not considered in the selection of these levels. Therefore, they should not necessarily be translated directly into regulations. The identified levels are given in Table 3-2. Note that these figures are "yearly averages." In view of that qualification, it should be understood that the identified levels were not intended to be taken directly from the levels document and made into law. The enforcement of "yearly average" sound levels would be technically difficult and administratively unsound. Also note the further qualification that the hearing-loss levels in the table are not only yearly averages, but they are to be considered over a 40-year period of persistent exposure.

In the table, $L_{eq}(24)$ refers to the L_{eq} for a 24-hour time interval and $L_{eq}(8)$ is for an 8-hour interval. L_{dn} is $L_{eq}(24)$ with a 10-dB penalty added to the sound levels that occur during the night, which extends from 2200 to 0700 hours. In the example that was given with L_{eq} = 70.6 dB, this is $L_{eq}(24)$ because the time period was 24 hours. If the 1-second 120-dBA sound occurred in the daytime, L_{dn} is also 70.6 dB; but, if the sound occurred at night, L_{dn} would be 80.6 dBA because of the 10-dB penalty for nighttime sound. It should be noted that the L_{dn} descriptor should not be used to assess sound levels with respect to their hazardous qualities for hearing health. The 10-dB penalty for nighttime sounds represents an adjustment for intrusiveness and would have no bearing on the damaging effect of a given sound. Therefore, damage-risk considerations can be assessed by using the L_{eq} descriptor, not L_{dn}.

TABLE 3-2 Yearly Average* Equivalent Sound Levels Identified as Requisite to Protect the Public Health and Welfare with an Adequate Margin of Safety.[3]

		Indoor		To Protect Against Both Effects[a]	Outdoor		To Protect Against Both Effects[a]
	Measure	Activity Interference	Hearing-Loss Consideration		Activity Interference	Hearing-Loss Consideration	
Residential with outside space and farm residences	L_{dn}	45		45	55		55
	L_{eq} (24)		70			70	
Residential with no outside space	L_{dn}	45		45			
	L_{eq} (24)		70				
Commercial	L_{eq} (24)	b	70	70[c]	b	70	70[c]
Inside transportation	L_{eq} (24)	b	70	b			
Industrial	L_{eq} (24)[d]	b	70	70[c]	b	70	70[c]
Hospitals	L_{dn}	45		45	55		55
	L_{eq} (24)		70			70	
Educational	L_{eq} (24)	45		45	55		55
	L_{eq} (24)[d]		70			70	
Recreational areas	L_{eq} (24)	b	70	70[c]	b	70	70[c]
Farm land and general unpopulated land	L_{eq} (24)				b	70	70[c]

*Refers to energy rather than arithmetic averages.
[a]Based on lowest level.
[b]Since different types of activities appear to be associated with different levels, identification of a maximum level for activity interference may be difficult except in those circumstances where speech communication is a critical activity.
[c]Based only on hearing loss.
[d]An L_{eq} (8) of 75 dB may be identified in these situations as long as the exposure over the remaining 16 hours per day is low enough to result in a negligible contribution to the 24-hour average, i.e., no greater than an L_{eq} of 60 dB.
Note: Explanation of identified level for hearing loss: The exposure period that results in hearing loss at the identified level is a period of 40 years.

PRIMER ON METHODS FOR MEASURING L_{eq} AND L_{dn}

Since the equivalent sound level, L_{eq}, has been selected by the EPA as the method of choice for the cumulative measurement of sound in the environment, it is logical to predict that this will encourage the design and manufacture of numerous sound-measuring devices that register the L_{eq} measure. With the increase in numbers of models and designs of such sound-analyzing units, the per-unit cost will undoubtedly be lower than is presently the case with the small number of commercial units available.

There is need at present for obtaining L_{eq} and L_{dn} values simply and economically. Methods cited in this section are intended to answer that need. This section is organized so that the progression of subject matter moves to more sophisticated equipment and techniques toward the end. No particular method will be suggested as being more desirable than another, for each one discussed herein has its appropriate applications. It should be emphasized, however, that the simpler the approach the better; thus, it is likely that the simplified method discussed first might be the most commonly used because of its simplicity and the use of less expensive sound-measuring equipment. Further, these methods are not without their limitations, which will be pointed out. The limitations must be considered before deciding if a given cumulative method is appropriate.

The subsequent discussion involves the methodology used to determine L_{eq}. Should one desire to measure L_{dn}, 10 dBA is added to each measure taken between the hours of 2200 and 0700 as a nighttime penalty. Aside from the 10-dBA penalty, the method of measurement and calculation of L_{eq} applies for L_{dn}.

Simplified Measurement of L_{eq} Using Minimum Equipment

Equations 3.2 and 3.3 have been given for the calculation of L_{eq}. Accordingly, two simplified methods will be given, each corresponding to one of the equations.

Option 1: Use of Equation 3-3. Measuring L_{eq} or L_{dn} using Equation 3-3 requires only a sound level meter and necessitates a small number of calculations, which are facilitated by the use of nomograms. An advantage of this technique is that one person can collect the data with a sound level meter mounted on a tripod, usually with a windscreen on the microphone. The procedure is as follows:

1. Be certain that the meter is calibrated.
2. Begin taking measures by turning on the sound level meter and adjusting its response characteristics to the A-weighting scale. Set the meter response on "fast" setting.
3. Data are recorded on a data form (such as that shown in Fig. 3-3) by noting the decibel reading on the meter every 5 or 10 seconds and making a tally mark adjacent to the appropriate decibel level on the form. The selection of sampling interval is an option, which can vary according to the sound source. In general, if there is little variation in sound level at the point of measurement, longer intervals (e.g., 10 seconds) can be accommodated. In the event the sound level changes rapidly, the accuracy of the measure is increased by shortening the sampling interval to 5 seconds. With some unusually rapid varying sound sources, the 5- or 10-second sampling interval will not suffice, and equipment providing more rapid sampling is required.

4. Terminate the measurement period after the desired number of samples are taken. As few as 100 samples can give quite accurate indications of the L_{eq} value in many cases. With increased sampling, however, the accuracy of the approximation is increased. The amount by which accuracy improves will vary according to the type of noise source and the application for which the measures are intended. Therefore, enforcement should be based on knowledge of the number of samples necessary to provide an accurate indication of the sound environment in question. This knowledge can be obtained by taking as many as 400 samples in the field, and then analyzing the first 100 samples, then adding the next 100, and so on, until the calculations indicate that the data have stabilized. This way, enforcement personnel will know how many samples must be taken for monitoring a given sound source or environmental region. The upper limit of samples is the endurance of the person taking the measures. It is tiring to read and mark the form every few seconds; thus, this factor should be taken into consideration.

5. In following this method, one accumulates the number of times that the sound level was found to be in any of a range of discrete intervals. Since the total number of samples is known, the fraction of the total time for each sound level is easily computed. These calculations can then be used to construct a histogram such as the one in Fig. 3-2. With these data, L_{eq} can be calculated by using either Equation 3.2 or 3.3 (see previous section).

6. Two nomograms are given in Figs. 3-4 and 3-5. The nomogram in Fig. 3-4 enables one to quickly obtain ($10 \log_{10} x_i$) when x_i (the fraction of time the sound is at a particular sound level) is known. This quantity is added to L_i (the sound level in question) to get the partial L_{eq}. Partial L_{eq}'s are obtained for each level in the measurement scheme.

7. Figure 3-5 is used to facilitate addition of the partial L_{eq} values. (Note: these values cannot be simply summed arithmetically.) This nomogram is based on the fact that when two simultaneous sounds of identical sound level are in the same vicinity, they combine to raise the overall sound level 3 dB higher than either sound level of the single unit. As the difference increases between the sound level of two sound sources, the combination effect is less until the increase to overall sound level is only about 0.1 dB when a second sound is 16 dB lower than the primary sound source (Note: this nomogram is based on the same calculations as in Fig. 1-14.)

The foregoing discussion can be clarified with the use of an example. In the sample data form in Fig. 3-3, the results of measurements taken on a busy city street are shown. Measures were taken every 5 seconds until a total of 200 readings were obtained over a period of approximately 17 minutes. Column 1 lists the decibel values with 1-dB increments. Column 2 contains the tally marks made every 5 seconds. After column 2 has been completed in the field, the

(1) Sound Pressure Level–dBA	(2) Field Measurements Counts–5 sec. intervals	(3) Counts per dB	(4) Midpoint of SPL interval L_i (dBA)	(5) Counts per Interval	(6) Fraction of total time (x_i)	(7) 10 log x	(8) Partial L_{eq}
100							
99							
98	|	1	97.5	6	0.03	$\frac{6}{200} = -15.0$	82.5
97	||	2					
96							
95	|||	3					
94	𝍸	5					
93	|||	3					
92	|	1	92.5	22	0.11	$\frac{22}{200} = -9.5$	83.0
91	𝍸	5					
90	𝍸 |||	8					
89	𝍸 𝍸 𝍸 |||	18					
88	𝍸 𝍸 𝍸 𝍸 ||||	24					
87	𝍸 𝍸 𝍸 𝍸 𝍸 |	26	87.5	106	0.53	$\frac{106}{200} = -2.7$	84.8
86	𝍸 𝍸 𝍸 𝍸 |	21					
85	𝍸 𝍸 𝍸 ||	17					

<table>
<tr><td>84</td><td>𝍸𝍸𝍸𝍸 ||</td><td>22</td><td rowspan="5">82.5</td><td rowspan="5">64</td><td rowspan="5">0.32</td><td rowspan="5">$\frac{64}{200} = -4.8$</td><td rowspan="5">77.7</td></tr>
<tr><td>83</td><td>𝍸𝍸𝍸𝍸</td><td>20</td></tr>
<tr><td>82</td><td>𝍸𝍸 ||</td><td>12</td></tr>
<tr><td>81</td><td>𝍸 |||</td><td>8</td></tr>
<tr><td>80</td><td>||</td><td>2</td></tr>
<tr><td>79</td><td></td><td></td><td rowspan="5">77.5</td><td rowspan="5">2</td><td rowspan="5">0.01</td><td rowspan="5">$\frac{2}{200} = -20.0$</td><td rowspan="5">57.5</td></tr>
<tr><td>78</td><td>|</td><td>1</td></tr>
<tr><td>77</td><td>|</td><td>1</td></tr>
<tr><td>76</td><td></td><td></td></tr>
<tr><td>75</td><td></td><td></td></tr>
</table>

Total Counts = 200

Fig. 3-3 Typical form for data collection and partial L_{eq} calculation.

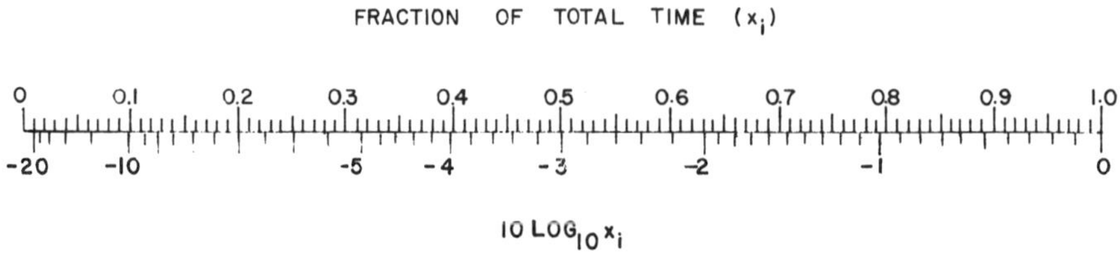

Fig. 3-4 Nomogram for the determination of 10 $\log_{10} x_i$.

remaining calculations can be completed later. The step-by-step calculation procedures are as follows:

1. Sum the tally marks for each discrete sound level, and enter the result in column 3.
2. Decide on the size of the sound level intervals to be used in the calculation. In this example, 5-dB intervals were chosen: 75–79, 80–84, 85-89, and so on. (If 1-dB intervals had been used, the final L_{eq} value will change by only a fraction of a decibel.) The midpoints of the intervals (L_i) are listed in column 4.
3. In column 5, the counts (tally marks) per total interval are summed.
4. Divide each number in column 5 by the total number of observations (in this example, there were 200). Place these fractions of the total time (x_i) in column 6.
5. Use the nomogram in Fig. 3-4, and place the numbers in column 7. The numbers are obtained by finding x_i (information in column 6) from the top of the nomogram and then reading the corresponding value from the bottom of the nomogram.
6. Calculate partial L_{eq} values by adding each entry in column 7 to its counterpart in column 4. Place this sum in column 8.
7. Add the partial L_{eq} values by using the nomogram in Fig. 3-5. This procedure is illustrated in detail in Fig. 3-6. The partial L_{eq} values are summed by combining two partial L_{eq} values, then by adding a third to the result of summing the first two. Next, add the fourth partial L_{eq} to the aggregate total of the first three, and so on. In the example, the summation process yields an L_{eq} value of 88.8 dB. This can be interpreted to mean that the sound during approxi-

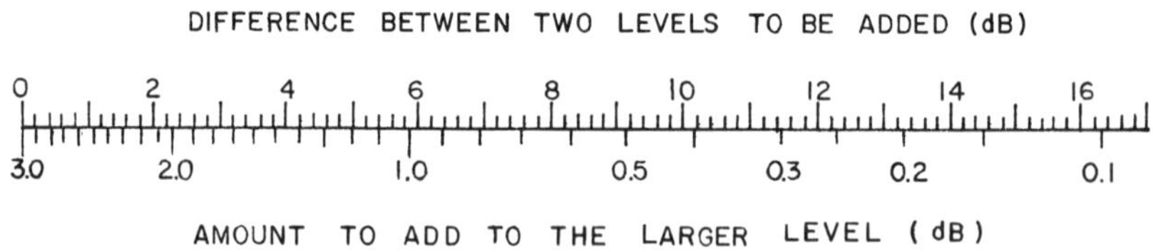

Fig. 3-5 Nomogram for addition of partial L_{eq}'s.

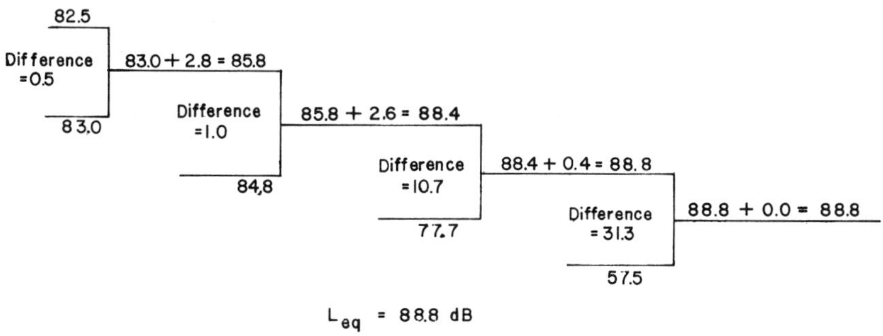

Fig. 3-6 Addition of partial L_{eq}'s.

mately 17 minutes of sound analysis delivered energy to the ear which would be equivalent to a steady-state sound of 88.8 dBA. An appropriate shortcut is to sum only those partial L_{eq} values within 20 dB of the highest one since the rest are negligible in their effect of the final L_{eq} value.

Option 2: Use of Equation 3-2. An alternate method of obtaining L_{eq} values is exemplified in Table 3-3 using Equation 3-2. Procedures for this method according to the example data are:

1. List the midpoints of the level intervals (L_i) and place in column 1. These are obtained in the same way as discussed for column 4 of Fig. 3-3.
2. Compute the fraction of total time (x_i) the sound level was in each of the discrete intervals. This too, appears in Fig. 3-3 (column 6).
3. Compute $10^{L_i/10}$, and place it in column 3. This is the number that has a logarithm of $L_i/10$ (antilogarithm of $L_i/10$). For example, for the first interval, $97.5 \div 10 = 9.75$. This is the logarithm of 0.563×10^{10}.
4. Calculate the product of x_i times $10^{L_i/10}$ for each interval, and place it in column 4.

TABLE 3-3 Calculation of L_{eq} by Equation 3-2.

(1) Sound Level L_i (dbA)	(2) Fraction of Total Time – x_i	(3) $10^{L_i/10}$	(4) $x_i\,10^{L_i/10}$
97.5	0.03	$.563 \times 10^{10}$	$.169 \times 10^9$
92.5	0.11	$.178 \times 10^{10}$	$.196 \times 10^9$
87.5	0.53	$.563 \times 10^9$	$.298 \times 10^9$
82.5	0.32	$.178 \times 10^9$	$.057 \times 10^9$
77.5	0.01	$.563 \times 10^8$	$.000 \times 10^9$

$L_{eq} = 10 \log_{10} 0.720 \times 10^9$ $\qquad \Sigma\, x_i\, 10^{L_i/10} = 0.720 \times 10^9$

$L_{eq} = 10\,(8.856) = 88.56$ dB or 88.6 dB

5. Sum the fractions in column 4.

6. Calculate L_{eq} by finding 10 times the logarithm of the sum of column 4. The L_{eq} in the example is 88.6 dB. This L_{eq} value compares closely to the one obtained using the first optional method. (Note: L_{eq} = 88.1 dB, when 1-dB intervals are used.)

In recent years, so-called electronic slide rules have been developed. These small calculators are particularly useful in obtaining partial L_{eq} values and for combining the partial L_{eq} values in order to obtain the total L_{eq} for a particular acoustic situation. The calculator can be used for computing L_{eq} using either Equation 3-2 or 3-3.

Tape Recordings with Laboratory Analysis

Field recordings of noise can be made using magnetic tape-recording equipment. The tapes are returned to a laboratory for analysis and calculation of L_{eq} values. Numerous laboratories are presently equipped to perform this analysis, and one facility could serve as a processing center for a sizable geographic region.

Many types of recorders are available that are suitable for fieldwork. The most important specification for recorders is that the frequency response be adequate to record the sound in the environment with a minimum of frequency distortion. Additionally, battery life for portable recorders should be sufficient to cover the time periods the unit is in use. Other specifications, such as wow and flutter due to variations in tape speed, are not of major consequence in this application since the desired measure is amplitude of the sound source. Wow and flutter would have to be very extreme to cause errors in amplitude recordings. The automatic gain control circuit *must* be deactivated for accurate level recordings.

Field recordings are primarily dependent on two factors: (1) the microphone must be of good quality and durable enough to withstand the elements (a windscreen is desirable to reduce the contribution of wind noise), and (2) calibration of the tape recorder is essential. As sounds are recorded in the environment, all levels are relative to each other. A calibration sound should be at the head of the tape to provide a reference level to which all other analysis of the tape is related. This is accomplished by creating a steady sound of known amplitude, with the recorder microphone and a sound level meter microphone close together. The level of the calibration sound is then marked on the tape box.

Instrumentation for laboratory analysis varies according to the facility. In some instances, a sound-analyzing system is used to provide a direct readout of the amount of time the sound was in any of several 5-dB ranges. Calculation of L_{eq} can then proceed as outlined in the optional methods earlier. Other laboratories have computerized sound analytic equipment that automatically calculates the L_{eq} values after the data are fed from the tape recorder to the computer.

It is advisable to explore the possibility of establishing sound-analysis centers that can serve an entire region. In this case, there is a saving on the total cost of monitoring environmental noise in a given region.

Short-Term Automatic Monitoring Instruments

Instruments in this class are those for which the measuring-time capability is less than 24 hours. Typically, the operator stays with the instrument as would be the case with a hand-held sound level meter. At the end of the allotted monitoring time period, the instrument gives a direct readout in L_{eq}.

At this writing, the selection of such devices is limited, but it is anticipated that a number of short-term monitoring instruments will be marketed in the future.

Inherent in the design of such equipment is the limitation of the duration of measurement. In addition, L_{dn} could be determined only by monitoring daytime or nighttime levels and completing the summarized L_{dn} calculations as cited in the simplified method after the addition of 10 dB to all measures taken between 2200 and 0700 hours.

The original cost of automatic monitoring devices will be higher than simple sound level meters, but the saving of staff time in calculating L_{eq} might well balance the cost over the long run.

THE SPECIAL CASE OF IMPULSE (IMPACT) NOISE

The major characteristics of impulse noise are: (1) the extremely fast rise time (time consumed in reaching maximum sound pressure level), and (2) the high peak levels attained for brief moments (less than 0.5 second). Typical impulse sounds, such as gunshots, reach peak levels of from 140 to 170 dB. Fortunately, the sound remains at the peak for a very brief time (only a few microseconds) before decaying to safe levels. If sound of the magnitude of impulse sound were maintained for one full second, it is certain that damage would occur in the ear. As an example, a sound peaking at 165 dB and lasting for 1 second delivers the same amount of energy to the ear as a 115-dB sound experienced continually for 24 hours (L_{eq} = 115). (Note that impulse sound is usually measured using linear meter response; therefore, the levels (in dB) and the L_{eq} values are not A-weighted.) Since impulse sounds do not linger at their peak for long periods, most damage-risk criteria (predictors of hearing hazard) for impulse sounds are based on the repetition of exposures to impulse sounds. If, for example, one gunshot peaking at 165 dB has a duration of peak level for only 10 microseconds, this would be equivalent to a 24-hour level of about 65 dB. If this sound is repeated often, the peak durations are accumulated so that 10,000 repetitions of the exposure would accumulate to a duration of combined discrete events totaling 0.1 second. This would be an L_{eq} value of 105 dB. In the example, the gun would have to be fired 100,000 times in one day to achieve the accumulated duration of exposure of 1 second (L_{eq} = 115 dB). This cumulative factor is cited in the discussion of impulse sound found in the levels document (Ref. 2, p. 25).

A special consideration regarding impulse sound has to do with the measuring equipment. A 50,000-Hz pure tone has a rise time of approximately 5 microseconds. This also is a very common rise time for impulse sounds. Therefore, to be able to respond to the rapid increase in sound energy, a measuring device would need a frequency response at the upper end of the spectrum of about 50,000 Hz. Most sound level meters are made to give accurate responses only up to about 18,000 Hz. Therefore, much of the sound pressure in an impulse sound is not sensed by conventional sound level measuring equipment. The peak level readings are lower than is really the case, even with the meter set for "fast" response. Measurement of impulse noises necessitates the use of special-purpose impact-noise measuring devices.

The data provided in the levels document show that a daily impulse noise exposure exceeding 145 dB for a period of 10 years should cause a 5-dB hearing threshold shift for the 4000-Hz test tone.[4] If the sound has a duration of less than 25 microseconds, a peak level of 167 dB would produce the same effect. Therefore, the duration of the impulses being regulated bear directly upon the allowable maximum levels for protection from hearing deficiency. In the levels document, a table and chart (Ref. 2, pp. 25–27) provide interim figures. Again, it must be stressed that these levels are the ideal and must necessarily be amended to accomplish impulse noise control regulations based on technical feasibility and economic impact.

USE OF THE IDENTIFIED LEVELS

The major use for the identified levels cited in the levels document will be for guidance in selecting permissible levels for land use. Even in our generally noisy society, for certain applications and in some locales, the identified levels might be adopted without change in a regulation in order to maintain the existing quiet environment. In most applications, however, for reasons of economics and/or technical feasibility, the levels of the levels document will necessarily be set higher initially.

One of the preferred ways to arrive at reasonable and enforceable allowable noise levels is to study the levels currently in use in other jurisdictions. For that purpose, Table 3-4 is given with the levels specified for residential, commercial, and industrial land use. In all cases, the levels are for the regulation of intruding noise sources and, therefore, are not true cumulative noise measures because they do not include all environmental sounds. In some of the cases cited in Table

TABLE 3-4 Examples of Existing Noise Level Limits (dBA).

	Residential		Commercial		Industrial	
Community	Day	Night	Day	Night	Day	Night
Baltimore, MD	55-70	50-55	58-70	58-65	61-75	56-70
San Francisco, CA	55-60	50-55	70	60	70-75	70-75
Boston, MA	60	50	65	55	70	60
Denver, CO	55	50	65	60	80	75
Hawthorne, CA	42	42	–	–	53	53
Loveland, CO	48	45	52	49	56	53
New York City, NY	65	45	65	–	80	–
Fountain Valley, CA	50-60	45-55	–	–	–	–
Santa Rosa, CA	55	45	60-65	55	70	70
Billings, MI	55	50	60-70	55-65	80	75
Missoula, MI	60	55	65	60	70-80	65-75
Coral Gables, FL	35-40	32-35	45	40	50	45
Helena, MI	55	50	60	55	80	75
Wheatridge, CO	37	37	–	–	–	–
Henrosa Beach, CA	45	45	55	54	65	65
Grand Rapids, MI	45-52	38-45	52-63	45-56	–	–
Albuquerque, NM	55-61	55-61	62-66	62-66	–	–
San Diego, CA	50-60	45-55	60	55	70-75	70-75
Colorado	55	50	60	55	70-80	65-75
Illinois	55-62	45-62	55-62	45-62	61-70	51-70
Chicago, IL	55-61	55-61	62-66	62-66	–	–
Dallas, TX	56	–	56-63	–	56-70	–
Minneapolis, MN	55	–	62	–	–	–
North Carolina	55-60	50-55	–	–	70	70
Lakewood, CO	55	50	60	55	80	75
Inglewood, CA	55	45	65	65	70	70
Salt Lake City, UT	65	55-60	70	65	75-80	75-80

3-4, the levels are maximum allowable levels for steady-state noise sources. In other cases, an allowance is made for sounds lasting less than an hour. Thus, they are probably quite close to an L_{10} level (i.e., the level exceeded 10 percent of the time). In still other cases, the levels are ambient levels with an allowance for noise intrusions. Thus, these levels are quite similar to L_{90} (i.e., the level exceeded 90 percent of the time). The most frequently used limits are: (1) residential areas, 55 dBA (day) and 45 or 50 dBA (night); (2) commercial districts, 60-65 dBA (day) and 55-60 dBA (night); and (3) industrial regions, 70 dBA for both day and night.

If true cumulative measures are to be used to assess and monitor land use characteristics of a circumscribed region with respect to the noise environment, L_{eq} or L_{dn} values can be used. There are two cases when the conversion to L_{eq} can be accomplished with the use of methods discussed earlier without field measurements of the actual levels. First, if the sound is steady-state with no significant variability and without large intrusive noises, then $L = L_{eq}$. Second, if the variation of sound levels with time is known to be normally distributed, the conversion from L_{90} can be made by:

$$L_{eq} - L_{90} = 1.28\,s + 0.115\,s^2 .$$

In this equation, s is the standard deviation of the normal distribution. The equation is plotted in Fig. 3-7 with standard deviations ranging from 0 to 12. In the case of highway noise, which often is approximately normally distributed with a standard deviation ranging from 2 to 5 dB, the difference between L_{eq} and L_{90} ranges from 3 to 9 dB. As seen in Fig. 3-7, the difference between L_{eq} and L_{90} is about 30 dB with a standard deviation of 12; however, a standard deviation of 12 is unusually large for most environmental noise conditions.

As an example, consider a community that has a noise ordinance with levels for L_{90} determined by land use. In residential areas, L_{90} cannot exceed 50 dB. Assume that most of the noise in the residential areas is normally distributed traffic noise with a standard deviation of 5 dB. By looking at Fig. 3-7, the difference between L_{90} and L_{eq} is found to be nearly 9 dB. Therefore, $L_{eq} = 59$ would be equivalent to $L_{90} = 50$ dB in that situation.

For those regulations that establish L_{10} levels, the conversion to L_{eq} is simple

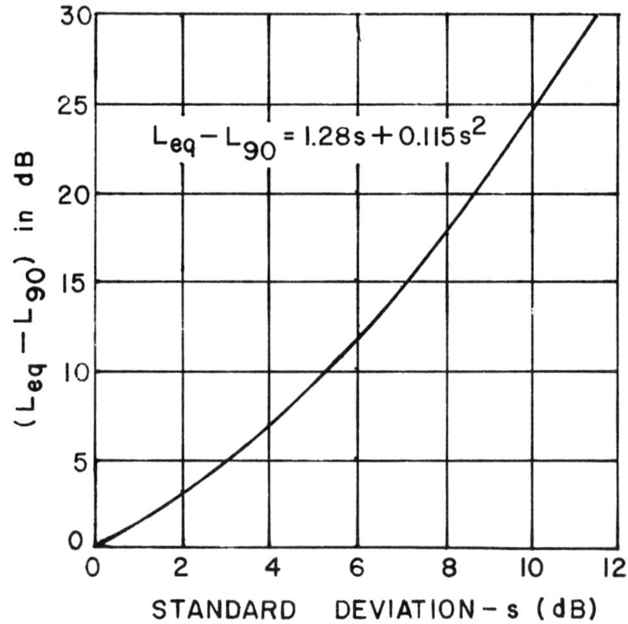

Fig. 3-7 Difference between L_{eq} and L_{90} for a normal distribution.

because L_{eq} and L_{10} are nearly equal. The relationship between L_{10} and L_{eq} for a normal distribution is given by:

$$L_{10} - L_{eq} = 1.28\,s - 0.115\,s^2.$$

This equation is plotted in Fig. 3-8 for standard deviations ranging from 0 to 14. It can readily be seen that $L_{eq} = L_{10} - 2$ dB within +2 dB for standard deviations smaller than 11 dB.

The best policy is to make preliminary sound measurements and analyses before setting new sound level limits to ensure that the levels are appropriate and enforceable. For example, a large metropolitan city has set a limit for stationary noise sources in residential districts at 55 dBA. Yet, measurements showed that the actual levels were never less than 59 dBA during the quietest part of the day. Thus, enforcement of the regulation is impossible.

As part of the total noise control scheme, a permit system may be instituted. Permits may be required for certain types of construction equipment and/or for mechanical and industrial equipment. They should also be required for parades or for sporting events such as automobile races. It is necessary to weigh the social value of each equipment usage or event against the contribution to the noise environment; this can best be determined by adding to the existing L_{eq} or L_{dn} value the L_{eq} value of the sound source for which the permit is requested.

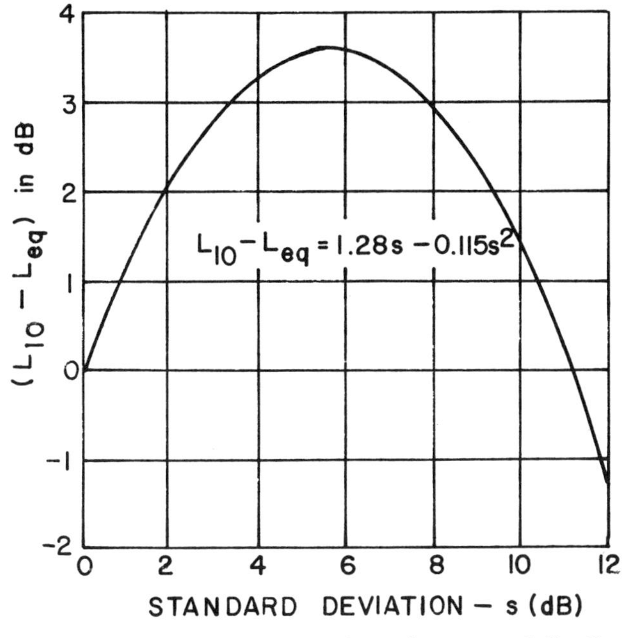

Fig. 3-8 Difference between L_{10} and L_{eq} for a normal distribution.[4]

As an example, a permit request is submitted to operate a certain type of construction equipment that produces 85 dBA sound for one hour a day in a commercial zone. In this zone, the existing L_{eq} (24) is 70 dB. First, the fraction of the total time, x_i is computed as $\frac{1}{24} = 0.0417$. The term 10 log x_i is determined from the nomogram in Fig. 3-4 to be 13.8 dB. So, the partial L_{eq} due to the construction equipment is 85 - 13.8 = 71.2 dB. When this is added to the existing L_{eq} of 70 dB by using the nomogram in Fig. 3-5, the result is 71.2 + 2.5 = 73.7 dB. So the effect of the 85-dBA device for 1 hour is to add 3.7 dB to the 24-hour L_{eq}. The significance of that rise in sound level must be determined by the local authorities in deciding whether to allow the permit.

A similar use for L_{eq} values will occur in the preparation of environmental impact statements and in site study requirements for new developments. In both cases, it will be necessary to measure the L_{eq} or L_{dn} before the development and then to project the effect of the new sound source(s) by computing partial L_{eq} values and then combining them to achieve a total L_{eq} or L_{dn}.

Some communities are adopting use restriction in high-noise areas as part of their noise control ordinances. This type of regulation restricts single-family development from areas where the L_{dn} is above some specified limit. For multiple-family residences, dormitories, and so on, the indoor L_{dn} is specified by requiring sufficient sound insulation from the outdoor high-level sound.

One of the most important aspects of any noise abatement program is follow-up sound monitoring, which can and should be done with L_{eq} and L_{dn} measures. As the federal government controls the manufacturing of various noise-producing products and as local communities apply their regulations to control local use of noisy devices, it is necessary to periodically monitor the sound to measure any progress in noise control. Only in this way can the effect of the various elements in a noise abatement program be assessed.

REFERENCES

1. Noise Control Act of 1972, Public Law 92-574, Ninety-second Congress, HR11021, October 27, 1972.

2. U.S. Environmental Protection Agency, Public Health and Welfare Criteria for Noise, EPA 550/9-73-002, July 27, 1973.

3. U.S. Environmental Protection Agency, Information on Levels of Environmental Noise Requisite to Protect Public Health and Welfare with an Adequate Margin of Safety, EPA 550/9-74-004. March 1974.

4. U.S. Environmental Protection Agency, Impact Characterization of Noise Including Implications of Identifying and Achieving Levels of Cumulative Noise Exposure, EPA NTID 73.4, 1973.

4

Environmental assessment of noise

D. J. Hagerty, Ph.D.

Associate Professor, Civil Engineering Department
University of Louisville
Louisville, Kentucky

and

J. E. Heer, Jr., Ph.D.

Assistant Dean of Engineering
Civil Engineering Department
University of Louisville
Louisville, Kentucky

INTRODUCTION

World population explosion, urban sprawl, blatant water and air pollution, shortage in energy supplies, depletion of mineral resources . . . . A virtually endless list of factors has created a growing awareness in the American public of the necessity of engineering in proper environmental planning. Never before have Americans, by general consensus, so completely agreed with the necessity to "look before we leap."

One of the most comprehensive measures of the legislative effort to secure adequate environmental planning was the passage, in 1969, of the National Environmental Policy Act (NEPA). This law, which requires the preparation of comprehensive environmental impact statements, has emerged as a major force in the study and abatement of the effects on individuals and communities of unabated noise in our technological society. Environmental assessments have become necessary in order to identify the consequences of unabated environmental contaminants, including noise. Such identification leads to the development of effective means for the prevention and abatement of specific noise problems.

History of the National Environmental Policy Act (NEPA)

The passage of NEPA was the culmination of almost 15 years of attempts by Congress to define a national policy that would guarantee the conservation and improvement of environmental quality. The impact statement requirement, which has become so important in the implementation of NEPA, was actually added to the bill at the last minute. At the time, the rider's far-reaching consequences were unforeseen by the bill's authors. This is somewhat ironic, since a considerable legislative history preceded the formulation and passage of NEPA.

In 1959, in the Eighty-sixth Congress, Senate Bill 2549, entitled the "Resources and Conservation Act," was proposed as a measure to coordinate the many federal programs aimed at environmental conservation. This bill failed in the Congress, and several years passed before another bill was introduced. In 1965, the "Ecological Resources and Surveys Bill," Senate Bill 2282 of the Eighty-ninth Congress, was proposed. This bill provided for the coordination of the then-existing environmental programs within various federal agencies and for the use and exchange of information on environmental quality among these agencies. This bill also failed to pass Congress. Finally, in 1969, Senate Bill 1075 was introduced by Senator Henry M. Jackson, who stated that the measure was intended to: (1) formulate a national environmental policy, (2) authorize research on the conservation and preservation of natural resources, and (3) establish a council of environmental advisers to the president.

The original draft of NEPA contained no required operational procedures designed to ensure implementation of the environmental policy formulated according to the philosophy of the law. The bill called for ". . . a national strategy for management of the human environment."[30] However, it provided no procedures for review, coordination, or control of continuing federal agency activities in management of the environment. After the bill was referred to the Senate Interior and Insular Affairs Committee, a provision requiring certain operational procedures was added. This action-forcing provision is the now-famous Section 102(2)C, which requires all federal agencies to prepare environmental impact statements for any proposed actions judged to be significant with respect to possible detrimental environmental effects.

The final form of the national environmental policy act as enacted represented a synthesis of Senate Bill 1075 and House of Representatives Bill 6750, which had been sponsored by Rep. John Dingell of Michigan. This synthesis had been prepared by a combined House and Senate conference committee. The bill was signed into law by President Richard M. Nixon on January 1, 1970.

The passage of NEPA signaled a major renovation in federal policy concerning environmental protection. Previous policy had been directed toward dealing with environmental crises as or after they occurred. The new legislation generally was directed toward anticipating and preventing environmental degradation. In essence, NEPA was seen by its initiators as a means of providing an interdisciplinary approach to decisionmaking by all agencies of the federal government concerned with management of the nation's environment.

NEPA in Brief

The complete text of NEPA can be found at the end of this chapter. The following discussion covers the more important features of the law.

Section 101. The general statement of a national environmental policy, which was a major goal of this law, is contained in Section 101. Senator Jackson and other members of the Senate and House committees who formulated this legislation have since indicated that this statement was not intended simply to be a reiteration of basic national beliefs. On the contrary, it was intended to establish *definite priorities* and to give expression to national goals and ideals with respect to environmental preservation. It was also intended to provide a legal foundation upon which federal administrators could base decisions that might significantly affect the environment. This statement of policy was intended to guide such administrators in evaluating environmental quality, especially when environmental values appear to be in conflict with other value parameters or measures. The policy statement, upon examination, can be seen as a national commitment to avoid or prevent actions that would endanger the health or the continued existence of humankind. In other words, it meant that Americans would no longer intentionally initiate any action that would irreparably harm the land, water, and air of their nation.

Section 102. The second section of Title I of NEPA, Section 102, provided the "action-forcing" procedures of the act, which, as mentioned, were added almost as an afterthought. This section requires (1) that all federal agencies proposing significant actions consult with all other appropriate agencies having jurisdiction over, or expertise in, environmental matters pertaining to air, water, noise, and so on concerning the contemplated action, and (2) that they prepare a comprehensive statement on the expected environmental impact of the proposed action. This impact statement must be prepared in accordance with detailed provisions spelled out in the law. These provisions are described in greater detail later.

The remaining three sections of Title I of the act—Sections 103, 104, and 105—were designed to require all federal agencies to review their agency procedures for compliance with the NEPA requirements. These sections also insured that the statutory obligations previously existing for any federal agency were not affected by the provisions of NEPA, which was to be considered supplementary to existing authorizations of federal agencies.

Section 201-207. Sections 201-207 in Title II of the act all pertain to the establishment of a Council on Environmental Quality (CEQ), as an advisory body in the Office of the President. The purpose of this council was to advise the president on the impact of various federal activities on the national environment and to maintain continued supervision of such activities. By law, the council is required to establish a system for monitoring indicators of environmental quality. Furthermore, it must maintain comprehensive records on the status of the environment, as well as ensure that a comprehensive data base will be maintained so that decisions on environmental problems can be made rationally and logically on the basis of sufficient data. The CEQ, by law, is composed of three members, appointed by the president to serve at his or her pleasure, by and with the advice and consent of the Senate. In addition to analyzing environmental information to predict trends in environmental quality, the CEQ is charged with formulating national policy to improve the quality of the environment.

The wording of NEPA seems to indicate that the members of Congress wished to establish a group of environmental experts who would view the national situation in a broad setting and who would establish environmental goals to be attained. However, this situation has not materialized. Because of the position of the CEQ within the Office of the President, its position as an independent group of advisers has been somewhat compromised. In some situations that have developed since the passage of NEPA, the CEQ has appeared to speak *for* the president rather than *to* the president.

The intentions of the federal lawmakers also seem to have been thwarted in regard to the requirements for an impact statement. This provision, for a comprehensive statement of expected impact, was inserted into the bill with the intention of developing a summary statement of two or three pages on the environmental effects of proposed federal actions. A review of the environmental impact statements currently being submitted to CEQ shows that present-day statements exceed these expectations more than a thousandfold.

Ramifications of NEPA

To date, the most important consequence of NEPA—perhaps contrary to the expectations of Senator Jackson and his colleagues—has been the preparation of numerous detailed assessments of the possible effects of actions undertaken by federal agencies. The environmental impact statement (EIS) has emerged as the most important step in comprehensive environmental planning. As such, it merits careful scrutiny. Why and how are such statements prepared? When is it necessary to prepare an EIS?

Section 102(2)C of NEPA, states that a detailed report must be prepared on "... proposals for legislation or other major federal actions significantly affecting the quality of the human environment ...." Considerable uncertainty existed in the days immediately following passage of NEPA concerning the interpretation of the words *major* and *significantly*. Federal court decisions have since clarified the uncertainties. Two major cases, *Hanley* v. *Mitchell*[39] and *Scrap* v. *United States*,[42] established the precedents for analysis of the significance of federal action with respect to environmental effects. The philosphy expressed in these decisions can best be summarized by saying that the term *major federal action* "... refers to the cost of the project, the amount of planning that preceded it, and the time required to complete it, but does not refer to the impact of the project on the environment." The environmental significance of a project could be determined in the following way:

> ... in deciding whether a major federal action will "significantly" affect the quality of the human environment, the agency in charge, although vested with broad discretion, should normally be required to review the proposed action in the light of at least two relevant factors: (1) the extent to which the action will cause adverse environmental effects in excess of those created by existing uses in the area affected by the action, and (2) the absolute quantitative environmental effects of the action itself, including the cumulative harm that results from its contribution to existing adverse conditions or uses in the affected areas.[39]

Also, in *Hanley* v. *Mitchell*, the court declared that close public interaction must be maintained throughout the evaluation process to determine the importance and significance of a specific action. This decision guarantees public participation in environmental planning as was envisioned by the initiators of the law.

Included in the actions for which federal agencies must prepare impact statements are (1) proposals or recommendations for legislation, (2) agency reports and evaluations on legislation initiated in other agencies, and (3) projects or continuing programs undertaken by an agency, if these programs have been judged to have significant environmental effects. In decisions to date, actions that have been considered to have major environmental significance have included those whose impact is highly controversial on environmental grounds, those that are precedents for continuing activity or for more sweeping activity, and those that involve a number of federal agencies (and thus are significant because of the level of agency effort involved).

In order to ensure sufficient time for review of proposed actions and their possible effects, agencies are required to prepare a draft environmental statement and to circulate it for evaluation for a period of at least 90 days. A final statement, including replies to questions and comments raised by the draft statement, must be circulated to the public at least 30 days before the implementation of any proposed action.

Each federal agency is required to obtain reaction from other agencies, interested parties, and the general public at the earliest possible time. To ensure that agency review of proposed actions is complete and appropriate, the CEQ has issued guidelines for use by federal agencies in implementing the requirements of NEPA.[5] Specifically, with respect to actions involving noise control and abatement, the following agencies must be consulted regarding any proposed activity:

Department of Health, Education and Welfare (health aspects)
Department of Commerce–National Bureau of Standards
Department of Transportation–Assistant Secretary for Systems Development and Technology, Federal Aviation Administration (Office of Noise Abatement)
Environmental Protection Agency (Office of Noise Abatement and Control)
Department of Housing and Urban Development (urban land-use aspects, building materials standards).

Other agencies which must be consulted on particular projects are identified by concern in areas of air quality and air pollution control, weather modification, environmental aspects of electric energy generation, toxic materials, pesticides, and so on.

The role of the general public in formulating environmental policy under the guidelines of NEPA is significant. By law, every draft EIS prepared by a federal agency must be made available to the general public at the same time that it is circulated to other federal and state agencies for review and comment. The 90-day delay between the time of circulation of the draft EIS and the date of implementation of any proposed action should allow ample time for citizen comment on draft statements. It is important to note that federal agencies are obligated to retain and, upon demand, make available all comments received from other agencies and from the public on draft environmental statements. In addition, the final EIS must include an evaluation of all comments received on the draft statement and must include an attempt to resolve any objections raised in the comments received. The final version of the impact statement also must be made available to the public 30 days before implementation of the proposed action. If a citizen believes that his or her comments and questions have not been adequately addressed in the final EIS, he or she may so notify the CEQ. The public availability of impact assessment reports and impact statements is perhaps one of the most significant advances in environmental planning produced by the passage of NEPA.

In addition to review by other governmental agencies and by interested parties in the general public, virtually all proposed actions covered by an EIS must be reviewed by the U.S. Environmental Protection Agency. Total review is not specifically required by law, but the EPA is required to review all statements concerned with air and water pollution, drinking-water supplies, solid waste, pesticides, radiation hazards, and noise. This comprehensive listing of projects ensures that the great majority of impact assessments filed with the CEQ will have been reviewed by the EPA. The EPA's obligation to review proposed federal actions exceeds the responsibilities of other agencies in this regard because of the EPA's unique position as the principal federal regulator of pollution-control activities.

After review by all concerned agencies, the initiating agency can forward a final impact statement to the CEQ. This pursuance of proposed action can take place even in the face of unfavorable comment and opposition from all reviewing agencies, even the EPA. In essence, none of the reviewing agencies, including the EPA, have any authority to stop any project sponsored by another federal agency; they simply function in advisory capacities to the CEQ and the president. As a matter of law, the CEQ does not possess statutory authority to stop or prevent the initiation of a project; it may, however, make recommendations to the president, who does have the appropriate authority to stop projects judged to be significantly detrimental to the nation's environmental quality. The role of the CEQ in many respects is that of a coordinator; it is not considered to be a review agency because its comments will not be appended to impact statements as are the comments of other agencies. Since the CEQ makes recommendations directly to the president, its failure to comment on either draft environmental statements or final environmental statements cannot be taken to indicate either approval or disapproval of any proposed action.

The ramifications of NEPA have been broadened and intensified in a number of states where state laws have been enacted to require an EIS on any proposed action to be undertaken by state agencies or to be supported by state or local funds. Environmental policy legislation requiring some form of impact assessments has been passed in California, Washington, Delaware, Montana, Wisconsin, North Carolina, and Indiana. At least a dozen other states have evaluated such environmental assessment legislation. The effect of such state and local legislation will be the requirement for comprehensive land-use planning at federal, state, and local levels. Such land-use planning can have great social and political importance as well as environmental and economic significance.

GROWTH OF THE IMPACT STATEMENT

As previously mentioned, the requirement for a comprehensive assessment and disclosure of the expected impacts from proposed federal actions has proven to be the most significant result of the passage of NEPA. The EIS has reached its present significance principally through judicial interpretation of NEPA and subsequent enforcement of the provisions of the act by the federal courts. The history of interpretation and enforcement has had great significance because judicial decisions have established the ground rules for adherence to the provisions of NEPA and, in some cases, have been extended to the specification of the contents of an EIS.

Enforcement of NEPA

During the committee hearings on Senate Bill 1075, several senatorial committee members and witnesses testifying expressed the opinion that the implementation of NEPA would rest principally with the Office of Management and Budget (OMB). The institutional power of OMB to regulate expenditure of funds in federal agencies seemed to be the principal qualification of this agency as the prime enforcer of the stipulations of NEPA.[6] However, the senators who formulated the legislation did not delegate this specific authority or responsibility to OMB. Since the passage of NEPA, OMB personnel have been exceedingly reluctant to assume any responsibility for enforcement of NEPA through the withholding or allocation of funds on the basis of agency compliance or noncompliance with NEPA requirements. In fact, OMB personnel have even been reluctant to create specific guidelines for their own utilization in assessing the impact of OMB policies, even though such guidelines have been called for by the CEQ.[2] In general, the implementation of NEPA has not taken place in any executive office or agency.

Since NEPA was a comprehensive reform of governmental procedures concerning environmental conservation, the task of enforcing the act fell rather naturally to the judicial branch of government through interpretation of the provisions of the act. The federal courts have responded to this allocation of responsibility with alacrity.

Federal judges apparently have considered NEPA a powerful tool to assist them in a general review of governmental policy. This judicial enthusiasm may stem from the fact that much of the judicial review of governmental action in recent years has been directed toward achieving systematic decisionmaking on the part of federal agencies. In several instances, judicial review has strongly stressed the need for full evaluation of alternate courses of action and more comprehensive assessment of the public interest, primarily through greater public participation in governmental decisionmaking. In light of these goals, which have been set apparently by the federal judiciary, NEPA has proven to be a convenient mechanism for judicial review of agency activities.

The initial involvement of federal judges in interpretation of the requirements of NEPA arose as a result of citizen challenges, in the form of lawsuits, to federal agency actions. Early court decisions (soon after the passage of NEPA) were based generally on matters of procedure, on considerations of whether the procedures called for in NEPA had been followed by an initiating agency. In later decisions, however, the courts have set aside agency actions on the basis that certain crucial factors have not been considered in environmental assessments, or that impermissible factors had entered into the decision.[35] In retrospect, several important trends can be seen in judicial interpretation of NEPA.

Federal courts have been zealous in establishing the lower limits of applicability of NEPA. Judicial decisions have emphasized the necessity of considering even projects of small or moderate size, on the basis that their long-term consequences and secondary impacts (which, in spite of relatively minor primary effects) may warrant environmental assessment. Second, the initial guidelines issued by the CEQ called for the formulation of specific guidelines by each federal agency. In several early court cases, the decisions rendered required the preparation of impact statements solely on the basis that formulated agency policy had explicitly required such preparation. In other words, federal agencies are being held to the letter of the guidelines that they have formulated.

Judicial reviews of impact assessments have endeavored to identify particularly unique or controversial projects. A set of typical projects has been so identified on the basis of either previous court decisions, agency requirements, or on the intentions stated by the committees that framed the legislation. All highly controversial projects have been included in this category as projects requiring impact statements. The courts have interpreted the word "controversial" to refer to "... size, nature or effect of the major Federal action rather than to the existence of opposition to a use, the effect of which is relatively undisputed."[38] However, if the guidelines of a particular agency have defined "controversial" as "meeting strong public opposition," then this strong public opposition to a particular project will inherently require the preparation of an impact statement.

The most important result of the early judicial reviews of NEPA has been the establishment of judicial authority to review impact assessments and statements, and the delineation of agency liability as specified in NEPA. Two landmark cases, *Environmental Defense Fund* v. *Hardin*[36] and *Scrap* v. *United States*,[42] decided the question of judicial authority with respect to NEPA. In these cases, the courts ruled that the wording and spirit of NEPA itself provided grounds for judicial review. After the authority of the courts to review and enforce NEPA had been formalized, the next most important consideration was the delineation of agency responsibilities in the implementation of NEPA. In this respect, the phrase "to the fullest extent possible" included in Section 102 of the act has been used by the courts as a yardstick to evaluate agency compliance with NEPA.

In the widely quoted case of *Calvert Cliffs' Coordinating Committee* v. *Atomic Energy Commission*,[34] the courts declared that the wording of NEPA was such that very important procedural guidelines were established by the act and that federal agencies must strictly comply with the provisions of the act. In this decision, Judge Skelly Wright stated,

> Section 102 duties are not inherently flexible. They must be complied with to the fullest extent, unless there is a clear conflict of *statutory* authority. Considerations of administrative difficulty, delay and economic cost will not suffice to strip the section of its fundamental importance.[34]

Thus, it becomes important to examine the requirements stated in the act for the contents of an EIS.

Contents of the EIS

The requirements for complying with the provisions of Section 102(2)C of NEPA, as interpreted in various court decisions, can be stated in three basic requirements:

1. the integration of the impact assessment into the environmental decision-making analysis
2. the thorough review of all feasible alternatives in the assessment
3. the use of appropriate detail in the impact statement.

In *Environmental Defense Fund* v. *Corps of Engineers*,[37] the court outlined the minimum requirements for an impact statement by stating that

> At the very least NEPA is an environmental full disclosure law .... The "detailed statement" required by Section 102(2)C should, at a minimum, contain such information as will alert the President, the Council on Environmental Quality, the public, and, indeed, the Congress, to all known possible environmental consequences of proposed agency action.

With respect to the evaluation of alternatives, the decisions handed down by the courts, in *Natural Resources Defense Council* v. *Morton*[40] and in subsequent cases, have indicated that agencies must evaluate all plausible alternatives to a proposed action. Included in this set of alternatives must be not only those actions within the purview of the agency but also those actions that are beyond the present scope of the initiating agency's jurisdiction.

This stipulation by the courts was apparently intended to increase agency interaction. However, the courts specifically mentioned that consideration of alternatives should not be limited to those plans offering universal solutions to the particular problem under consideration, since alternatives that offer only a partial solution could in some cases be more advantageous than the proposed action because of decreased environmental damage. With regard to the detailed compliance of agencies with the letter of the law, the courts have been very strict in evaluating the overall adequacy of filed impact statements. In the case

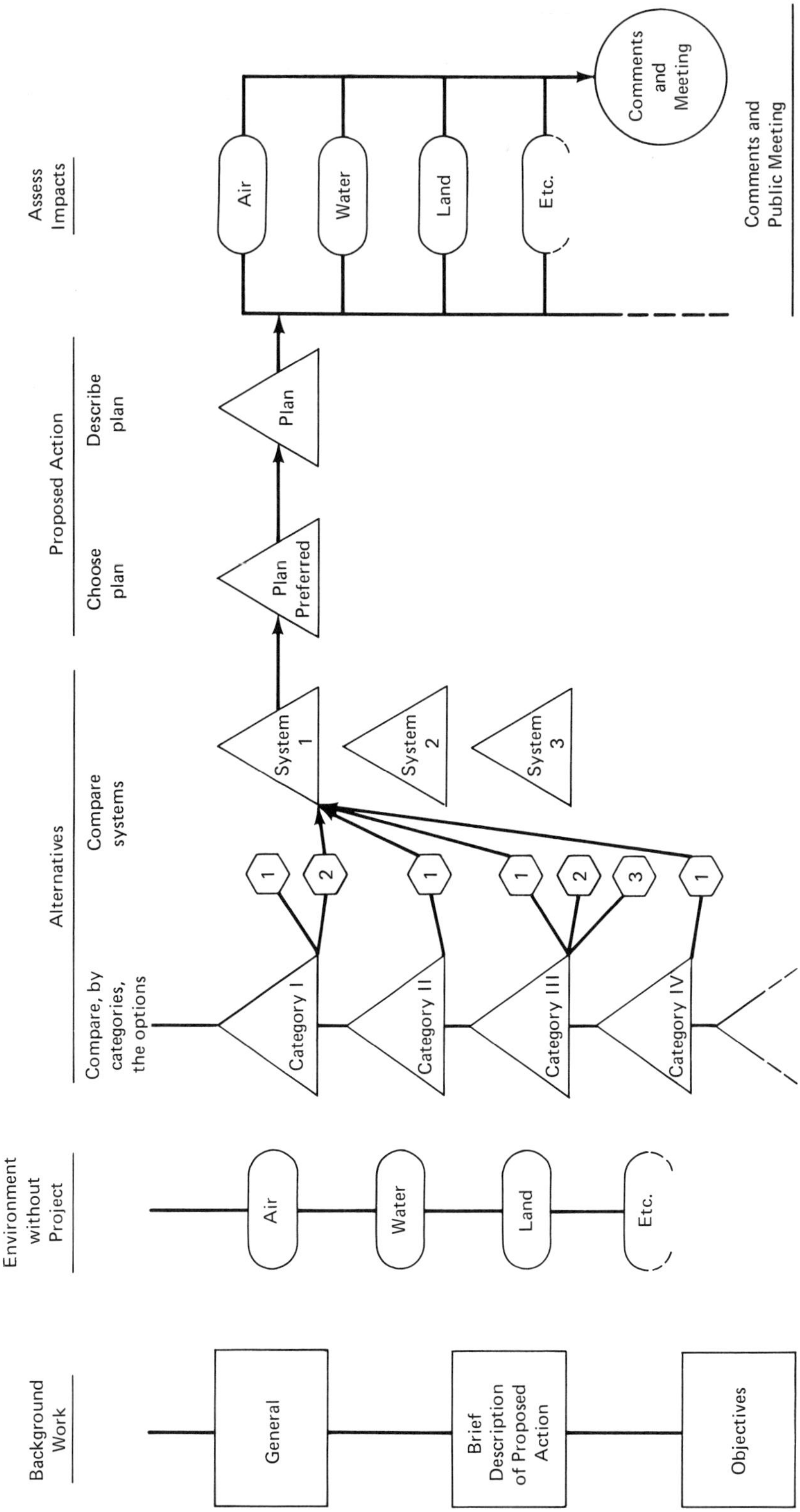

Fig. 4-1 Flowchart for EIS preparation. (After EPA.[26])

of *Environmental Defense Fund* v. *Corps of Engineers*, the court ruled that "the EIS must be written in language that is understandable to non-technical minds and yet contain enough scientific reasoning to alert specialists to particular problems within the field of their expertise."[21] In the *Calvert Cliffs* decision mentioned previously,[34] the court decided that all alternatives to the proposed action, including the null or do-nothing alternative, must be described in sufficient detail so that a reviewer could make an independent, reasonable decision concerning the advisability of proceeding with the proposed action and the relative merits of each alternative to the proposed action. In this decision, the courts ruled that not only should the environmental statement enumerate all alternatives, but that it should also contain an evaluation of the unique impacts of each alternative with regard to the proposed course of action. Generally, the courts have held that an evaluation and balancing among feasible alternatives must be performed at every major stage in the decisionmaking process. Figure 4-1 shows a typical flowchart for evaluation of alternatives in such a federal agency operation.

Finally, the courts have examined environmental impact statements very critically for compliance with the five requirements that are specifically stated in Section 102 (2) C of the act. In the case of *Scrap* v. *United States*,[41] the court stated, "It should go without saying that the mere filing of a document labelled 'Final Impact Statement' is insufficient to shield an agency from judicial review if the document fails to comply with standards outlined in NEPA." Table 4-1 shows, in outline, the specific requirements incorporated in Section 102 of NEPA.

TABLE 4-1 Outline for EIS Content. (From U.S. Department of the Army, 1974.)

1. PROJECT DESCRIPTION
   a. Purpose of action
   b. Description of action
      (1) Name
      (2) Summary of activities
   c. Environmental setting
      (1) Environment prior to proposed action
      (2) Other related Federal activities
2. LAND-USE RELATIONSHIPS
   a. Conformity or conflict with other land-use plans, policies and controls
      (1) Federal, state and local
      (2) Clean Air Act and Federal Water Pollution Control Act Amendments of 1972
   b. Conflicts and/or inconsistent land-use plans
      (1) Extent of reconciliation
      (2) Reasons for proceeding with action

TABLE 4-1 *(Continued)*

3. PROBABLE IMPACT OF THE PROPOSED ACTION ON THE ENVIRONMENT
   a. Positive and negative effects
      (1) National and international environment
      (2) Environmental factors
      (3) Impact of proposed action
   b. Direct and indirect consequences
      (1) Primary effects
      (2) Secondary effects
4. ALTERNATIVES TO THE PROPOSED ACTION
   a. Reasonable alternative actions
      (1) Those that might enhance environmental quality
      (2) Those that might avoid some or all adverse effects
   b. Analysis of alternatives
      (1) Benefits
      (2) Costs
      (3) Risks
5. PROBABLE ADVERSE ENVIRONMENTAL EFFECTS WHICH CANNOT BE AVOIDED
   a. Adverse and unavoidable impacts
   b. How avoidable adverse impacts will be mitigated
6. RELATIONSHIP BETWEEN LOCAL SHORT-TERM USES OF MAN'S ENVIRONMENT AND THE MAINTENANCE AND ENHANCEMENT OF LONG-TERM PRODUCTIVITY
   a. Trade-off between short-term environmental gains at expense of long-term losses
   b. Trade-off between long-term environmental gains at expense of short-term losses
   c. Extent to which proposed action forecloses future options
7. IRREVERSIBLE AND IRRETRIEVABLE COMMITMENTS OF RESOURCES
   a. Unavoidable impacts irreversibly curtailing the range of potential use of the environment
      (1) Labor
      (2) Materials
      (3) Natural
      (4) Cultural
8. OTHER INTERESTS AND CONSIDERATIONS OF FEDERAL POLICY THAT OFFSET THE ADVERSE ENVIRONMENTAL EFFECTS OF THE PROPOSED ACTION
   a. Countervailing benefits of proposed action
   b. Countervailing benefits of alternatives

PREPARING AN ENVIRONMENTAL ASSESSMENT

If preliminary evaluations indicate, on the basis of an environmental assessment, that it is necessary to prepare an EIS, then certain procedures will be required by law. The general sequence of preparation and of time periods for review are specified in the act itself. However, the methodology to be used in accomplishing environmental assessment, and the organization of an assessment team have not been addressed in the act.

General Sequence of Preparation

The general sequence of events in preparing an EIS within a federal agency is illustrated in Fig. 4-2 where the various steps in preparing a statement are illustrated. In step 1, the applicant agency prepares an environmental assessment of the proposed action together with all collected baseline environmental data. In step 2, the agency itself performs an environmental review of the actions at a very early stage in the overall project development in order to decide whether

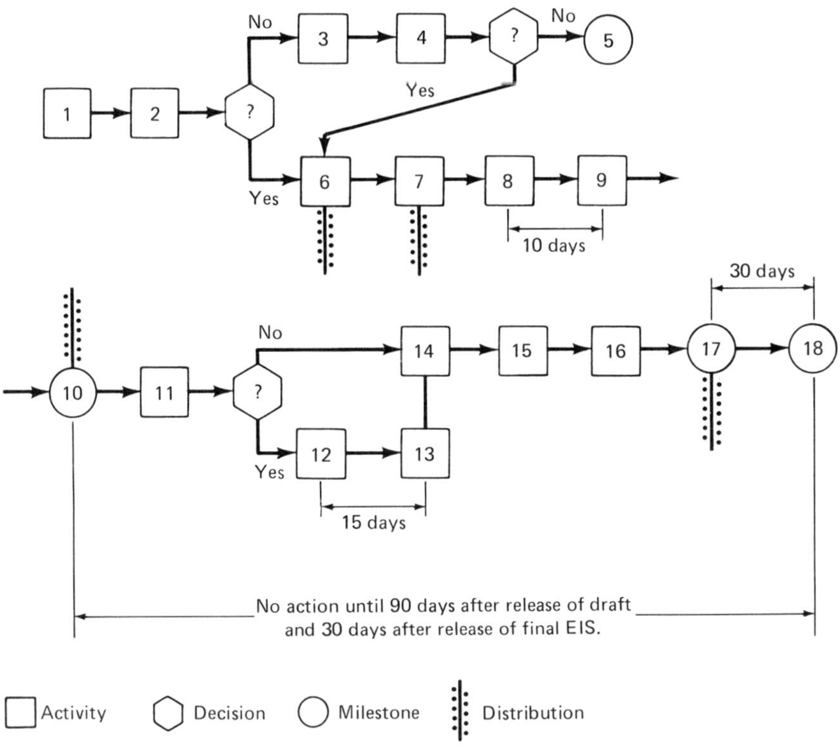

Fig. 4-2 Sequence of EIS preparation in federal agency.[25]

an EIS is required. If the project is or will be highly controversial, or if it is likely that the project will have a significant impact, an EIS will be prepared.

If no EIS is deemed necessary, an environmental impact appraisal, a rather brief document, is prepared (in step 3). Also in step 3, a negative declaration (that no EIS is required) is prepared and circulated to pertinent federal, state, and local agencies; interested persons; and local newspapers and other media. The impact appraisal, negative declaration, and other baseline supporting information are maintained available for public inspection. If a negative declaration is made and circulated for comments, the agency receives and evaluates these comments in step 4. If the agency reevaluates the situation and decides that an EIS is necessary, a full review of the project is made. If, after reviewing comments received on the negative declaration, the agency maintains that no EIS is required, it proceeds with administrative action to implement the project (step 5).

If an impact statement is considered necessary, the agency proceeds to step 6. This is a preliminary step—the agency simply prepares a notice of intent to prepare an EIS and circulates this notice of intent to appropriate federal agencies, regional groups, local agencies, and interested persons. After this notice of intent is circulated, agency personnel proceed, in step 7, to prepare a preliminary draft EIS. This statement must address all of the issues outlined in Table 4-1. As step 8 in the overall sequence of events, the agency personnel coordinate an internal review of the preliminary draft statement and solicit comments from within the agency hierarchy. After in-house comments have been received and the draft statement has been appropriately revised, it is submitted to the Office of Federal Activities for further review.

In step 9, agency personnel receive the comments from the Office of Federal Activities, revise the draft statement accordingly, and make it available to various news media. In step 10, the draft environmental statement is distributed to the CEQ, appropriate federal agencies, state and local agencies, and interested persons. At this time, agency personnel proceed to step 11 in which they determine the need for a public hearing. If it is deemed necessary to hold a public hearing, appropriate public notices are circulated as step 12, and, as step 13, a public hearing is conducted. After the hearing, if one is held, all suggestions, criticisms, and comments that have been received on the draft environmental statement are reviewed and evaluated, in step 14. Comments generated at the public hearing, if one was held, are included in this review and evaluation.

In step 15, the agency personnel prepare the final EIS on the basis of previous preparations and reactions. As step 16, a news release is submitted to local news media to announce the completion of a final EIS. As step 17, the finalized EIS is distributed to the CEQ and the various agencies and individuals mentioned previously.

After a 30-day period of review and consideration, if the initiating agency maintains that all comments have been answered and the proposed action is appropriate, administrative action to implement the proposed project may be initiated. This sequence illustrates the general procedures used by most federal agencies. These procedures are usually described in comprehensive detail in agency guidelines. However, the methodology used in evaluating environmental impacts and in assessing alternative effects may not be specified, in all cases, in agency guidelines. The methodology to be used in the environmental impact assessment is extremely important and also bears some examination.

Methodology

Because of the complex nature of environmental impact assessment investigations, several attempts have been made to develop standardized methods for such assessment. Most of these methods have not proven to be adequate for all types of assessment and for all categories of projects. Traditional economic methods (equivalent uniform annual costs, present worth, equivalent uniform annual net return, net present value, benefit/cost ratio, rate of return, etc.) generally have been found to be inadequate because of the difficulty in evaluating most environmental characteristics in monetary terms. Likewise, a variety of graphical comparison techniques have been developed to visually represent the location and degree of impact in a proposed activity. These techniques include the use of map overlays,[15] computer map analysis, value profiles,[16] shaded matrices,[4] relative merit analyses,[1] physical models,[9] and contour mapping of severity.[17] These methods are useful primarily as communication tools; they are not inherently quantitative, nor are they sufficiently precise in a quantitative fashion. In order to provide a more quantitative analysis, certain numerical systems also have been developed for environmental analysis, such as ranking systems,[4] rating systems,[7] expected value methods,[19] value matrices,[18] utility theory applications,[4] and decision models.[10] These numerical methods generally are not sufficiently discriminatory and are by necessity somewhat oversimplified.[2,4,6,11]

Although none of these methods are sufficiently comprehensive for total applicability in environmental assessment, some of the methods are useful in accomplishing portions of the assessment task. For example, matrices can be utilized profitably to identify proposed activities and to identify anticipated impacts. The matrix developed by the U.S. Geological Survey (USGS) is quite useful in preliminary investigations.[14] However, this system has not proved to be sufficiently quantitative for total application. The critical operation in obtaining a quantitative estimate of impact, according to the procedure recommended by the developers of the matrix, is not inherently flexible; it is not possible to assign proper weighting to the various aspects of the location, construction, and operation of particular facilities. In the USGS procedure, the matrix consists of 100 possible actions listed along the horizontal matrix axis and 88 identified impacts listed along the vertical axis, yielding 8800 possible interactions. This matrix was intended only as a guide; only first-order, direct impacts would be studied, and probably no more than a dozen activities would be significant for any one project.[14] Each element (box) in the matrix consists of an upper half in which the degree of impact of a particular action is expressed as a number from 1 (least) to 10 (greatest), and a lower half in which the overall importance of that particular impact on the project under study is expressed in a similar fashion. Since a highly subjective series of many value judgments must be incorporated in the use of this matrix, it is subject to some criticism. No criteria have been given by the developers of the matrix for making these highly subjective judgments. No systematic analysis of interactions among possible actions is considered, and secondary effects are ignored. Thus, matrix methods are useful primarily for preliminary studies and for communicating with the public concerning identified impacts.

Another method that has been utilized for environmental assessment can be classified as a "checklist."[13,22,23,27,28] In this method, a long list of possible areas of impact is compiled, and the investigator compares the list with the possibilities inherent in the proposed action. He or she simply proceeds through the checklist in order to uncover any possible areas of impact. This method is useful primarily as an aid to the memory of the investigator. It certainly is not a quantitative investigation tool. However, some checklists do have worthwhile features. For example, the AEC guidelines for power plant assessments are rather comprehensive, as they require a number of descriptive evaluations of a proposed plant: purpose; site character; plant character; environmental effects and construction; impacts of normal operation; effluent control and monitoring; effects of accidents; socioeconomic effects; alternative energy sources and sites; plant design alternatives; benefit/cost analysis; a list of environmental approvals from and consultations with other agencies; and a list of documentary references. Obviously, it is not possible to quantify all of these factors or even to quantify environmental factors in a summary benefit/cost ratio. Using such a comprehensive checklist usually results in the amassing of an unwieldy compilation of baseline data. These inventories are of little use without some selective criteria to identify the significant impact areas. On the other hand, guidelines developed by the Department of Housing and Urban Development[27] incorporate a valuable concept of "threshold" values of environmental parameters for impact assessment. However, these guidelines are not sufficiently comprehensive. Another checklist method has been developed by personnel of the Army Construction Engineering Research Laboratory at Champaign, Illinois.[13] It is not possible to obtain quantitative evaluations of impact through the use of this method but the method does incorporate index matrices for identification of impacts and provides descriptive material on the evaluation of such impacts in many phases of the environment. Also, it features good reference lists for each area of impact assessment.

In addition, several highly quantitative methods have been developed. For example, Battelle Memorial Institute, Columbus Laboratories developed a system for use in the evaluation of water resources projects.[3] This system includes a complicated weighting procedure in which numerical values are given for a long list of possible impact areas. In this way, it is claimed that a quantitative evaluation can be obtained, and a comparison of alternatives can be accomplished quantitatively. Four basic areas are included in this system: ecology, environmental pollution, aesthetics, and human interests. A total of 1000 parameter importance units (PIU) are distributed among 78 different environmental parameters. The "environmental quality" of each parameter is predicted for a given project, on the basis of a value from 0 (worst) to 1 (best). This is then multiplied by the PIU value for that parameter to yield a number of environmental importance units (EIU). The sum of all environmental importance units for all parameters is assumed to be a measure of total impact for a proposed action. In the same way, values of environmental importance units can be obtained for all alternatives to a proposed action, and thus the alternatives can be compared quantitatively on the basis of the EIU total scores. Also, in this system, especially delicate or controversial environmental elements are singled out for particular attention.

Although this system was designed to furnish a quantitative evaluation of impact, the method is inflexible and contains no provision for giving appropriate weight to parameters that are more important in particular situations, especially situations other than water resources systems. Other quantitative systems have been developed, but they are subject to much the same criticism.

A review of the systems for impact assessment developed to date indicates that no one given system is generally applicable to all types of projects in all areas of the world. In order to improve this situation, especially with respect to the evaluation of the environmental assessments of noise effects, it is suggested that types of projects be identified and grouped into major categories. These categories of activity could be analyzed in an effort to develop generalized impact assessments and avoid the duplication associated with individual assessments performed for particular reasons. In performing the analysis of types of activities, it is suggested that a matrix be developed as a preliminary analysis tool and vehicle for communication.

Second, a comprehensive checklist could be assembled by experts in noise abatement and control for use in assessing noise impacts from various types of activities. This development of a matrix and a checklist for analysis of classes of projects could greatly simplify environmental assessment of noise effects. One type of activity, for example, could be the construction or improvement of a major highway. A working committee could identify the major noise-producing elements in highway construction and use, and could greatly facilitate the assessment of noise effects from particular highways by performing a generalized assessment of all highways. By assessing general situations and identifying general impacts, much effort that has been expended in duplicate investigations from one project to another could be saved.[11]

Organization of Personnel

A very important step in any environmental impact assessment is the organization of the investigative team assigned the task of impact evaluation. One of the most practicable methods of organizing such a team is to define the investigative tasks to be performed and then to secure appropriate individuals to perform the tasks for which they are most suited. The investigative team should be structured so that it is possible to take a systematic, interdisciplinary approach, which should include integration of natural and social sciences with the environmental design arts in both the planning and decisionmaking stages. Interdisciplinary investigations are not identical with multidisciplinary studies; assessment-team members must be willing to recognize values and constraints pertinent to fields outside their own areas of expertise. To achieve an appreciation for other disciplines among the members of an investigative team, it may be helpful to seed the team with some individuals who have expertise in more than one discipline. Experience has shown that an interdisciplinary approach is preferable to all others.[29] The assessment should consist of an attempt to bring together individuals possessing specialized knowledge in a variety of appropriate disciplines in order to best determine societal needs, identify environmental impacts, and search for the best possible alternative solutions.[31]

In addition to formulating an interdisciplinary team, the selection and appointment of an overall coordinator is also important. This individual will oversee the efforts of the many specialists involved in the assessment. A wide range of specialist disciplines, including noise control, will be required for a comprehensive assessment. Thus, the coordinator must be capable of developing a working knowledge of, and a familiarity with, each discipline included in the study team. This familiarization must be carried out within the constraints of time and money associated with a particular project; however, in any major assessment effort, the coordinator must have sufficient time to establish at least a minimal working knowledge of most of the disciplines represented on the study team. Thus the coordinator must be able to communicate well with the other team members.

It is essential that effective communication be established among all members of the investigative team at the outset of the assessment study. One of the most neglected—yet perhaps one of the most worthwhile—undertakings in any assessment is a comprehensive description of the proposed action by the personnel of the initiating agency. All too often, members of the investigative team are left to discover their own facts about the proposed action, in a misguided effort to maintain objectivity. The team should be adequately informed about any physical constraints on the project location and operation. Most of the alternatives would have been considered during the initial planning of the proposed facility. These considerations must be discussed with the personnel of the initiating agency so that the investigative team gains an appreciation of the practical difficulties faced by the project designers. Conversely, the personnel from the initiating agency must make every effort to develop an appreciation for the difficulties associated with environmental impact assessment. Frequent conferences between the project initiators and the project evaluators may be highly effective in furthering communication and in producing a more exact and realistic evaluation of impact.

Finally, in organizing an assessment team and in supervising an assessment study, the coordinator must establish communication, at a very early stage in the assessment operation, with the public. It is advisable to employ professional writers and communications experts to present detailed information on project characteristics to the general public. One of the chief failings of early impact assessment studies and impact statements was a virtually complete lack of communication with the public during the assessment process. If the public does not understand the characteristics of a proposed facility (e.g., the operation of a power plant), their ignorance may lead to hostility, or at best apathy, toward the proposed activity. This can be eliminated to a great extent through educational programs designed to inform the public about the proposed action and to elicit suggestions and comments from them for possible alternative activities.

ENVIRONMENTAL ASSESSMENTS OF NOISE

Because of the relatively high state of development in water- and air-pollution-control technology and because of the attention focused upon natural ecosystems, much of the investigative work in impact assessments to date has centered on the water and air pollution problems associated with particular projects and the effects of these pollutants on ecosystems. With increasing capabilities to quantify noise effects and with increasing awareness of the importance of noise as a possible health hazard, environmental assessments of noise are attaining increased significance.

Planning Implications

The intent of NEPA included the utilization of systematic, interdisciplinary studies by federal agencies in developing environmental plans. This planning ideal was to be implemented on all levels of federal agency activity. The implementation of an interdisciplinary approach is significant for environmental planning in several distinct categories:

1. Policy planning: the formulation of agency policies and procedures that could affect environmental quality
2. Program planning: the application of available technology in specific types of locations such as urban areas and coastlines
3. Functional planning: the development and construction activities carried out by an agency within a limited territory to serve a specific need and
4. Project planning: the implementation of plans for individual activities or projects such as the construction and operation of a particular facility.

The goals of NEPA included the correlation of these levels of planning in order to identify possible environmental consequences at each level in order to minimize any detrimental effects. With respect to noise control and abatement, federal policy, especially as established through legislative actions, should be subjected to environmental assessment to determine the possible impacts in the noise environment of contemplated changes, and to determine the possible alternatives to the proposed policies.

Of course, the bulk of the environmental assessments which have been performed to date in the United States have dealt with the effects and ramifications of specific projects. In all likelihood, the most significant assessment of noise impacts in the future also will be accomplished with respect to noise created during the construction, operation, and maintenance of particular facilities or activities.

Specific Project Considerations

With respect to the evaluation of noise impacts associated with specific projects, assessment activities fall into several general categories, including: the monitoring of noise in an existing environment in order to characterize that environment without the proposed project or activity; the prediction of impacts from a proposed action on the basis of modeling that action and its noise generation; and the evaluation of impacts predicted for a proposed action on the basis of direct and indirect effects, short-term and long-term impacts, secondary effects, and effects on the individual as well as on the community.

Required in the environmental assessment process is a full quantitative description of the existing environment without the proposed project or activity. Thus, it is necessary to characterize the sound environment in the area to be affected by the proposed action. In some cases, this might be a rather small geographic area, that would be affected by the project, or it could be a rather extensive area such as the land adjacent to a new highway or a new airport. In other cases, the noise environment may consist of the individual's working space in a facility or factory. Such would be the case, for instance, if the impact assessment is directed toward evaluating the effects of a regulatory move that would create a change in working conditions throughout a general industry. In monitoring the existing noise environment, it will be necessary to characterize the general outdoor environment in a rather large geographic area for most projects. The means of quantitatively describing the noise environment, and the parameters that can be used for such description, have been described in previous chapters. However, it is useful to consider several factors that could affect the choice of a parameter or descriptor for use in environmental impact assessments.

First, because of the rather large land area involved in many projects considered to have significant environmental effects, it is highly desirable to develop a parameter that can be established through automated monitoring, data acquisition, and information processing systems in order to minimize the cost of the monitoring program. In addition, because of the necessity to characterize a very large area, it may be advantageous to establish a descriptor (such as L_{eq} or L_{dn}) that can be measured with fairly simple, lightweight, and rugged equipment. Since the overall purpose of the impact-assessment procedure is to preserve the human environment, the parameters selected to describe the noise environment should correspond to parameters characteristic of human response to noise, and should be relatively easy to use; that is, they should be based upon relatively simple measurement and data-reduction procedures. Furthermore, the parameters used to characterize the noise environment should be useful in conjunction with existing regulations, such as federal regulations, local noise ordinances, and local building codes. Finally, the parameters used to describe the noise environment should be understandable to the lay public, since a major emphasis in environmental impact analysis is on the communication of information to the public and their involvement in the decisionmaking process.

Since the evaluation procedure is to a large degree a prediction process, it is necessary to be able to predict expected levels of noise from given proposed actions. The prediction of noise levels that would be generated by given activities has been studied in great detail for certain activities, such as highway construction and use,[8] whereas only rudimentary predictive techniques exist for other proposed actions. Where possible, the parameters selected for characterizing the noise environment should be amenable to use in mathematical models that will be used to simulate the noise environment and to predict noise levels created because of a proposed activity at some future date. It is mandatory that a standard measure of noise effects be employed when one is to use mathematical models, together with computer hardware, to evaluate a number of exceedingly complex alternative plans or modes of construction or operation.[20]

In regard to the prediction of noise levels to be generated by proposed actions, very little information exists for the calibration and evaluation of prediction techniques. In other words, very few impact studies and investigations have been carried out in "before-and-after" situations. The limited empirical data that are available suggest that some unexpected results may be obtained when major facility changes are made that affect the noise generated in a particular area.

As an example of a before-and-after study, the authors completed a study using systematic and periodic measurements of the noise generated in a suburban area in order to investigate the effects of the construction of an interstate highway through the area.[12] In this study, a simple descriptor of noise was utilized: decibel readings on the A-weighted scale of a standard sound level meter. This method was employed because of the similarity of the response of the meter and the response of the human ear, and because the device is so simple to use. Prior to the construction of the highway, noise levels were determined at a large number of locations distributed throughout the suburban area. Sound-pressure levels (L_{max}) were measured during morning rush hours, the noon hour, and the evening rush hours. Measurements were taken a number of times on different dates at each location. After the data had been collected, arithmetic average sound pressure level contours were drawn for the entire area.

Several years later, after completion of construction of the interstate highway through the suburban area, the monitoring was repeated. The results indicated that vehicular traffic was the predominate noise source. Throughout most of the daylight hours, noise levels in the entire suburban area far exceeded the acceptable noise level criteria established by national and international health agencies. Prior to construction, the highest sound pressure level reading (97 dBA) was obtained at the intersection of two residential/commercial streets at the noon hour. The construction of the interstate highway removed most of the trucks and many of the automobiles from the residential and commercial streets in the study area and concentrated them instead on the interstate highway. Exceedingly high sound pressure levels were created within the immediate vicinity of the interstate highway. However, sound pressure levels were significantly reduced throughout other portions of the suburban areas, due to the redistribution of traffic after completion of the interstate highway. An interesting corollary of this study was that apparently little official notice was taken of the changes in noise levels or of the possible detrimental effects of high sound pressure levels on human beings in the study area, since two hospitals were located, planned, and constructed in close proximity to the new interstate highway—after the highway's completion.

In the assessment of noise effects on particular projects or associated with particular actions, it is necessary to distinguish between the direct and indirect effects of generated noise. In this regard, direct effects could be classified as those that would occur simultaneously with the generation of a noise, or as a direct result of exposure to noise. On the other hand, indirect effects would be those that occur *after exposure* to the noise and as a result of the cumulative direct effects of the generated sound pressure levels.

In agreement with the philosophy of NEPA, it is necessary to consider short-time noise impacts created during construction of a facility or during implementation of a particular plan. However, it is also necessary to consider the long-term effects. In this regard, a significant level of noise may be created during the construction of a facility such as a power plant, whereas continued long-term operation of that plant may only create relatively insignificant noise levels. On the other hand, the noise levels associated with construction of an airport facility may be insignificant when compared to the long-term noise impacts associated with the operation of the airport and the generation of traffic on roadways connecting urban centers with the airport.

In addition, the secondary effects of noise created by a particular action must be considered—for example, the overall effects upon a community of the noise created by a highway that bisects that community. Such noise may lead to the necessary renovation of such public buildings as schools, to include air conditioning, baffles, screens, and other means of reducing noise exposure in the classroom. These measures may have undesirable side effects. The location of a noisy facility in a residential area may discourage informal community social activities that would take place outdoors. In some instances, the entire sense of community well-being within a given area has been altered drastically by the generation of noise in nearby facilities. It is important in any impact assessment to include community effects as well as the effects upon individuals. To illustrate some of these considerations, an example of an impact assessment that includes significant studies of noise impacts is presented in the following section.

Example of a Highway Impact Study

As an example of a comprehensive evaluation of environmental impact, including a significant study of noise impact, the study of the location, design, and use of Interstate 66 in the Washington, D.C., area is valuable.[32, 33] The evaluation of the impact of the proposed interstate highway included a comprehensive study plan, with the EIS being only a portion of the study plan. The complete study was divided into seven distinct phases:

Phase I: Inventory
Phase II: Evaluation Criteria
Phase III: Identification of Alternatives
Phase IV: Evaluation and Selection of Feasible Alternatives
Phase V: Detailed Evaluation of Feasible Alternatives
Phase VI: Refinements of Selected Alternatives
Phase VII: Conclusions.

In the inventory phase, the limits of the study area were determined by the boundaries of impact. The geographic limits varied according to the particular area of impact being studied. For example, the area of noise impact was not considered to be nearly as large as the area of impact for land-use changes or for secondary economic impacts. In phase I, an initial community attitude survey was conducted to identify and evaluate local desires regarding transportation needs, neighborhood character and development, noise, and other environmental factors.

In phase II, the goals and objectives for the region as a whole, and for individual communities, were identified. This identification included local and regional environmental values as well as transportation objectives. These goals and objectives were used by the study team in establishing criteria for evaluation of alternative transportation plans. In phase III, all possible transportation systems that appeared to be at all feasible for meeting the requirements of moving people, goods, and services in the study area were considered. New technology, including monorails, air-cushion vehicles, and other innovative techniques, was studied in addition to conventional transportation facilities. Included in this phase was an evaluation of the noise predicted to be generated by each mode of transportation as selected. In phase IV, further refined and preliminary system layouts were sketched. In phase V, selected feasible alternatives underwent detailed evaluation.

These detailed evaluations included community/sociological, ecological/natural environment, and economic/cost impact data. At this stage of the investigation, noise contours for the various transportation corridors were calculated on the basis of standard predictive models and the anticipated volumes of traffic on the proposed facilities. In phase VI, the selected alternative schemes were further refined, and in phase VII the preferred alternative was selected. In addition, in phase VII, the need for any additional economic, social, and environmental studies was analyzed. Adverse effects in all of these areas were identified, as were appropriate measures designed to eliminate or minimize these detrimental effects. This final phase in the study culminated in the preparation of a complete EIS.

In connection with the noise survey for the I-66 study, a firm of consulting engineers assessed the effects of the predicted generated noise. Considerable attention was given to the psychological as well as the physiological effects of the noise predicted to be generated by the new transportation system. The physiological effects evaluated included sleep intrusion, constriction of the veins, and loss of hearing. Psychological effects were judged to depend upon the individual involved and the characteristics of the sound. Individual studies were made to a limited extent. The noise in the I-66 study area was evaluated on the basis of procedures established by the Federal Highway Administration, which has established a standard (L_{10}) for exterior noise in residential areas at 70 dBA. The most significant measures of highway noise, in terms of individual annoyance, were taken to be the peak sound-pressure levels rather than the average sound-pressure levels associated with the highway traffic. The peak noise levels in turn were associated with the acceleration of vehicles.

Included in the assessment study for I-66 was an evaluation of various noise-abatement procedures. The noise abatement procedures and structures that were selected were chosen on the basis of predictions of noise levels from alternate transportation systems and adjacent land use in the I-66 corridor. Noise levels were predicted with the use of a highway noise computer model calibrated for specific use in the I-66 corridor. This model was validated with data obtained on ambient sound levels taken earlier at key points in the study area. Additional sound pressure level readings were taken during the preparation of the environmental assessment. Ambient noise data were collected at critical receptors such as hospitals and schools. Additional data were collected at points near the alternatives in order to analyze possible changes in the noise levels created from changing the traffic patterns on other arteries.

The predicted community noise levels for each of the alternative systems were then compared with the existing levels of ambient noise. Abatement measures were evaluated for those parts of each alternative system where an adverse change in noise level was anticipated. The alternative transportation systems were compared on the basis of noise-evaluation criteria, which included the number of critical receptors exposed to various noise levels, the cost of noise-abatement devices, and the net change in the number of persons exposed to noise levels in excess of recommended standards. The noise impact in this study was classified by land use type, such as residential, school, commercial, park, and open space. After the noise levels in the communities were predicted for different times of the day for each alternative system, the cumulative noise dosage for a particular area from a proposed transportation system was estimated. This dosage was computed as person-hours of exposure above given thresholds of noise, such as 70 dBA. Consideration was given to different types of land use; noise levels and exposures were then determined at those times when the land was in maximum use. For example, residential occupancy is ordinarily low during peak traffic hours, while noise that affects schools occurs during the midday hours between 8:00 A.M. and 4:00 P.M.

Contour maps of expected noise levels were produced for each proposed transportation system design and community landscape. Various design features including noise abatement techniques were taken into account through the use of the computer model. Noise abatement strategies were included in the refinement of selected alternatives with regard to both physical design of the facilities and traffic operation. These measures for noise abatement included possible construction of noise barriers or earth berms, constructing the transportation systems at a lower elevation than the level of the surrounding land, or establishing a buffer zone between the transportation corridor and the community. With respect to the use of highways in the corridor, noise abatement methods included the possible use of special surface materials for the roadway, the location of interchanges at points of lowest noise impact, the grading of the highway and ramps for least noise impact, and the regulation of the vehicle mix (trucks and autos), speed, and volume of traffic on the roadways.

SUMMARY

The National Environmental Policy Act of 1969 has emerged as the most important legislative tool to date for environmental protection and land use planning in the United States. This law includes requirements for comprehensive evaluation of proposed actions and all feasible alternatives including the null or do-nothing alternative. The enforcement of NEPA requirements has rested with the judiciary system in the United States. In enforcing this law, the courts have adhered to the letter of the law and have required strict compliance with its provisions on the part of all federal agencies.

In preparing an EIS, certain procedures are required by law, such as the provision of sufficient time for agency and citizen reaction to the draft and final versions of impact statements. Other items, which are not covered in the law itself, include the methodologies for impact assessment and the organization of an assessment team.

No one single methodology is sufficiently flexible to be applicable to all systems and all projects. A combination of the use of matrices to identify areas of impact, checklists to ensure a comprehensive review, and the assessment of classes of action and classes of impact offers perhaps the most promising means of accomplishing meaningful environmental assessment.

With respect to noise control and abatement, the EIS has great significance with respect to future environmental planning. Noise-related impact statements will be required for legislative actions, for development of new regulations, and for the design, construction, and operation of specific projects. With respect to specific projects, it will be necessary to characterize the noise in the existing environment and to predict future generated noise levels on the basis of feasible alternatives. The impact prediction must include direct and indirect noise effects for both short-term and long-term impacts. Secondary effects of the noise must not be neglected. Finally, the effects on the community as a whole as well as individual effects must be adequately considered.

THE NATIONAL ENVIRONMENTAL POLICY ACT

To establish a national policy for the environment, to provide for the establishment of a Council on Environmental Quality, and for other purposes.

Be it enacted by the Senate and House of Representatives of the United States of America in Congress assembled, That this Act may be cited as the "National Environmental Policy Act of 1969."

PURPOSE

Sec. 2. The purposes of this Act are: To declare a national policy which will encourage productive and enjoyable harmony between man and his environment; to promote efforts which will prevent or eliminate damage to the environment and biosphere and stimulate the health and welfare of man; to enrich the understanding of the ecological systems and natural resources important to the Nation, and to establish a Council on Environmental Quality.

TITLE I

DECLARATION OF NATIONAL ENVIRONMENTAL POLICY

Sec. 101. (a) The Congress, recognizing the profound impact of man's activity on the interrelations of all components of the natural environment, particularly the profound influences of population growth, high-density urbanization, industrial expansion, resource exploitation, and new and expanding technological advances and recognizing further the critical importance of restoring and maintaining environmental quality to the overall welfare and development of man, declares that it is the continuing policy of the Federal Government, in cooperation with State and local governments, and other concerned public and private organizations, to use all practicable means and measures, including financial and technical assistance, in a manner calculated to foster and promote the general welfare, to create and maintain conditions under which man and nature can exist in productive harmony, and fulfill the social, economic, and other requirements of present and future generations of Americans.

(b) In order to carry out the policy set forth in this Act, it is the continuing responsibility of the Federal Government to use all practicable means, consistent with other essential considerations of national policy, to improve and coordinate Federal plans, functions, programs, and resources to the end that the Nation may—

(1) fulfill the responsibilities of each generation as trustee of the environment for succeeding generations;

(2) assure for all Americans safe, healthful, productive, and esthetically and culturally pleasing surroundings;

(3) attain the widest range of beneficial uses of the environment without degradation, risk to health or safety, or other undesirable and unintended consequences;

(4) preserve important historic, cultural, and natural aspects of our national heritage, and maintain, wherever possible, an environment which supports diversity and variety of individual choice;

(5) achieve a balance between population and resource use which will permit high standards of living and a wide sharing of life's amenities; and

(6) enhance the quality of renewable resources and approach the maximum attainable recycling of depletable resources.

(c) The Congress recognizes that each person should enjoy a healthful environment and that each person has a responsibility to contribute to the preservation and enhancement of the environment.

Sec. 102. The Congress authorizes and directs that, to the fullest extent possible: (1) the policies, regulations, and public laws of the United States shall be interpreted and administered in accordance with the policies set forth in this Act, and (2) all agencies of the Federal Government shall–

(A) utilize a systematic, interdisciplinary approach which will insure the integrated use of the natural and social sciences and the environmental design arts in planning and in decisionmaking which may have an impact on man's environment;

(B) identify and develop methods and procedures, in consultation with the Council on Environmental Quality established by title II of this Act, which will insure that presently unquantified environmental amenities and values may be given appropriate consideration in decisionmaking along with economic and technical considerations;

(C) include in every recommendation or report on proposals for legislation and other major Federal actions significantly affecting the quality of the human environment, a detailed statement by the responsible official on–

(i) the environmental impact of the proposed action,
(ii) any adverse environmental effects which cannot be avoided should the proposal be implemented,
(iii) alternatives to the proposed action,
(iv) the relationship between local short-term uses of man's environment and and the maintenance and enhancement of long-term productivity, and
(v) any irreversible and irretrievable commitments of resources which would be involved in the proposed action should it be implemented.

Prior to making any detailed statement, the responsible Federal official shall consult with and obtain the comments of any Federal agency which has jurisdiction by law or special expertise with respect to any environmental impact involved. Copies of such statement and the comments and views of the appropriate Federal, State, and local agencies, which are authorized to develop and enforce environmental standards, shall be made available to the President, the Council on Environmental Quality and to the public as provided by section 552 of title 5, United States Code, and shall accompany the proposal through the existing agency review processes;

(D) study, develop, and describe appropriate alternatives to recommended courses of action in any proposal which involves unresolved conflicts concerning alternative uses of available resources;

(E) recognize the worldwide and long-range character of environmental problems, and, where consistent with the foreign policy of the United States, lend appropriate support to initiatives, resolutions, and programs designed to maximize international cooperation in anticipating and preventing a decline in the quality of mankind's world environment;

(F) make available to States, counties, municipalities, institutions, and individuals, advice and information useful in restoring, maintaining, and enhancing the quality of the environment;

(G) initiate and utilize ecological information in the planning and development of resource-oriented projects; and

(H) assist the Council on Environmental Quality established by title II of this Act.

Sec. 103. All agencies of the Federal Government shall review their present statutory authority, administrative regulations, and current policies and procedures for the purpose of determining whether there are any deficiencies or inconsistencies therein which prohibit full compliance with the purposes and provisions of this Act and shall propose to the President not later than July 1, 1971, such measures as may be necessary to bring their authority and policies into conformity with the intent, purposes, and procedures set forth in this Act.

Sec. 104. Nothing in Section 102 or 103 shall in any way affect the specific statutory obligations of any Federal agency (1) to comply with criteria or standards of environmental quality, (2) to coordinate or consult with any other Federal or State agency, or (3) to act, or refrain from acting contingent upon the recommendations or certification of any other Federal or State agency.

Sec. 105. The policies and goals set forth in this Act are supplementary to those set forth in existing authorizations of Federal agencies.

TITLE II

COUNCIL ON ENVIRONMENTAL QUALITY

Sec. 201. The President shall transmit to the Congress annually beginning July 1, 1970, an Environmental Quality Report (hereinafter referred to as the "report") which shall set forth (1) the status and condition of the major natural, manmade, or altered environmental classes of the Nation, including, but not limited to, the air, the aquatic, including marine, estuarine, and fresh water, and the terrestrial environment, including, but not limited to, the forest dryland, wetland, range, urban, suburban, and rural environment; (2) current and forseeable trends in the quality, management and utilization of such environments and the effects of those trends on the social, economic, and other requirements of the

Nation; (3) the adequacy of available natural resources for fulfilling human and economic requirements of the Nation in the light of expected population pressures; (4) a review of the programs and activities (including regulatory activities) of the Federal Government, the State and local governments, and nongovernmental entities or individuals, with particular reference to their effect on the environment and on the conservation, development and utilization of natural resources; and (5) a program for remedying the deficiencies of existing programs and activities, together with recommendations for legislation.

Sec. 202. There is created in the Executive Office of the President a Council on Environmental Quality (hereinafter referred to as the "Council"). The Council shall be composed of three members who shall be appointed by the President to serve at his pleasure, by and with the advice and consent of the Senate. The President shall designate one of the members of the Council to serve as Chairman. Each member shall be a person who, as a result of his training, experience, and attainments, is exceptionally well qualified to analyze and interpret environmental trends and information of all kinds; to appraise programs and activities of the Federal Government in the light of the policy set forth in title I of this Act; to be conscious of and responsive to the scientific, economic, social, esthetic, and cultural needs and interests of the Nation, and to formulate and recommend national policies to promote the improvement of the quality of the environment.

Sec. 203. The Council may employ such officers and employees as may be necessary to carry out its functions under this Act. In addition, the Council may employ and fix the compensation of such experts and consultants as may be necessary for the carrying out of its functions under this Act, in accordance with section 3109 of title 5, United States Code (but without regard to the last sentence thereof).

Sec. 204. It shall be the duty and function of the Council–

(1) to assist and advise the President in the preparation of the Environmental Quality Report required by section 201;

(2) to gather timely and authorative information concerning the conditions and trends in the quality of the environment both current and prospective, to analyze and interpret such information for the purpose of determining whether such conditions and trends are interfering, or are likely to interfere, with the achievement of the policy set forth in title I of this Act, and to compile and submit to the President studies relating to such conditions and trends;

(3) to review and appraise the various programs and activities of the Federal Government in the light of the policy set forth in title I of this Act for the purpose of determining the extent to which such programs and activities are con-

tributing to the achievement of such policy, and to make recommendations to the President with respect thereto;

(4) to develop and recommend to the President national policies to foster and promote the improvement of environmental quality to meet the conservation, social, economic, health, and other requirements and goals of the Nation;

(5) to conduct investigations, studies, surveys, research, and analyses relating to ecological systems and environmental quality;

(6) to document and define changes in the natural environment, including the plant and animal systems, and to accumulate necessary data and other information for a continuing analysis of these changes or trends and an interpretation of their underlying causes;

(7) to report at least once each year to the President on the state and condition of the environment; and

(8) to make and furnish such studies, reports thereon, and recommendations with respect to matters of policy and legislation as the President may request.

Sec. 205. In exercising its powers, functions, and duties under this Act, the Council shall:

(1) consult with the Citizens' Advisory Committee on Environmental Quality established by Executive Order numbered 11472, dated May 29, 1969, and with such representatives of science, industry, agriculture, labor, conservation organizations, State and local governments, and other groups, as it deems advisable; and

(2) utilize, to the fullest extent possible, the services, facilities, and information (including statistical information) of public and private agencies and organizations, and individuals, in order that duplication of effort and expense may be avoided, thus assuring that the Council's activities will not unnecessarily overlap or conflict with similar activities authorized by law and performed by established agencies.

Sec. 206. Members of the Council shall serve full time and the Chairman of the Council shall be compensated at the rate provided for Level II of the Executive Schedule Pay Rates (5 U.S.C. 5313). The other members of the Council shall be compensated at the rate provided for Level IV of the Executive Schedule Pay Rates (5 U. S.C. 5315).

Sec. 207. There are authorized to be appropriated to carry out the provisions of this Act not to exceed $300,000 for fiscal year 1970, $700,000 for fiscal year 1971, and $1,000,000 for each fiscal year thereafter.

REFERENCES

1. Adkins, W. G., *et al.*, "Social, Economic, and Environmental Factors in Highway Making," *148 Progress Report*, Study 2-1-71-148, Texas Highway Department and Federal Highway Administration, Washington, 1971.

2. Amlung, D. K., and Eith, A. W., "An Assessment of Methodologies of Environmental Planning," M.E. Thesis, University of Louisville, Louisville, May 1975.

3. Battelle Memorial Institute, Columbus Laboratories, "Final Report on an Environmental Evaluation System for Water Resource Planning to the Bureau of Reclamation, Department of the Interior," January 1972.

4. Carter, Everett C., Haefner, Lonnie E., and Hall, Jerome W., "An Informational Report on Techniques for Evaluating Factors Relevant to Decision Making on Highway Locations," Department of Transportation, Federal Highway Administration, Office of Environmental Policy, Washington, November 1972.

5. Council on Environmental Quality, "Preparation of Environmental Impact Statements: Guidelines, " 38 *Federal Register*, 20550, August 1, 1973.

6. Dries, David A., "Environmental Impact Assessment Methodology," M.E. Thesis, Speed Scientific School, University of Louisville, May 1974.

7. Engelen, R. E., and Stuart, D. G., "Transport Technologies and Urban Structure: A Framework for Evaluation," Presented at the ASCE-ASME National Meeting on Transportation Engineering, Seattle, July 1971.

8. Galloway, W. J., Clark, W. E., and Kerrick, J. S., "Highway Noise: Measurement, Simulation and Mixed Reactions," National Coop. Highway Research Program Report 78, Highway Research Board, Washington, 1969.

9. Girand, J., Technical movie of Papago Freeway, *Transportation Engineering Journal of ASCE*, **97**, (TE4): 619–626 (November 1971).

10. Haefner, L. E., and Redding, M. J., "An Analytical Structure of Community Public Works Decision Process," *Papers and Proceedings*, Joint Conf. Env. Des. Res. Assoc. and Amer. Inst. Arch., Irvine, California, January 1972.

11. Hagerty, D. J., and Heer, J. E. Jr., Impact assessment of power plants, *Journal of the Power Division*, ASCE, **102** (PO1): 81–94 (January 1976).

12. Heer, J. E., Hagerty, D. J., and Knadler, D. J., Effect of an interstate highway on urban area noise levels, *Public Works* **105** (1), January 1974.

13. Jain, R. K., Urban, L. V., and Stacey, G. S., *Handbook for Environmental Impact Analysis*, Dept. of Army, CERL, Champaign, Illinois, (Draft), April, 1974.

14. Leopold, Luna B., Clarke, Frank E., Hanshaw, Bruce B., and Balsley, James R., "A Procedure for Evaluating Environmental Impact," Geological Survey Circular 645, U.S. Government Printing Office, Washington, 1971.

15. McHarg, Ian L., *Design With Nature*, Natural History Press, Garden Press, Garden City, New York, 1969.

16. Oglesby, C. H., Bishop, B., and Willeke, C. E., A method for decisions among freeway location alternatives based on user and community consequences," *Hwy. Res. Record* 305, Highway Research Board, Washington, 1970.

17. Redding, M. J., "The Quality of Residential Environments, Preference for Accessibility to Residential Opportunities," Ph.D. Thesis, Northwestern University Technological Institute, Evanston, June 1970.

18. Schimpeler, C. C., and Grecco, W. L. Systems evaluation: An approach based on community structure and values, *Hwy. Res. Record* 238, Highway Reseach Board, Washington, 1968.

19. Schlager, K., "The Rank-Based Expected Value Method of Plan Evaluation," *Hwy. Res Record* 238, Highway Research Board, Washington, 1968.

20. Sutherland, L. C., Braden, M. H., and R. Colman, "A Program for the Measurement of Environmental Noise in the Community and Its Associated Human Response," Wyle Research Laboratories (for the Department of Transportation), El Segundo, California, December 1973.

21. Twiss, Robert H., Linking the EIS to the planning process, *Environmental Impact Assessment: Guidelines and Commentary*, ed. Thomas G. Dickert, University Extension, University of California, Berkeley, 1974.

22. United States Atomic Energy Commission, "Preparation of Environmental Reports for Nuclear Power Plants," Regulatory Guide 4.2, U.S. Government Printing Office, Washington, March 1973.

23. United States Department of Agriculture, Environmental Impact Statements: Proposed Guidelines, 38 *Federal Register*, 31904, November 19, 1973.

24. U.S. Department of the Army, *Handbook For Environmental Impact Analysis* Construction Engineering Research Laboratory, Champaign, April 1974.

25. U.S. Environmental Protection Agency, Preparation of Environmental Impact Statements-Interim Regulation, Washington, D.C., 38 *Federal Register,* 1696 (Part II, January 17, 1973).

26. U.S. Environmental Protection Agency, Office of Federal Activities, "Manual for Preparation of Environmental Impact Statements for Wastewater Treatment Works, Facilities Plans, and 208 Areawide Waste Treatment Management Plans," Washington, July 1974.

27. United States Department of Housing and Urban Development, "Handbook of Departmental Policies, Responsibilities and Procedures for Protection and

Enhancement of Environmental Quality, 38 *Federal Register*, 19182, July 18, 1973.

28. Werner, Robert R., "Environmental Impact and the U.S. Army Corps of Engineers," Presented at Ecological Society of America Symposium, Amer. Assoc. Adv. Sci., Washington, December 1972.

29. "Environmental Quality as an Objective of Water Resources Planning," Special Printing, Comm. Int. and Ins. Affairs, U.S. Senate, U.S. Govt. Printing Office, Washington, 1972.

30. Hearings on S. 1075, S. 237 and S. 1752 Before the Senate Committee on Interior and Insular Affairs, 91st Congress, 1st Session, p. 117.

31. *Proceedings*, Panel Discussion on the Systematic Interdisciplinary Approach in Highway Planning and Design, ed. MIT Trans. and Comm. Values Proj., Urban Sys. Lab. with Fed. Hwy. Adm., Cambridge, March, 1973.

32. Virginia Department of Highways, *The I-66 Question*, Virginia Department of Highways, Alexandria, Virginia, Vol. 1, No. 1, October 1972.

33. Virginia Department of Highways, *The I-66 Question*, Virginia Department of Highways, Alexandria, Virginia, Vol. 2, No. 4 April 1973.

34. *Calvert Cliffs' Coordinating Committee* v. *Atomic Energy Commission*, 449 F. 2d 1109, 1 ELR 20346 (D.C. Cir. 1971). *Cert denied*, 404 U.S. 942 (1972).

35. *Environmental Defense Fund* v. *Ruckelshaus*, 439 F. 2d at 597, 1 ELR at 20064 (D. C. Cir. 1971).

36. *Environmental Defense Fund* v. *Hardin*, 325 F. Supp. 1401, 1 ELR 20207 (D.D.C. 1971).

37. *Environmental Defense Fund* v. *Corps of Engineers* 325 F. Supp. 728, 1 ELR 20130 (E.D. Ark. 1970–71), 4 ERC 1097, 2 ELR 20260 (E.D. Ark.) 342 F. Supp. 1211, 2 ELR 20353 (E.D. Ark.), *Aff'd.* 470 F. 2d 289, 2 ELR 20740 (8th Cir. 1972).

38. *Hanly* v. *Kleindienst*, F. 2d, 2 ELR 20717 (2d. Cir. 1972).

39. *Hanly* v. *Mitchell*, 460 F. Supp. 640, 2 ELR 20181 (S.D. N.Y.), 460 F. 2d. 640, 2 ELR 20216 (2d Cir.), *cert. denied*, 41 U.S.L.W. 3247 (Nov. 7, 1972), sub. nom. *Hanly* v. *Kleindienst*, 471 F. 2d 822, 4 ERC 1785 (2nd Cir. Dec. 5, 1972), *cert. denied*, 484 F. 2d 448, 5 ETC 1416 (May 21, 1973).

40. *Natural Resources Defense Council* v. *Morton*, 337 F. Supp. 165, 2 ELR 20028 (D.D.C. 1971), 337 F. Supp. 167, 2 ELR 20089 (D.D.C. 1971), *motion for summary reversal denied*, 458 F. 2d 827, 2 ELR 20029 (D.C. Cir.), *dismissed as moot*, 337 F. Supp. 170, 2 ELR 20071 (D.D.C. 1972).

41. *SCRAP* v. *United States*, 346 F. Supp. 189, 195 note 8, 2 ELR 20486 20488 note 8 (D.D.C.), *application for stay pending appeal denied sub nom. Aberdeen & Rockish Ry. Co.* v. *SCRAP*, –U.S.–, 2 ELR 20491 (1972) (Burger, Circuit Judge), *cert. granted* 41 U.S.L.W. 3339 (Dec. 19, 1972).

42. *SCRAP* v. *United States*, 346 F. Supp. 189, 2 ELR 20486 (D.D.C. 1972), *rev'd on other grounds*, 5 ERC 1449 (S. Ct. June 18, 1973).

5

Introduction to basic principles of noise control

Mancil W. Milligan
Professor and Head
Department of Mechanical and Aerospace Engineering
The University of Tennessee
Knoxville, Tennessee

All noise control problems can be fundamentally represented as shown in Fig. 5-1. In each problem, a source or sources generate the noise, and both direct and indirect paths exist by which the generated noise is transmitted to one or more receivers. In general, the conditions of noise level and exposure time for the receiver are fixed by occupational safety and health standards for industrial situations or by some other type of standard for nonindustrial situations.

RECEIVER

There is little opportunity for noise control at the receiver. Normally, the permissible noise levels are set for the receiver, and engineering techniques must be used on the source and/or paths in order to limit the exposure of the receiver. Current OSHA (Occupational Safety & Health Administration) standards permit the use of personal hearing protection (earplugs

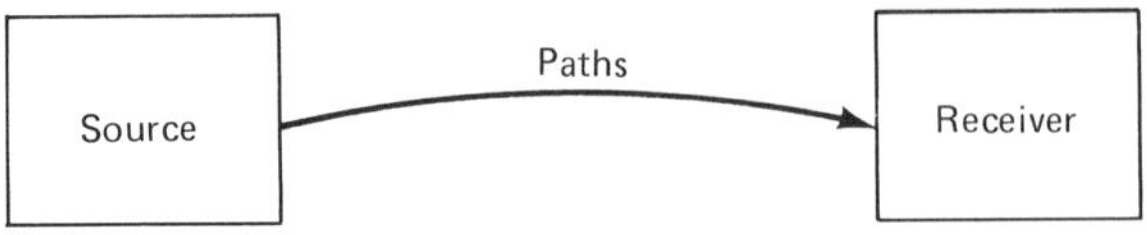

Fig. 5-1 Fundamental elements in noise control.

or earmuffs) if feasible engineering and/or administrative control techniques cannot reduce the exposure to an acceptable level. Personal hearing protection devices are in reality a treatment of the transmission path from source to receiver but, since they are worn by the receiver, it makes some sense to treat them as part of the receiver. Personal ear protection and hearing conservation procedures are discussed in more detail in Chapter 7.

One other aspect relating to the receiver is associated with controlling noise exposure by so-called administrative techniques. In other words, we can control both the exposure time and level for workers by arranging their work pattern in a way that limits the amount of time that they spend in certain noise level environments. Theoretically, this appears to be a powerful concept but in practice it has limited potential. Due to worker skill requirements and union job descriptions, it is often very difficult to arrange an individual's work pattern so that it consists of activities in significantly different noise level environments. Another complication is the difficulty of assessing the cost. We would normally expect some loss in productivity when the range of required skills is increased. Presently, there appears to be no basis for assessing the cost of this type of noise control even when the work pattern could be arranged to limit exposure.

Table 5-1 indicates the permissible noise exposures currently associated with most occupational safety and health standards.

Table 5-1 Permissible Noise Exposures.[a]

Duration (per day in hours)	Sound Level (dBA, slow response)
8	90
6	92
4	95
3	97
2	100
1.5	102
1	105
0.5	110
0.25 or less	115

[a]When the daily noise exposure is composed of two or more periods of different levels, the combined effect should be considered, rather than the individual effect of each. That is, if the sum of the following fractions exceeds unity (1), then the mixed exposure is considered to exceed the limit value.

$$\frac{C_1}{T_1} + \frac{C_2}{T_2} \cdots \frac{C_n}{T_n} \tag{5-1}$$

where

C_n = total time of exposure at a specified noise level
T_n = permissible item at that exposure.

An example will illustrate how to determine compliance. Suppose an employee works for 6 hours in a noise level of 90 dBA, 1 hour in a level of 100 dBA, and 0.5 hour in a level of 105 dBA. Let us calculate the cumulative exposure for this working day;

$$\left.\frac{4}{8}\right|_{90\text{ dBA}} + \left.\frac{1}{2}\right|_{100\text{ dBA}} + \left.\frac{0.5}{1}\right|_{105\text{ dBA}} = 1.5 \tag{5-2}$$

Since the cumulative exposure fraction exceeds unity, this employee's sound exposure is not in compliance. Now, if by administrative control we could arrange this employee's work pattern so that he spends 2 hours at 90 dBA, 0.5 hour at 100 dBA, 0.5 hour at 105 dBA, and the remainder of the work shift at less than 90 dBA, then the employee would be in compliance since

$$\left.\frac{2}{8}\right|_{90\text{ dBA}} + \left.\frac{0.5}{2}\right|_{100\text{ dBA}} + \left.\frac{0.5}{1}\right|_{105\text{ dBA}} = 1.0 \tag{5-3}$$

In summary, the opportunities for noise control at the receiver are worthy of consideration but are usually very limited in practice. The real challenges in noise control are associated with interrupting the path and abating sound at the source.

THE PATHS

We will be concentrating our attention in this chapter on the reduction of noise at the source. However, in addition to source control, it is possible to obtain significant noise reduction at the receiver by controlling the noise that "builds up" from the indirect paths typified in Fig. 5-1. This phenomenon, termed *reverberation*, is caused by the lack of absorbing surface in the region around the source. The reflected noise level will increase until the rate of sound absorption is equal to the rate of sound generation. Walls, floor, and ceilings in industrial plants are typically hard and acoustically reflective; thus, reverberation or indirect path transmission may significantly affect the noise level at a specific location.

If each surface of a room were highly sound absorbent, the intensity of the sound would be diminished each time the sound is reflected. Therefore, the sound energy would decrease, at a rate that depends largely upon the amount of absorption in the room. The rate of decay is usually characterized as reverberation time and is defined as *that amount of time required for the average sound pressure level to decrease 60 dB after the sound source is silenced.* The reverberation time (t) can be roughly approximated as:

$$t = \frac{0.161\ V}{a}\ (\text{sec}) \tag{5-4}$$

where

V = room volume (m^3)
a = total room absorption in sabins (m^2).

Often, it is more convenient to calculate the total absorption, a, rather than measure the reverberation time and use Equation 5-4 to calculate a. The total absorption, a, can be calculated by summing the product of area, s, and corresponding acoustical absorption coefficient, α, for each surface.

$$a = S_1\alpha_1 + S_2\alpha_2 + S_3\alpha_3 + \cdots + S_k\alpha_k \tag{5-5}$$

where

S_k = surface area of surface k in square meters
α_k = absorption coefficient associated with surface k.

The average absorption coefficient, $\bar{\alpha}$, for a room is defined as the total absorption, a, divided by the total surface area, S.

$$\bar{\alpha} = \frac{a}{S} \tag{5-6}$$

See Chapter 6 for additional discussion of these principles.

With this background, we can now look at some of the characteristics associated with the semireverberant fields found in most industrial situations. If we were to walk away from a noisy machine in a typical industrial room, we would observe that the noise level decreases for some distance at a rate similar to an outdoor environment or free field. At some point, however, depending on room acoustics, we would observe little or no additional reduction in sound level as we moved farther away from the source. The parameters that describe this semireverberant field can be related in an approximate manner by the following relationship:

$$SPL = PWL + 10 \log_{10}\left(\frac{1}{4\pi r^2} + \frac{4}{R}\right)(\text{dB}) \tag{5-7}$$

where

SPL = sound pressure level referenced to 0.0002 dyne/cm^2
PWL = sound power level referenced to 10^{-12} W
r = distance from source in meters
R = room constant $= \dfrac{s\bar{\alpha}}{1-\bar{\alpha}}$.

By examining a plot of Equation 5-7, it is possible to gain insight into the importance of the direct and indirect transmission paths. Figure 5-2 graphically depicts the relative sound pressure level in decibels as a function of distance

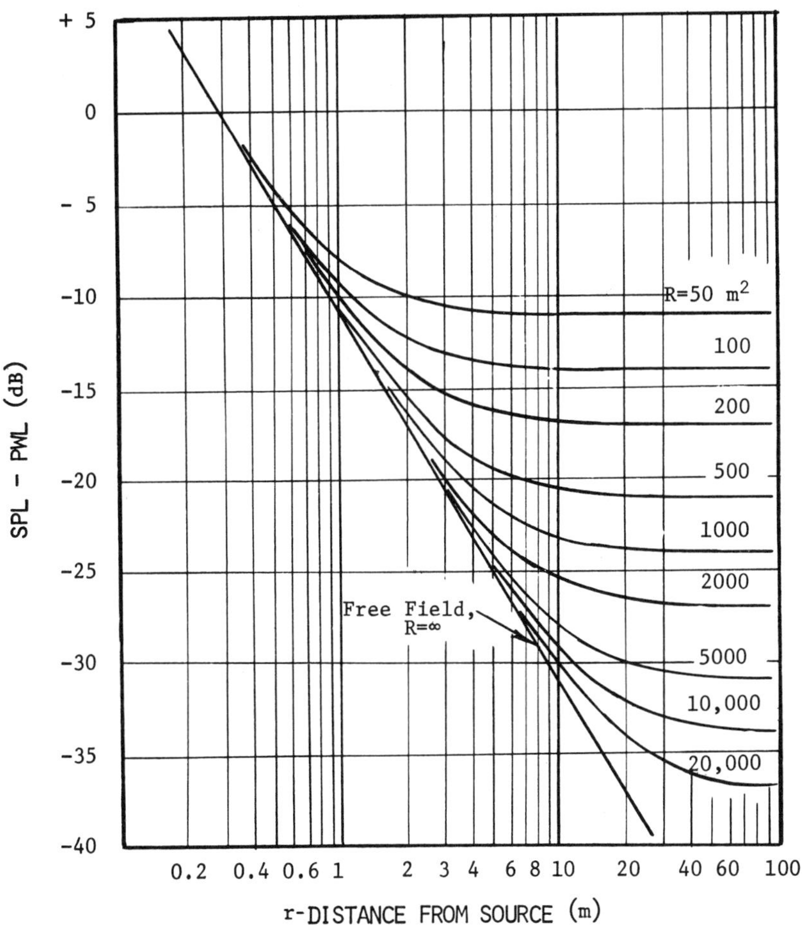

Fig. 5-2 Sound pressure level behavior in a large enclosure.

from the source for various values of room constant. The straight line is for free-field, where $R = \infty$. This condition exists when all sound waves are absorbed and no reflected sound is present. The free-field condition is approximated outdoors in the absence of large reflective surfaces and may be approximated in an anechoic room. It can be seen from Equation 5-7 and Fig. 5-2 than when the room constant is large, the decrease in sound pressure level with distance from the source remains close to the free-field values longer than when the room constant is small. In the reverberant part of the field, where the curves are almost parallel to the x-axis, sound pressure level changes little with distance and depends primarily upon the room constant.

Since the room constant, R, is proportional to the total absorption, a, it is obvious that we can change the room constant in a workspace by adding absorbing materials to the reflective surfaces. Now, the true value of the simple model presented in Fig. 5-2 becomes apparent. If the employee in question is in the near field (the sound-level plots as the free-field straight line), then adding absorptive materials in the room will not change the noise level for the individual. Many plant managers have been misled by fast-talking or uninformed sellers of acoustical absorbing panels, surfaces, and so on. Either the salespeople are ignorant (they do not understand the model in Fig. 5-2) or dishonest (they do understand the model and know *a priori* than no amount of acoustical treatment in the form of increased absorption will decrease the noise level at the location in question). In either case, many dollars have been wasted on the installation of absorbing surfaces when there was no potential for a solution to the noise problem.

In those situations where the worker is located in the reverberant field (on the flat portion of the plot in Fig. 5-2), it is possible to lower the noise level at the worker by increasing the room constant (i.e., increasing total absorption of the room). For example, if the employee were 10 meters from the noise source in a room with a room constant of 100, the noise level at the employee could be reduced by 9 dB if the room constant could be increased to 1000.

A detailed discussion of the treatment of sound paths including enclosures and their application is presented in Chapter 6. If one always investigates noise problems from the standpoint of Fig. 5-2, then the potential of path treatment will be obvious before any expenditure of resources is necessary.

THE SOURCE

In the previous sections, we have briefly explored some fundamental concepts associated with noise control as it relates to the path and receiver. There are at least three primary areas in which control of noise generated by a source may be initiated: proper design, proper equipment operation, and equipment maintenance.

The *proper design* of equipment to minimize noise generation is a rather complex engineering problem requiring a strong background in the fundamentals of vibrations, fluid mechanics, dynamics, materials, and machine design. It should be apparent that in a treatment such as this, it is impossible to develop the background required to effectively design for noise control. In a very general way, some of the things that may result from good noise design are:

1. Using shock-absorbing techniques to absorb impact energy (for example, the use of nonmetallic gears to reduce the noise generated by the metal-to-metal impact associated with metallic gears; the use of flexible mounts to support wood planer knives to reduce the noise generated by the impact of the knives on the board).
2. Using efficient flow techniques to reduce noise associated with high-fluid velocities and turbulence (for example, the new quiet hydraulic pumps in which the flow paths have been redesigned to give less turbulence; the reduction in noise generated by control valves in both gas and liquid systems achieved by simply reducing system pressure to as low as possible).
3. Reducing fluid jet velocities since jet noise is proportional to the eight power of jet velocity. Often the air jets from pneumatic control valve exhaust ports are a major source of industrial noise. By reducing such jet velocities, much quieter operation is achieved.
4. Reducing sound-radiating areas. For example, as much as 5 dBA reduction has been achieved at the operator of industrial sewing machines by detaching the metal cloth table from the machine and mounting the table separately on a wooden workbench. Also, in some situations it is possible to divide a large radiating surface into smaller areas, thus changing both the frequency and magnitude of the noise emitted.
5. Reducing peak accelerations by utilizing the maximum time to produce required velocity or displacement changes. This is critical in the design of cams and cam followers. In many industrial machines, such as those used in making straight pins, the major noise sources are the cams and followers.
6. Significantly separating operating speeds and resonant speeds. Since large-amplitude vibrations are associated with resonant speeds, noise generation is reduced by keeping operating speeds well removed from the resonant speeds to prevent excitation at resonance.

Although this list is incomplete, one general conclusion can be drawn: the optimum time to incorporate engineering design considerations relating to noise is the initial design phase. Many of these considerations offer potential for abating the noise generated by existing sources; however, they would usually require extensive and expensive redesign and modification.

The second primary area is *proper operation*. All equipment should be operated at the design conditions. Operating equipment at design pressure and speed should result in minimum noise generation. Several areas relating to proper operation are:

1. If acoustic guards, covers, or enclosures are used, make certain that they are in place and that all openings are acoustically sealed. Noise is a lot like water—it will rush out through any cracks or openings.
2. If the equipment is equipped with a muffler system, make sure that it is performing according to design. Since many mufflers depend upon absorption within the case, significant changes in insertion loss can occur with age and use.
3. Install machinery on adequate mountings and foundations to reduce structureborne sound and vibrations. There are many good isolation pads commercially available. Make sure your choice isolates both sound and mechanical vibrations. An isolator that is good for mechanical vibrations may pass the higher frequency sound energy without significant reduction.
4. Use proper cutting speeds and feed-rates to control tool chatter in machining operations.
5. Provide good support for tool-bits and workpiece to reduce vibrations. Often, an increase in constraint will decrease these vibrations and eliminate the noise generation.
6. Apply additional sound control devices. Inlet and discharge silencers or mufflers can often be added with a small investment. Effective acoustic enclosures are often easy to design and construct.

The important point to be made here is that the manner in which a piece of equipment is operated can significantly influence the magnitude of the noise generated.

The third primary area is *equipment maintenance*. There are many maintenance techniques that can affect noise. Some of these could just as well have been classified under the operation area. The following list illustrates potential maintenance activities:

1. Maintain good dynamic balance. This decreases rotating forces and normally reduces noise generated by secondary sources, such as shaking guards, shells, or vibrating enclosures.
2. When purchasing replacement components, such as gears, motors, and pumps, do not ignore noise specifications. Most vendors can quote sound power level data for their products. Using these data, one can accurately predict the noise impact.
3. When installing fluid pipes or lines, use gradual rather than abrupt transitions from one part of a flow passage to another.
4. Improve lubrication. Inadequate lubrication due to poor design, degradation due to age, or incorrect lubricants are often the cause of bearing noise and also cutting noise associated with machining.
5. Install bearings correctly. Improper installation is often the reason for bearing-noise problems.

6. Reduce mechanical runout of shafts. This improves both initial static balance and dynamic balance. By reducing this source of vibration excitation, many components such as bearings, gears, and cams may generate less noise and have greater life.

It should be apparent that maintenance is an important element in controlling the noise generated by equipment. Maintenance departments should be instructed that noise is a very important consideration in all of their activities.

CASE HISTORIES

To illustrate some of the ideas presented in this chapter, two actual case histories follow. The first case highlights the importance of careful identification of the sound path. The second case illustrates the importance of enclosure design and source identification.

The Zipper Machine

A plant manager had invested a good deal of money in installing sound-absorbing material on all walls and ceilings, yet there was no improvement in the noise level at the worker location near the zipper machine. The manager called in a knowledgeable noise control engineer. The engineer spent 10 minutes observing the workroom, 10 minutes making eight noise measurements near the worker location, 5 minutes plotting the measured noise levels on a plot such as Fig. 5-2, and 5 minutes interpreting the results. Within 30 minutes after arrival, the engineer concluded that

1. the worker was in the near-field of the zipper machine (the noise level data plotted along the straight-line, free-field part of Fig. 5-2)
2. no amount of added noise absorption (increased room constant) could change the noise level at the worker location
3. the plant manager had been misled by the advice to install acoustical absorbing material.

The case was concluded by the design and installation of a simple enclosure for the zipper machine, which resulted in noise levels at the worker location that were in compliance with health and safety standards.

The Sewing Machine

Noise exposure at a sewing-machine operator was too high; thus, the plant "safety man" attempted some corrective action. The noise level at the operator location was 92 dBA, and so the safety man decided that he should be able to handle a 2-dBA reduction. Since the operator had to have access to the foot of the machine, the safety man decided that a good rigid partial enclosure covering the back, top, and one end of the machine should solve the problem. Since there was a plant in the same town that manufactured steel sinks and coated the backside of the sinks with "sound deadener," he decided to coat his fine partial enclosure with the same material. After completing the enclosure, it was placed into position with much anticipation. With the noise enclosure in place, however, the noise level at the operator rose from 92 to 94 dBA.

The safety man called in an engineer who explained the behavior of the partial enclosure. The type of enclosure did not absorb any significant amount of noise but reflected it back to the operator location. As for the sound deadener, the engineer explained that it was simply a damping material used to keep the thin sink walls from "ringing."

In examining the problem, the engineer discovered a large stack of used bearings in a corner near the sewing machines. He was informed that they were no good because it was necessary to replace them every two weeks in the sewing machines. The engineer wondered if there was a connection between high noise and short bearing life. Upon careful examination, it became obvious that the wrong lubricant was being used in the bearings. The engineer determined the correct lubricant, which was tried in the sewing machines. Bearing life was extended to more than 16 months, and the noise level at the operator was reduced to 90 dBA or below. The plant manager liked what he saw—he saved money and solved his noise problem at the same time.

STEPS TO FOLLOW IN INDUSTRIAL NOISE CONTROL

This section presents a logical series of actions, which, if followed, should lead to effective noise control. If the individual conducting these actions comes to a step that he or she does not understand, then it is imperative that professional engineering assistance be obtained.

Step 1. Identify the noise problem areas in the plant. This can be done by using a sound level meter and a plant floorplan to conduct a noise survey. For this first step, it is necessary to identify all those areas in the plant where the noise level is 90 dBA and above if compliance with the current Occupational Safety & Health Standards is the goal.

Step 2. Establish the noise exposure of those employees in the potential problem areas identified in Step 1. This can be determined by timing the exposure at each noise level for the employee. Another technique is to have the employee wear an audiodosimeter on his or her normal work shift. By using the criteria associated with Table 5-1, you have now identified your noise exposure problems.

Step 3. Get to know the problem. Use your ears and eyes as first indicators of the nature of the problem. At this point, we would like to identify predominant noise sources since in most industrial problems there are many. Some qualitative indication of predominant frequencies can be obtained by A-, B-, and C-weighted sound pressure level measurements. By comparing readings on the three scales, it is possible to learn something about frequency content. A higher reading on C-scale than on A-scale would indicate that frequencies below 500 Hz constitute an appreciable portion of the total noise level. A difference between C- and B-scale readings indicate that frequencies below 250 Hz are important. When all three scale readings are essentially the same, the major noise frequencies are above 500 Hz. This is very limited information, and normally octave-band, 1/3-octave-band, or narrow-band analysis must be used to distinguish peaks indicating predominant noise frequencies. The instrumentation associated with these types of analysis is discussed in Chapter 2. Once you have identified the major noise frequencies, you have obtained the first diagnostic information that can help identify the source or sources generating the major portion of the noise.

Step 4. Using the data obtained in Step 3, you are ready to use proper operation and maintenance to decrease noise generation as discussed earlier.

Step 5. If careful attention to Step 4 has not resulted in the desired noise levels, then the next consideration is to the transmission path. In a previous section, we have explored some of the ideas necessary to understand and effectively treat the sound path. Added absorption in the workspace, enclosures, and/or barriers may provide effective control depending upon the situation.

Step 6. If all engineering efforts fail to reduce sound to the required noise level, then attention must be given to administrative control. Determine if a change in the work assignment of the employee will effectively limit his or her noise exposure.

Step 7. If the preceding six activities fail to lower noise levels to values that give exposures that meet safety and health standards, then the only alternative is to provide personal hearing protection for the employee. As an integral part of the personal-hearing-protection program, you are required to develop an audiometric testing program. A complete discussion of such a hearing conservation program is given in Chapter 7.

SUMMARY

The intent of this chapter has been to introduce the reader to some of the fundamental principles of industrial noise control. By the very nature of this presentation, it is notably incomplete; subjects such as vibration isolation are not even discussed. It is hoped that the material presented here has helped to develop a greater appreciation for the complexity of effective industrial noise control. To assist those with the background for, and the desire to increase their mastery of, industrial noise control fundamentals, a list of references follows.

Since there is some confusion about the current state of noise control technology, we offer the following observation. The current state of engineering capability is such that any noise level can be achieved for any process or application. For example, we can build a very quiet airplane, but the cost per seat mile would go up by such a large factor that air travel would disappear since no one could afford it. The real challenge is to achieve desired noise control at a cost that allows the process or application to remain competitive.

REFERENCES

Bell, L. H. *Fundamentals of Industrial Noise Control.* Trumbull, Connecticut: Harmony Publications, 1973.

Beranek, L. L. *Noise and Vibration Control.* New York: McGraw-Hill, 1971.

Beranek, L. L. *Acoustic Measurements.* New York: Wiley & Sons, 1949.

Crocker, M. J. *Tutorial Papers on Noise Control.* International Conference on Noise Control Engineering, Institute of Noise Control Engineering, 1972.

Diehl, G. M. *Machinery Acoustics,* New York: Wiley & Sons, 1973.

Diehl, G. M. *Think Quiet.* Compressed Air Magazine Company, 1971.

Harris, C. M. *Handbook of Noise Control.* New York: McGraw-Hill, 1957.

Hosey, A. D., and Powell, C. H. *Industrial Noise–A Guide to Its Evaluation and Control.* U.S. Department of Public Health, Publication No. 1572, 1967.

Morse, P. M., and Ingard, K. U. *Theoretical Acoustics.* New York: McGraw-Hill, 1968.

Peterson, A. P. G., and Cross, E. E. *Handbook of Noise Measurement.* General Radio Company, 1972.

Thuman, A., and Miller, R. K. *Secrets of Noise Control.* Atlanta: Fairmont Press, 1974.

Yerges, Y. F. *Sound, Noise and Vibration Control.* New York: Van Nostrand Reinhold Co., 1969.

6

Noise control by acoustical treatment

Anna K. Nábělek, Ph.D.
and
Igor V. Nábělek, Ph.D.
Department of Audiology and Speech Pathology
The University of Tennessee
Knoxville, Tennessee

The noise level in a building is a combination of noise coming from the outside and noise produced inside. The first section of this chapter deals with methods of controlling the ingress of outdoor noise, and later sections deal with noise control inside buildings.

OUTDOOR NOISE

Acceptable Noise Levels for New Construction

The best approach to limit the effects of outside noise is to select appropriate sites for new construction as recommended by the U.S. Department of Housing and Urban Development (HUD).[1] HUD recognizes noise as a major source of environmental pollution and, in its efforts to provide suitable living environments, has developed guidelines for the construction of new buildings.

HUD recommends that sites for future construction be

TABLE 6-1 External noise exposure standards for new construction sites. (From U.S. Department of Housing and Urban Development, 1971.)

No.	Noise Level	Category
1	Exceeds 80 dBA for 60 min/24 hr or 75 dBA for 8 hr/24 hr	Unacceptable
2	Exceeds 65 dBA for 8 hr/24 hr	Normally unacceptable
3	Does not exceed 65 dBA for more than 8 hr/24 hr	Normally acceptable
4	Does not exceed 45 dBA for more than 30 min/24 hr	Acceptable

carefully studied, particularly with respect to the location of potential sources of noises, such as airfields, industrial plants, railroads, and traffic arteries. With respect to noise levels, such sites have been divided into four categories, as shown in Table 6-1. HUD discourages the construction of new dwelling units on sites that have, or are projected to have, unacceptable noise levels.

Barriers

Often, new buildings can be located relatively close to sources of noise if the external noise levels are effectively reduced at the site of the building by the use of natural shielding of rolling terrain, shielding walls, thick growth of trees, or shielding by existing high buildings. Examples of good and bad solutions are shown in Fig. 6-1.

When barriers are situated between the source of noise (for example, a busy street) and the building, they interrupt the direct pathway of the sound. The barrier effect of a wall would be perfect if sound waves would not bend, and the sound would not go directly through the wall. A sound wave striking the wall would be reflected back toward the source, and while passing the wall edge, would continue directly upward. A barrier is usually less effective at low frequencies (with longer wavelength) because the sound waves bend around the wall edges when the wavelength becomes comparable with the size of the wall.

The effectiveness of the barrier depends on its dimensions and location relative to the source and receiver positions. The height of a constructed wall cannot usually exceed 6 meters (20 ft) for economical reasons. Empirical data indicate that the amount of noise reduction for such a wall will vary from 10 to 15 dB. The greatest reduction is obtained for the wall located close to the noise source or to the shielded object. For detailed computations, see Rettinger (p. 332).[2]

Hedges and single rows of trees have a minor effect on reducing traffic noise,

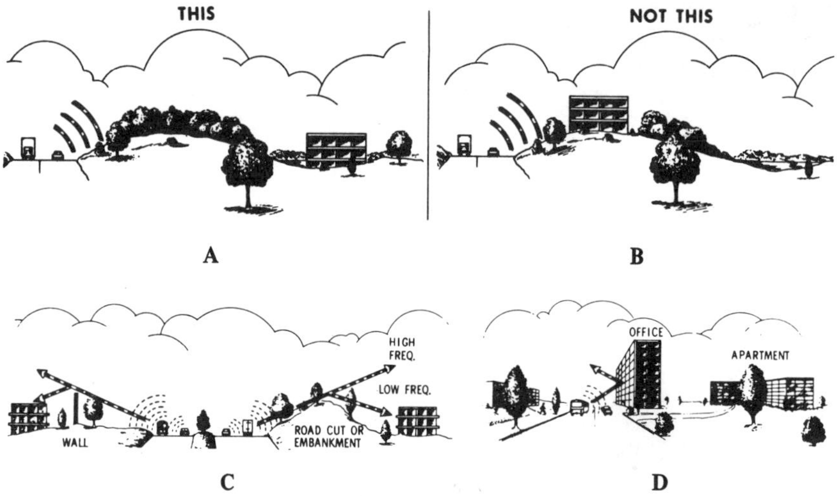

Fig. 6-1 Examples of noise barriers. A, good use of natural barriers. B, bad use of natural barriers. C, use of combination of barriers. D, use of buildings as barriers. (From Berendt et al., 1967.)

regardless of their location or proximity to the noise source. However, thinner hedges of trees can sometimes be used to provide an illusion of noise reduction. Thick, wooded patches are somewhat more effective. A wooded area of leafy trees with underbrush 30 meters (100 ft) thick will provide 3- to 4-dB noise reduction at low frequencies and 10- to 12-dB noise reduction at higher frequencies. Of course, in winter, when the leaves are gone, the sound reduction becomes negligible.

The hills along highways or depressions of the highways below the ground level produce noise reduction similar to other constructed barriers. The best noise control solution is to locate highways in tunnels from which the noise spreads at each end only.

Location of Buildings on Sites

The noise coming to a complex of buildings or originating within the complex, should not be "trapped" in the area by bouncing to and back from surrounding walls. Some examples of better or poorer orientations of buildings on sites are shown in Fig. 6-2. For a cluster of buildings, a random layout is preferred. Parallel building arrangements should be avoided because repetitive sound reflections occur easily between them. For the same reason, U-shaped courtyards should not be oriented toward noise sources such as highways.

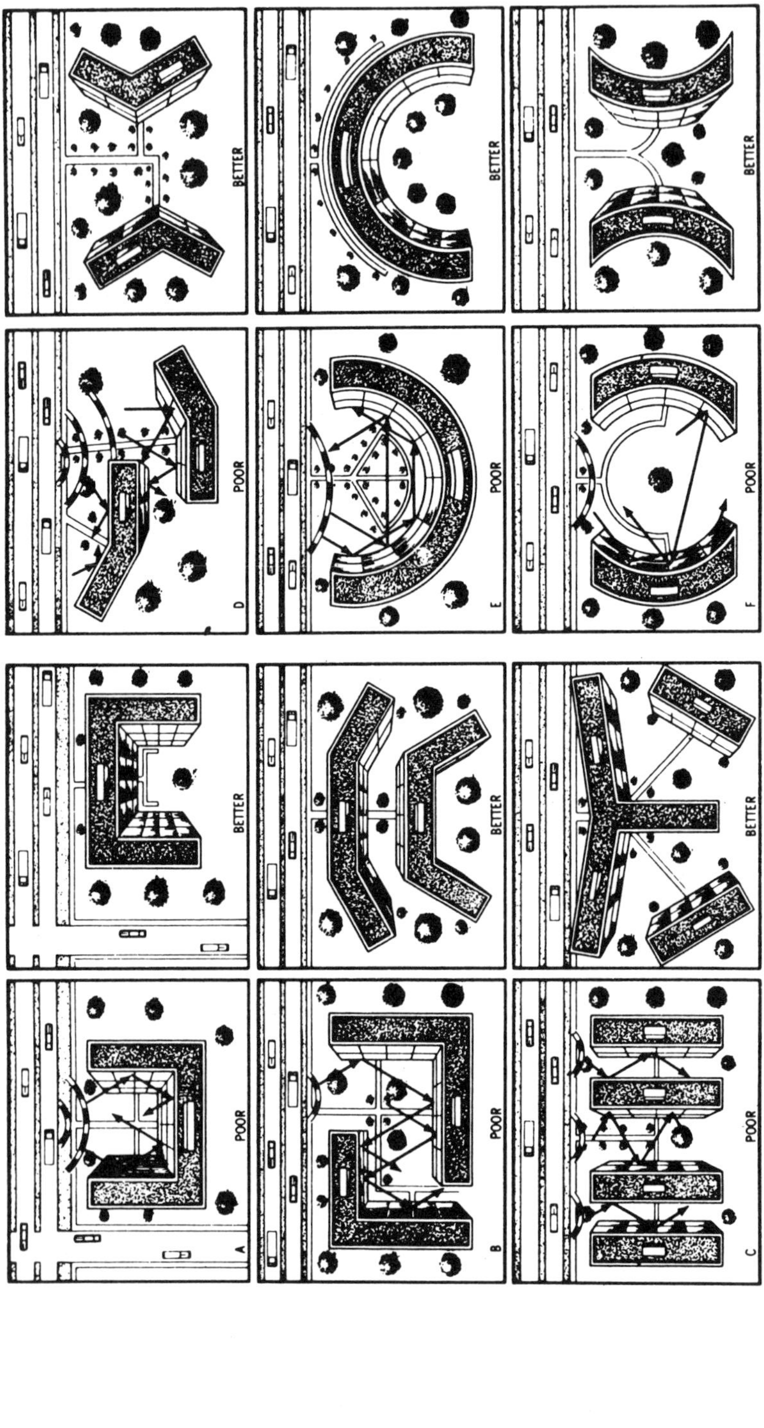

Fig. 6-2 Orientation of buildings on sites. (From Berendt et al., 1967.)

ENCLOSURE INSULATION PROPERTIES

As noise-reducing elements, enclosures are used (1) as shells surrounding the source of noise, or (2) to prevent the entrance of noise into the inside space. This section deals with the buildings.

Airborne and Structureborne Noise

Generally speaking, when the medium in which the noise starts to spread from the source is air, the noise is called *airborne;* when it starts with vibration within a structure, it is called *structureborne.* Airborne noise is produced by sources radiating directly to the air, such as traffic, music coming from an apartment, or a school or office activity. The structureborne noise occurs when wall, floor, or other building elements are set into vibrations by direct mechanical contact with the source, such as mechanical equipment or footsteps. The building materials should be effective in reducing both types of noises entering the enclosures.

Pathways of Airborne Sound through Partitions

Walls and other partitions should prevent the transmission of sound from one space to another. Airborne sound may pass through a partition in the following ways:

1. When a partition is rigid and nonporous, such as a concrete wall, the incident sound penetrates from the air into the partition, propagates through the material, and radiates from the other side. This type of energy transmission for walls and roofs is generally negligible for airborne sound, but may be considerable for impact (structureborne) noise.
2. When a partition is flexible, such as a windowpane or a plasterboard panel, it acts like an oscillating diaphragm and generates acoustic waves on the other side. This is how sound usually travels from one side of the partition to the other.
3. When a partition is made from a porous material, some sound energy is absorbed and changed into heat by friction between vibrating air particles in the material, but most of the energy passes through the pores to the space behind the partition. These materials are generally poor sound insulators, especially at low frequencies.
4. Sound may pass through a partition by "weak" spots in the partition. These weak spots may, for example, be joints in panels, common electric outlets, common pipes, or heating ducts.

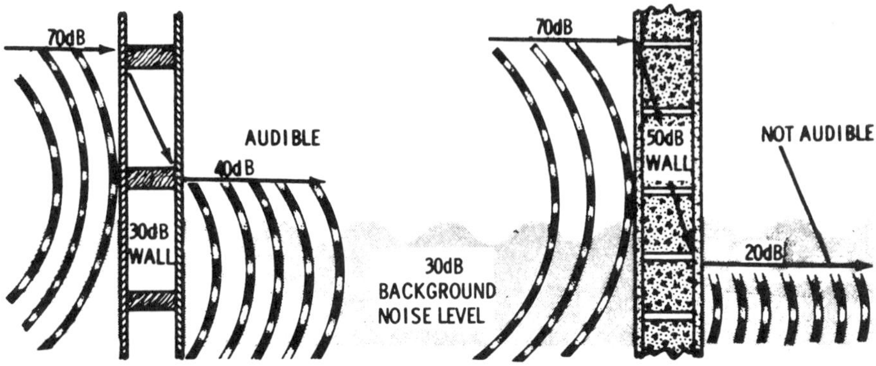

Fig. 6-3 Examples of noise reduction by two walls with different transmission loss. (From Berendt et al., 1967.)

Transmission Loss (TL)

The transmission loss (TL) of a partition (wall, ceiling, floor) is the ratio of airborne sound intensity on the source side of an infinitely large wall, to the sound intensity on the other side of the wall. It is expressed as the difference of sound pressure levels (in dB) in front of and behind the wall, respectively. The space behind the wall is supposed to be open. The larger the value of TL, the better are the sound-insulating properties of the partition.

The TL of a partition varies with frequency. Therefore, the TL is measured and tabulated for various frequencies, usually between 125 and 4000 Hz. Table 6-2 shows the TLs for some commonly used constructions of building partitions. STC (sound transmission class) and IIC (impact insulation class) are also included; both are explained in later sections. Additional TL data* of various structures can be found in Berendt et al.,[3] Bishop and Hirtle,[4] and Sabine et al.[5] The new constructions of partitions usually have manufacturer's TL data.

The TL data taken from the tables have to be treated as an approximation of field values. Some changes in physical properties of the partition, such as weight, density, thickness, and mounting, can change the TL considerably.

An example of how a wall reduces sound transmission is shown in Fig. 6-3. The sound pressure levels in front of and behind two different walls are shown schematically. In the left figure, a wall with 30-dB TL reduces the 70-dB noise level already present from other sources. In the right figure, a better wall, with 50-dB TL, reduces the same noise to a 20-dB level, which is 10 dB below the

*The publications by Berendt et al. and Sabine et al. can be purchased from the Superintendent of Documents, U.S. Government Printing Office, Washington, D.C. 20402. Price: $2.50 and $2.35, respectively.

TABLE 6-2 Transmission loss in decibels for common building constructions.

No. Material	Total Thickness (inch)	(cm)	Frequency (Hz) 125	250	500	1K	2K	4K	STC Rating	IIC Rating
Walls[a,b]										
1. Solid concrete	3	8	35	40	44	52	59	64	47	–[c]
2. Concrete (6, 15), layers of plaster	7	18	39	42	50	58	64	66	53	–
3. Concrete wall (6, 15), wood-wool + plaster on furring	10	25	31	40	52	58	60	60	52	–
4. Solid concrete blocks, layers of plaster	16	41	50	54	59	65	71	68	63	–
5. Brick (4.5, 11), layers of plaster	5.5	14	34	34	41	50	56	58	42	–
6. Brick (9, 23), layers of plaster	10	25	41	43	49	55	57	59	52	–
7. Brick (12, 30) without plaster	12	31	45	44	52	58	60	61	56	–
8. Stone (24, 61), layers of plaster	25	64	50	53	52	58	61	68	56	–
9. Hollow concrete block wall	12	34	47	43	45	52	54	56	48	–
10. Hollow concrete block	6	15	32	33	40	48	51	48	43	–
11. Cinder block (4, 10), layer of plaster	5.25	13	36	37	44	51	55	62	46	–
12. Cement block painted at both sides	3.75	10	40	40	40	48	55	56	44	–
13. Hollow gypsum block (3, 8), layers of plaster	4	10	39	34	38	43	48	46	40	–
14. Hollow gypsum block (3, 8), resilient one side, plaster	5	13	38	37	44	51	56	59	46	–
15. Hollow gypsum block (4, 10), gypsum lath + resilient clips	6	15	25	37	46	53	56	63	47	–
16. Hollow concrete (9.25, 23), layers of fiberboard	10.25	27	41	42	47	51	52	39	43	–
17. Double brick (4.5, 11) wall, cavity (2, 5), layer of plaster	12	31	37	41	48	60	60	61	49	–
18. Double brick (4.5, 11) wall, cavity (6, 15), layer of plaster	18	46	48	54	58	64	69	75	62	–
19. Wooden studs (4, 10), gypsum wallboards	5	13	22	30	35	40	41	40	39	–

TABLE 6-2 (*Continued*)

No. Material	Total Thickness (inch)	(cm)	Frequency (Hz) 125	250	500	1K	2K	4K	STC Rating	IIC Rating
20. Wooden studs (4, 10), gypsum laths + plaster layers	5.75	15	32	36	42	48	48	62	46	-
21. Steel truss studs (3.75, 8), metal lath, plaster	5.25	13	30	28	35	40	43	53	39	-
22. Steel truss studs (3.75, 8), double layers gypsum boards	4.75	12	35	38	44	50	50	51	48	-
23. Metal channel studs (1.625, 4), gypsum wallboards	2.625	7	20	30	38	47	48	45	39	-
24. Solid sanded gypsum plaster	2	5	36	28	35	39	48	52	36	-
25. Solid sanded gypsum plaster with metal channels	2	5	35	25	32	38	47	54	36	-
26. Solid gypsum core movable partition	2.25	6	34	34	37	38	39	45	36	-
Floors-ceilings										
1. Reinforced concrete slab	4	10	48	42	45	55	57	66	44	25
2. Reinforced concrete as above + carpeting and pad	4.5	11	48	42	45	55	57	66	44	80
3. Reinforced concrete slab (6, 15), layer of plaster	7.5	19	42	39	44	49	54	60	47	31
4. Concrete (4.5, 11), wood flooring, layer of plaster	7	18	35	37	42	49	58	62	46	47
5. Concrete (4.375, 11), screed, suspended plaster ceiling	10	25	38	41	45	52	57	59	48	47
6. Concrete (6, 15), wood, battens floating on glass wool, layer of plaster	9.5	24	38	44	52	55	60	65	55	57
7. Concrete (6, 15) with hollow blocks, wood on screed, plaster	8.165	21	40	42	46	52	58	60	50	48
8. Concrete (5.5, 14), floating floor, suspended ceiling	15.25	38	40	46	54	59	62	68	55	53

9. Hollow tile beam (5, 13), screed + linoleum, plaster	6.75	17	36	40	43	49	55	60	47	40
10. Hollow concrete slab (6, 15), cement mortar + cement finish, plaster	7.5	19	39	38	43	49	54	57	46	30
11. Concrete channel slab, sand cement finish	6.25	16	34	34	38	46	55	61	42	32
12. Ribbed concrete (7.25, 18) screed, wooden lath + plaster	9.50	24	33	37	42	52	58	62	46	42
13. Hollow concrete beam (6, 15), screed + floor, plaster	7.50	19	32	41	43	44	54	64	45	31
14. Wooden joists (8, 20), floor, gypsum wallboard	9.50	24	19	24	31	35	45	42	34	32
15. Wooden joists (7, 18), wood + linoleum, reeds + plaster	9.50	24	24	27	35	44	52	58	39	40
16. Wooden joists (10, 25), paperboard + pad + carpet, wallboard	12.5	32	27	32	38	44	49	60	38	57
Windows										
1. Sliding, aluminum frame	0.09375[d]	0.2	10	14	17	18	18	20	-	-
2. Projected (awning), aluminum frame, exceptionally good weather strip	0.09375[d]	0.2	16	22	25	28	32	28	-	-
3. Double window 0.09375[d], 0.2), air-space (0.1875, 0.5)	0.375	1.0	18	21	19	24	27	18	-	-
4. As (3) with a storm window, air-space (2, 5)	2.375	6.0	16	24	27	33	37	29	-	-
Doors										
1. Hollow-core flush, 30 lb (6.7 kg)	1.75	4	11	16	16	16	21	23	-	-
2. Solid-core flush, 95 lb (21.1 kg)	1.75	4	20	25	23	25	25	28	-	-

[a]Numbers in parentheses indicate thickness of layer, in inches and centimeters, respectively.
[b]Data for walls and floors-ceilings from Berendt et al., 1967. Data for windows and doors from Bishop and Hirtle, 1968.
[c]Not applicable.
[d]Conventionally described as $^3/_{32}$ inch.

background noise. The first wall would not provide satisfactory sound insulation whereas, in the second case, adequate insulation would be achieved.

Thus, the satisfactory performance of the wall in a given situation depends on three factors: (1) noise level on the noisy side, (2) the TL of the wall, and (3) interior background noise level. The inside noise levels are given by the inside activity. When walls are well designed, noise coming from outside should not increase the inside noise level appreciably. On the other hand, an enclosure does not have to attenuate the outside noise to levels much below the background level. It would be unnecessary and uneconomical.

Sound Transmission Class (STC)

The frequency dependence of the TL complicates its use. A single average value of TL over frequency would be handy but could correspond poorly to the subjective impression of sound insulation. Using reference tables, the TL values can be appropriately weighted and the insulating properties of building materials described by another single number called a *sound transmission class* (STC). To determine the STC of a given specimen, the curve representing its

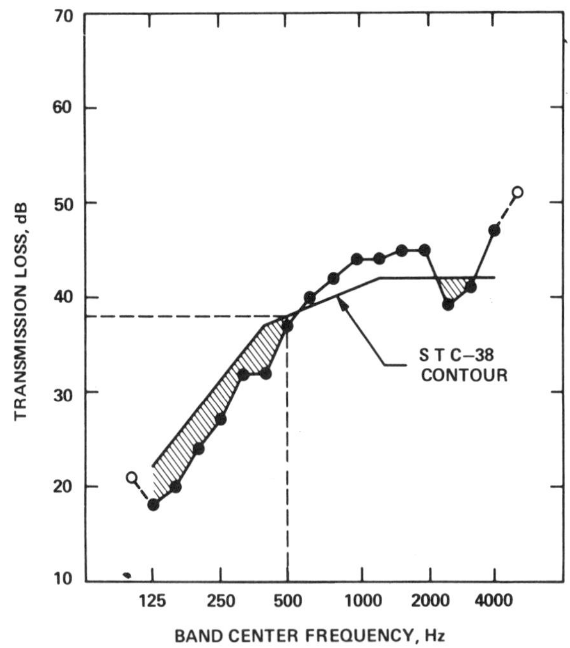

Fig. 6-4 An example of a sound transmission class contour fitted to a sound-transmission-loss curve. (From Sabine et al., 1975.)

transmission loss as a function of frequency is compared with a standardized reference frequency curve, the so-called STC contour (Fig. 6-4). The STC contour is moved vertically until the sum of deviations of the test curve from the STC contour *below* the contour at 16 test frequencies is not more than 32 dB and none of those deviations exceeds 8 dB. When the test curve is so adjusted, the STC value is read as the TL value corresponding to the intersection of the STC contour and the 500-Hz frequency line. An example is shown in Fig. 6-4. The cross-hatched area indicates the frequencies at which deviations below the contour occur; the sum of the deficiencies is about 29 dB, and the largest single deviations are 5 dB.

Noise Reduction (NR)

As already mentioned, the TL is the property of a partition of indefinitely large size. In practice, the partition has a limited size, and the space on the quiet side is completely or partially closed. The amount of noise reduced by the partition depends then, not only on the TL, but also on the size of the partition and the acoustic properties of walls in the receiving room. If these walls are reflective, the noise level is higher; if absorptive, the level is lower. The number taking all these factors into account is called *noise reduction* (NR), which can be calculated from the formula:

$$\mathrm{NR} = \mathrm{TL} + 10 \log a/S$$

where

NR = the noise reduction in dB
a = the total absorption in the receiving room
S = the area of the wall transmitting sound in ft^2 or m^2.

TABLE 6-3 Preferred noise criteria for various indoor spaces. (After Beranek et al., 1971.)

Type of Space	Noise Level in dBA
Concert halls, opera houses	20 to 30
Large auditoriums, drama theaters	Max. 30
Small auditoriums, churches, conference rooms	Max. 42
Bedrooms, hospitals	34 to 47
Large offices, stores, cafeterias, restaurants	42 to 52
Lobbies, small shops, laboratories	47 to 56
Kitchens, laundries, computer rooms	52 to 61
Work spaces without risk of hearing damage	66 to 80

The total room absorption a is the sum of all room surface areas in ft^2 or m^2, respectively, times their respective sound absorption coefficients, as in Table 6-4. The unit of absorption is the sabin. For examples of computations, see Egan (p. 65).[6]

The noise reduction will be smaller than the TL when the absorption of the receiving room is small and the wall is large. The NR may be larger than the TL when the absorption in the receiving room is large and the transmitting wall is small.

The insulating properties of partitions can be improved by increasing the weight (mass) of the partition, by increasing the number of insulating layers, and by eliminating all acoustic leaks.

Weight-Mass Law

The more massive and solid the structure, the greater its opposition to vibration, and consequently the higher the TL. This is the principle of the so-called mass law. According to this law, the TL of such a wall increases with frequency usually 3 to 6 dB per octave. Theoretically, the TL should increase 6 dB per doubling of the partition weight, but, in practice, the STC rating (a value averaged over frequency) increases about 4 to 5 dB for each doubling of weight.

Partitions from some materials have smaller TL values in the midfrequency region than predicted by the mass law. That is caused by various resonant effects of the partition.

Between the lowest natural resonant frequency of the wall and the frequency at which the wavelength of the bending of the wall is equal to the wavelength of the incident sound, the approximate TL values can be calculated as

$$\mathrm{TL} = 14.5 \log m + 14.5 \log f - 16$$

or

$$\mathrm{TL} = 14.5 \log m' + 14.5 \log f - 26$$

where

m = the mass in lb/ft^2 of the partition surface
m' = the mass in kg/m^2 of the partition surface
f = the frequency in Hz.

Other related formulas and graphs can be found in Rettinger (p. 130).[2]

Multiple-Layered Structures

The TL of walls, doors, or windows can be greatly improved if they are made up of two or more layers separated by an airspace. If the layers (panels) are sepa-

TABLE 6-4 Sound absorption coefficients for common materials. (By permission of American Board Products Association.[8,9])

No.	Material	Frequency (Hz) 125	250	500	1K	2K	4K	NCR rating
	Walls							
1	Brick	0.03	0.03	0.03	0.04	0.05	0.07	0.05
2	Concrete painted	0.10	0.05	0.06	0.07	0.09	0.08	0.05
3	Window glass	0.35	0.25	0.18	0.12	0.07	0.04	0.15
4	Marble	0.01	0.01	0.01	0.01	0.02	0.02	0.00
5	Plaster on concrete	0.12	0.09	0.07	0.05	0.05	0.04	0.05
6	Plywood	0.28	0.22	0.17	0.09	0.10	0.11	0.15
7	Concrete block, coarse	0.36	0.44	0.31	0.29	0.39	0.25	0.35
8	Heavyweight drapery	0.14	0.35	0.55	0.72	0.70	0.65	0.60
9	Fiberglass wall treatment, 1 in (2.5 cm)	0.08	0.32	0.99	0.76	0.34	0.12	0.60
10	Fiberglass wall treatment, 7 in (17.8 cm)	0.86	0.99	0.99	0.99	0.99	0.99	0.95
11	Wood paneling on glass fiber blanket	0.40	0.90	0.80	0.50	0.40	0.30	0.65
	Floors							
1	Wood parquet on concrete	0.04	0.04	0.07	0.06	0.06	0.07	0.05
2	Linoleum	0.02	0.03	0.03	0.03	0.03	0.02	0.05
3	Carpet on concrete	0.02	0.06	0.14	0.37	0.60	0.65	0.30
4	Carpet on foam rubber padding	0.08	0.24	0.57	0.69	0.71	0.73	0.55
	Ceilings							
1	Plaster, gypsum, or lime on lath	0.14	0.10	0.06	0.05	0.04	0.03	0.05
2	Acoustic tiles 0.625 in (1.6 cm), suspended 16 in (40.6 cm) from ceiling	0.25	0.28	0.46	0.71	0.86	0.93	0.60
3	Acoustic tiles 0.5 in (1.2 cm), suspended 16 in (40.6 cm) from ceiling	0.52	0.37	0.50	0.69	0.79	0.78	0.60
4	The same as (3), but cemented directly to ceiling	0.10	0.22	0.61	0.66	0.74	0.72	0.55
5	Highly absorptive panels, 1 in (2.5 cm), suspended 16 in (40.6 cm)	0.58	0.88	0.75	0.99	1.00	0.96	0.90
	Others							
1	Upholstered seats	0.19	0.37	0.56	0.67	0.61	0.59	0.55
2	Audience in upholstered seats	0.39	0.57	0.80	0.94	0.92	0.87	0.80
3	Grass	0.11	0.26	0.60	0.69	0.92	0.99	0.61
4	Soil	0.15	0.25	0.40	0.55	0.60	0.60	0.45
5	Water surface	0.01	0.01	0.01	0.02	0.02	0.03	0.00

rated by a considerable distance, they in effect form two independent partitions, and the total TL can approach the sum of the individual panel TLs. For most practical situations, however, the separation is relatively small; 25 cm or less for wall layers and 5 to 10 cm for windowpanes are more common, and the TL of the double partition is far less than the sum of individual panel TLs. In all double-wall construction, rigid ties between panels must be avoided.

Porous absorbers in the cavity between two layers contribute somewhat to sound insulation, but their density has little effect on the partition's STC rating. The use of a fiberglass blanket between two walls can improve the rating from 4 to 6 dB due to absorption of acoustic energy passing through them. Changing the density three times does not improve the rating, but the thickness of such material is important.

The TL of porous materials, as mentioned before, is relatively low. However, these materials are used as sound insulators in special constructions requiring light weight, such as aircraft cabins or folding or movable office partitions. Such lightweight dividers have usually poor TL at low frequencies but improve at higher frequencies. The porous material in connection with unporous layers has higher TL than predicted by the mass law, except at frequencies below 500 Hz. Porous materials tend to exhibit TLs proportional to their thickness and not to their mass.

Composite Constructions

Very often a partition is a combination of various materials. It can be brick wall with a glass window or a wall with a door, each of which, has a different TL. The composite TL has to be calculated from the formula

$$\mathrm{TL} = 10 \log \frac{S}{(\tau_1 s_1 + \tau_2 s_2)}$$

where

TL = the transmission loss of the composite partition in dB
S = the surface area of the partition in ft^2 or m^2
τ_1, τ_2 = the transmission coefficients of the components
s_1, s_2 = their corresponding surface areas in ft^2 or m^2.

The transmission coefficient τ can be calculated from the TL by using the equation

$$\mathrm{TL} = 10 \log 1/\tau$$

or

$$\tau = \frac{1}{\text{antilog TL}/10}$$

A good rule of thumb is that the composite TL is closer to the TL of the material with lower values of the TL, i.e., with poorer insulation properties. In that way, most of the noise transmitted to a building is coming through windows and doors, which usually have poorer insulation than the walls.

Decrease in NR by Special Phenomena

The effective NR in a room might be much less than estimated from the TLs of the surrounding partitions if the sound will come through via leaks and flanking paths. Probable paths are over and under partitions, around the edges of a floor-ceiling system, along a continuous wall past several partitions, and through doors, windows, electrical outlets, air-conditioning ducts, built-in cabinets, and other partition penetrations as shown in Fig. 6-5. Therefore, all the leaks should be sealed, and direct connection of spaces by common sound carriers, such as ducts, pipes, and others, should be avoided wherever possible. Noise

Fig. 6-5 Flanking transmission of airborne noise. Flanking noise paths: F1, open plenums over walls, false ceilings; F2, unbaffled duct runs; F3, outdoor path, window to window; F4, continuous unbaffled inductor units; F5, hall path, open vents; F6, hall path, louvered doors; F7, hall path, openings under doors; F8, open troughs in floor-ceiling structure. Noise leaks: L1, poor seal at ceiling edges; L2, poor seal around duct penetrations; L3, poor mortar joints, porous masonry blocks; L4, poor seal at sidewall, filler panel, etc.: L5, back-to-back cabinets, poor workmanship; L6, holes, gaps at wall penetrations; L7, poor seal at floor edges; L8, back-to-back electrical outlets; L9, holes, gaps at floor penetrations. (From Berendt et al., 1967.)

traveling inside ducts can be reduced by using acoustical treatment of the inside walls of the ducts.

Impact Noise

Impact noise occurs when an object makes an impact, strikes a partition, and sets it into mechanical vibrations; it is a structureborne noise. In most cases, impact noise is generated in floor-ceiling systems of multistory buildings, by footsteps, dropped objects, moved furniture, and so on. An impact sound transmitted from adjoining spaces is usually more annoying than a continuous noise because of its intrusive character. It can also cause moving of lighting fixtures and other suspended objects.

All ceiling-floor constructions have to be evaluated in terms of reducing both the airborne noise and the structureborne impact vibrations. While airborne noise can be reduced by heavy, rigid materials, the structureborne noise is best reduced by something soft such as cork or springs.

Impact Insulation Class (IIC)

The impact insulation class (IIC) is a single-number rating of floor-ceiling's impact-sound-reducing performance. The higher the IIC rating, the more efficient the floor-ceiling for reducing impact-sound transmission. To establish the IIC, noise is generated by a standard tapping machine, operating on the floor of the tested construction. The sound-pressure level is measured in the room below the floor-ceiling system at 16 test frequencies. The IIC is determined by comparing the performance curve with a reference IIC contour, a procedure similar for determination of the STC for a material.

The IIC rating for floor-ceiling structures is shown in Table 6-2. Note, for example, that constructions (1) and (2) have virtually the same STC, but the addition of a carpet to the construction (1) increases the IIC from 25 to 80. Efficient reduction of impact-noise transmission can be achieved by "floating" floors. A layer of concrete or similar material is placed on a porous sound absorber, which separates it from underlying layers. The concrete should not touch the walls.

The Federal Housing Administration recommends minimum IIC values of 48, 45, and 52 for buildings with an interior noise level of 45, 40, and 35 dBA, respectively.

PLANNING OF INTERIORS

When the penetration of outside noise is properly controlled, the inside noise depends primarily on the noise generated within the building.

Recommended Noise Criteria for Rooms

As a rule, we can tolerate, and even prefer to have, a certain amount of background noise. An acceptable level should neither disturb room occupants nor interfere with speech communication.

Examples of approximate A-scale sound pressure levels corresponding to recommended preferred noise criteria (PNC) curves[7] are shown in Table 6-3. These values should not be exceeded for rooms with indicated designation. They represent results of extensive studies based on human responses to sound exposure. The levels shown in Table 6-3 should not be considered valid if the noise spectrum does not have the same shape as a PNC curve. The use of PNC curves for calculation of TLs at different frequencies is shown, for example, in Rettinger[2] (p. 153) and Beranek et al.[7]

Note that in noise spaces, like factories, the noise criteria are a compromise between comfort and economy. The allowed levels of noise are higher than comfortable, but further reduction might be costly or even impossible.

Interior Plan

Quiet areas should be separated from noisy areas. Usually, one side of a building is exposed to higher external noise than the other. Noisy areas such as equipment rooms, recreation rooms, and kitchens should be located on the noisy side of the building. Garages, elevators, shops, and equipment and laundry rooms should be located at an end of the building far removed from quiet areas, rather than in some central location, which is often chosen for accessibility.

In apartment or office buildings, the layout of adjacent units should be planned so that the partitions separate similar functional spaces. For example, walls or floors should separate the bedrooms of one apartment from the bedrooms of the adjacent apartment, rather than separate a bedroom of one apartment from the living room or recreation room of another apartment. Examples of well-planned dwelling interiors are shown in Fig. 6-6.

To reduce the amount of noise produced by mechanical equipment, the following should be observed:

1. Equipment with the lowest sound power output rating should be selected.
2. It should be located as far as possible from the quiet areas in the building, or in a separate building.
3. It should be properly installed to isolate the vibrations from the structure of the building.

Sound Absorption Coefficient

When a sound wave strikes the surface of a material, only part of the incident energy is reflected back into the original medium. The other part penetrates

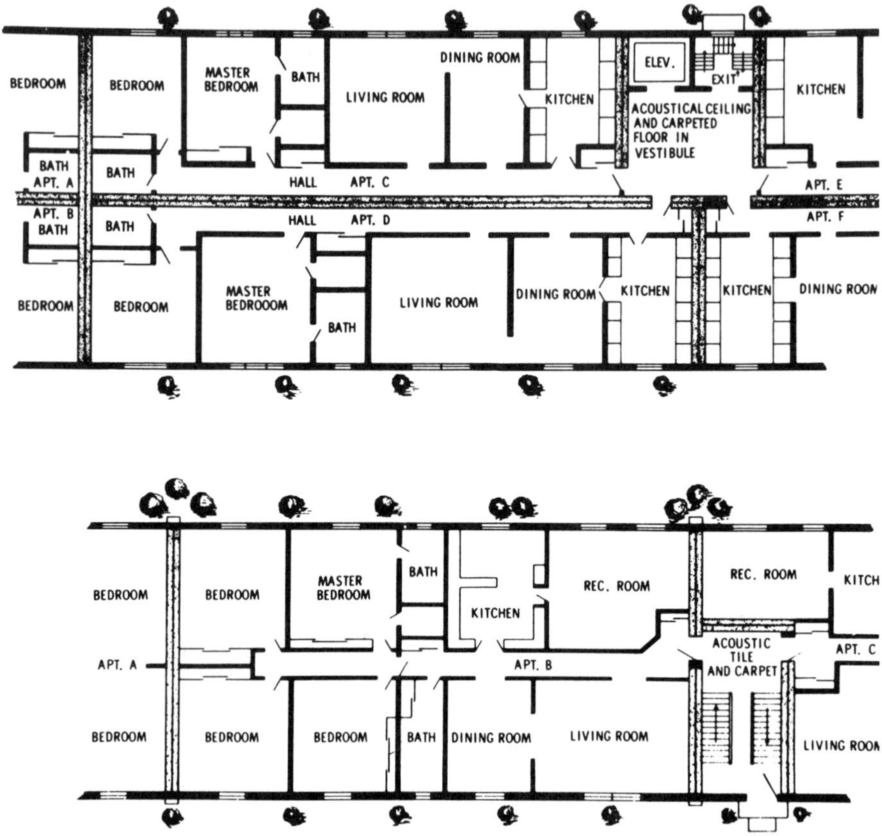

Fig. 6-6 Examples of well-planned buildings in which quiet areas are separated from noise areas. (From Berendt et al., 1967.)

into the material, where it is either converted into heat or passes through the material and comes out from the other side. For observers on the side of the material where the sound source is located, it does not matter if the sound energy is changed into heat or goes out on the other side. For them, the energy is not reflected—that is, it is absorbed by the material.

As a measure of absorption by a given material, the sound absorption coefficient is used. This coefficient is the ratio of energy not reflected by the surface to the incident energy.

All materials absorb some energy. Materials with large or appreciable sound absorption coefficients (usually greater than about 0.20) are referred to as *sound absorbers*, whereas those with small coefficients are called *sound reflectors.*

The absorption coefficient for a given material usually varies with frequency. Therefore, for practical purposes, it has to be specified for a wide range of frequencies—as a rule, from 125 to 4000 Hz. Table 6-4 gives absorption coefficients for some commonly used materials.*

Noise Reduction Coefficient (NRC)

For the convenience of builders, the absorptive properties of materials are described by a single number, a noise reduction coefficient (NRC). The term is rather confusing because "noise reduction" has been adopted for the noise-level difference between two adjacent enclosures. This coefficient is the arithmetic average of absorption coefficients for 250, 500, 1000, and 2000 Hz expressed to the nearest multiple of 0.05. The mean value does not provide information about the absorption at different frequencies, as the value of the coefficient may change considerably from low to high frequencies.

Sound-absorptive materials are often called *acoustic materials*, which may cause misunderstanding because absorptive properties of these materials are then sometimes confused with sound-insulating properties. The insulating properties of sound-absorptive materials are rather poor.

Absorption in Controlling Reverberation

In an enclosure, the sound reaching a listener's ears is a combination of the sound coming directly from the source and of innumerous reflections from walls, floor, and ceiling. These reflections are strong when the surfaces are reflective, and weak when the surfaces are absorptive. Rooms without reflections are subjectively judged as "dead," and rooms with reflections as "live." In a live room, sounds do not cease immediately after the source stops; they last for a while. This prolongation of sound is called *reverberation*. A measure of reverberation is the reverberation time, T, in seconds. This is the time interval between stopping the source and the time when the sound-pressure level has decreased 60 dB from the value when the source stopped.

The reverberation time can be calculated from the formula

$$T = 0.049\ V/a$$

where

T = the reverberation time in sec
V = the room volume in ft^3
a = the total room absorption in sabins.

*A list of absorption coefficients for most commercial acoustic tiles can be purchased from the American Board Product Association, 205 West Touhy Ave., Park Ridge, Illinois.

The total absorption is the sum of all room surface areas in square feet times their respective sound absorption coefficients.

The reverberation time T is also calculated by

$$T = 0.163\ V/a$$

when V and a are expressed in m^3 and m^2, respectively. For computation examples, see Egan (p. 41).[6]

A certain degree of liveness is acceptable and even preferable. The most preferred value is referred to in the literature as "the optimum reverberation time."[2] Its value changes with room volume and designation of the enclosure. To shorten the reverberation time to its optimum value, absorptive materials are introduced. In small rooms, the precise amount and distribution of absorptive materials is not critical. The absorptive ceiling, carpet, and perhaps drapes on the windows usually provide pleasant environment and eliminate overly long reverberation. To obtain a desirable reverberation in large auditoriums and concert halls is rather complicated, and detailed computations have to be made by a room acoustician.

Absorption in Controlling Noise

Similar to the wanted sound, noise in an enclosure is composed of a mixture of the direct and reflected noise. The absorption in the enclosure can reduce the reflected noise but not the direct noise. The direct noise has to be reduced in its source, or the direct path of noise between the source and the listener has to be interrupted. Therefore, the value of absorptive materials for noise control is limited.

The noise level in a room with highly reflective surfaces (for example, bare concrete with absorption coefficient of 0.02) will be reduced about 8 dB when the ceiling is treated with a highly absorptive material with an absorption coefficient of 0.7. The noise reduction will be about 11 dB when the ceiling and one wall are treated, and about 15 dB when all surfaces are treated.

The value of 15 dB is about the maximum possible NR by absorption in a room. The NR will be smaller if the original room already has absorption better than 0.02.

The noise reduction by the increased absorption can be calculated from the formula

$$\text{NR} = 10 \log a_2/a_1$$

where

NR = the noise reduction in dB
a_2 = the total room absorption after treatment in sabins
a_1 = the total room absorption before treatment in sabins.

(See Egan, p. 43, for a computation example.[6])

In spite of the limited power of the absorptive materials to control noise level, their role is very important. The 10-dB reduction in noise level can mean a change from hazardous working conditions to acceptable noise exposures; in speech communication, it means a change from difficult to easy communication.

Placement of Absorbers

The ceiling is usually the largest and most readily available area for absorbers; in most cases, it is the best place for such material. Sound-absorptive materials that are commercially available have a factory-applied surface finish, which is reasonably durable for ceiling applications. Typically, the ceiling application requires redesigning of light fixtures and/or the ventilation system. If the material cannot cover the surface of the ceiling, sound-absorbing units hanging from the ceiling can be used; however, they generally provide less absorption than the surface cover.

In places exposed to heavy pedestrian traffic, the floors should be covered by carpets, which are sound-absorptive at high frequencies. Carpets are typically used in office buildings, schools, corridors, and staircases. The primary acoustical benefit of carpet is not to absorb sound but rather to reduce the surface-generated noise of footfalls, furniture moving, and so on. Therefore, the variations in carpet weight and thickness and the carpet's sound absorption coefficient are not as important as practicality and durability.

Walls of narrow, long, and very high rooms should be covered by absorptive materials. Parallel reflective surfaces may create a problem called *flutter echo.* The sound is reflected back and forth between them and is distractive and unpleasant. This effect can be avoided by putting some absorptive material on one or both walls or by constructing walls so that they are not parallel. The materials used for wall treatment have to be carefully selected for abuse resistance. It is usually practical to limit the treatment to the area above chair backs or even above the reach of hands. Wall treatment may be used when other surfaces cannot be treated, or when they are already covered with absorbers yet more absorption is needed.

Generally, the construction of the building determines the mounting method. The actual method is important because it will control absorption efficiency. Tables of absorptive materials usually show absorption coefficients for various types of mountings. For example, sound-absorptive materials applied with adhesive directly to the surface are poor absorbers at low frequencies. However, when applied to furring systems, they will give somewhat better low frequency absorption; when used in suspended ceiling systems, they will provide absorption balanced over a large frequency range.

Noise Reduction in an Existing Environment

Reduction of noise in an existent situation has to be approached individually. The starting point is the analysis of noise sources, which can be outside or inside a room. Outside noise probably comes through partition components which have small transmission loss, as do doors and windows. When the pathways of the intruding noise are identified, the insulation of the poor partitions has to be improved. Single windows should be replaced by double windows with well-sealed borders. Light doors should be replaced with solid-core or double doors. Special construction doors with gaskets are also available. If a whole wall transmits excessive noise, its insulation can be improved by adding another full wall at a small distance from the original one.

Some outside noises can be attenuated at the sources before they reach the room. Acoustic connection between rooms by heating ducts may be avoided by separate ducts for each room. The noise of the airflow can be attenuated by absorptive materials placed on the inside walls of the duct.

The noises of the plumbing system should also be attenuated at their origin by appropriate mountings reducing the spreading of the vibrations along the system as well as into the walls.

The most difficult noises to deal with are those originating inside the room. If the source of the noise cannot be separated from the rest of the room by an enclosure or a barrier, the best solution is the application of sound-absorptive materials and carpeting for the elimination of reflections and footstep noise, respectively.

Disturbing noises can be masked sometimes by additional, controlled noise. This is a rather clumsy solution since it elevates the existing background noise level. It should be considered mainly in rather quiet enclosures where the background-noise level is lower than the recommended level, and the additional masker will not raise the level above the recommended one. The role of masking noise will be discussed in more detail in the following section of this chapter.

OPEN VERSUS CLOSED PLAN

The concept of *open plan* is not exactly new—it dates back to the earliest days of architecture. During the last decade, however, we have seen a radical change in its application. Previously, a large number of routine workers were accommodated in one place of business or industry. Today, unpartitioned office and school spaces house a multiplicity of activities. The current trend for open plan is on the rise. The open plan is treated as a modern environment for learning or working, having positive impact on interhuman relationships.

Contrary to the early ideas, the open plan is not cheaper than the conventional closed plan. It is not a building without walls for which construction money ran short; it is an enclosure with partitions, with specially designed walls and ceilings, with special design in acoustic absorption, and often with larger floor area per person than in the closed plan. The plan should be carefully designed by an acoustician; otherwise, the acoustics may be disastrous. There already exist open-plan schools with acoustical deficiencies and with strong teacher dissatisfaction with the teaching spaces. Various parent groups threaten legal action if the deficiencies are not corrected.

Since the requirements for offices and schools are quite different, these two types of open-plan facilities will be treated separately.

Open-Plan Offices

Two factors are the most important in an open-plan office: (1) good speech intelligibility and (2) good speech privacy. Good speech intelligibility is achieved when conversation can be carried on easily within a given work zone, at normal conversational voice level. Good speech privacy is achieved when speech is unintelligible or at least barely intelligible beyond the work zone where it originated.

Both speech intelligibility and speech privacy are evaluated in terms of an articulation index (AI). The AI is a fraction, representing, for a given listening condition, the effective proportion of the normal speech signal that is available to a listener for conveying speech intelligibly. This index is highly correlated with the intelligibility of speech as evaluated by speech-perception tests.

The speech-perception tests consist of a number of sentences, words, or syllables. A list of test items (for example, 10, 25, or 50 words) is presented to a listener or a group of listeners, and the percent of correct responses is scored. The score will be a function of many parameters, including the difficulty of the test and the difficulty of the listening conditions.

An AI value of 1.0 means perfect communication, and the intelligibility measured by any speech test is then 100 percent. An AI of 0.5 means that only half of the information comes to the listener. The intelligibility score varies depending on the difficulty of the test. For example, in good listening conditions, sentences, the easiest of all the tests, will produce 98 percent correct responses (98 percent intelligibility), while words and syllables can produce 95 percent and 70 percent correct responses, respectively. An AI of 0.5 is considered the lowest acceptable value for adequate communication.

Speech privacy is the inverse of intelligibility. An AI of 1.0 means zero privacy; an AI of 0 means perfect privacy. The minimum adequate privacy corresponds to the AI equal to 0.2. At an AI of 0.2, the intelligibility of sentences and words is about 50 percent; of syllables, it is 18 percent.

The goal of successful open-plan design is to achieve a minimum AI of 0.5 for intelligibility and a maximum of 0.2 for privacy.

There are five factors that can modify the AI in an open-plan situation:

1. *The distance between the speaker and the listener.* The further we go from a speaker, the softer is his voice, the lower the intelligibility, and the better the privacy. A 5-dB loss of sound intensity per doubling of distance is a reasonable and practical estimate in open-plan space with absorptive ceiling and walls. How-

ever, exploitation of this factor alone will result in an unpractical sparse occupancy of the office space.

2. *The level of the background noise.* The intelligibility of speech depends greatly on the level of background noise. The higher the noise, the lower the intelligibility and the better the privacy. Fortunately for the designers, the decrease in intelligibility with masking noise is negligible for high AI values but substantial for low AI. In other words, if speech is fully intelligible, it is resistant to deterioration by background noise until the noise level approaches the level of speech. But the overheard speech, which destroys privacy and already has low intelligibility, will be much less intelligible with an increase of background noise. How much of the masking noise has to be added is established by rather complex computations.

The masking noise, to be effective as a masker in increasing speech privacy and simultaneously not distracting in other ways, should be steady and continuous broad-band noise, controllable both in level and frequency spectrum. In practice, such background-noise sources as office activity, air conditioning, and music, do not fulfill these requirements and are not as effective as desired. Recommended is electronic noise, which can be fed into an array of loudspeakers above the suspended sound-absorptive panels.

The concept of additional masking noise is not readily accepted by the public. It has been found, however, that when the noise is applied appropriately and carefully and does not exceed recommended levels, it goes unnoticed by the office users if they are not informed of its presence.

3. *The loudness of the speaker's voice.* The louder the speaker's voice, the better his intelligibility but the poorer the speech privacy. Open-plan offices are designed for normal voice. This means some restrictions for the users, and it rules out the use of electroacoustic amplifying devices. The speaker's voice will usually remain at normal sound level unless the background noise is too high and the talker tries to overcome the disturbance. This creates another limitation for the background noise level.

4. *The speaker's orientation.* In most situations, the speaker and the listener are sitting facing each other, and most designs are made for that fixed relationship. Some privacy might be lost when the speaker is standing, and his voice carries over a barrier (as shown in Figure 6-7), or if he turns away from the listener and raises his voice to compensate for loudness. The reflections over the barrier should be minimized by a highly absorptive ceiling.

5. *The efficiency of barriers dividing work zones.* Barriers used in open-plan offices do not reach the ceiling. If they did, they would change the open plan to a conventional closed space. The barriers serve first of all as visual separation of working zones. Acoustic separation by barriers is very limited, and therefore barriers can secure privacy only if the initial privacy, without barriers, is good. To be effective, the barriers have to be at least 1.5 meters (5 ft) high and 2.4 meters

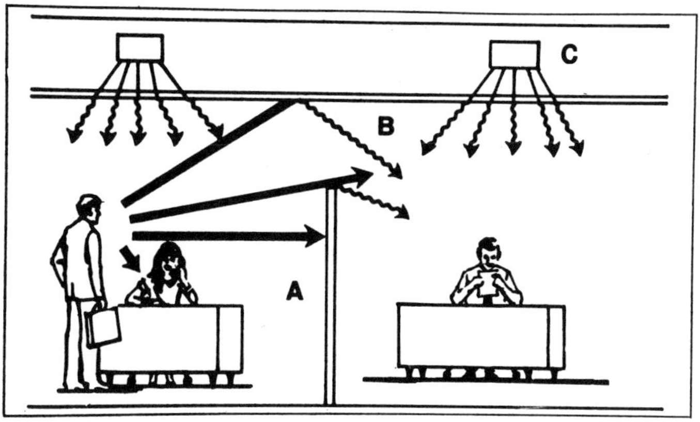

Fig. 6-7 Factors controlling speech privacy in open-plan office. A, barrier. B, absorptive ceiling. C, masking noise. (From Sabine, 1973.)

(8 ft) wide. They should extend to the floor and should be tied to the walls or other dividers to reduce vertical leaks.

For practical reasons and high cost of good privacy in the open plan, offices requiring very high privacy (for example, lawyer's offices, IRS auditing offices) should be separated from the open plan and accommodated in conventional closed rooms.

Open-Plan Schools

Originally, open-plan schools were designed similarly to the offices. However, some recent work has demonstrated that both the environment and the requirements of the schools are different from those of the offices.

While speech intelligibility is still essential, speech privacy is not critical. The major complaint about open-plan schools is usually the high noise levels with the resulting annoyance and distraction that can affect both teachers and students.

Contrary to the office space, the physical relationships between a teacher and variable size groups of students tend to change from hour to hour. Therefore, working zones cannot be designated as they are for the more static situation of an office.

There are no standards for maximum noise levels at schools. The noise level in some schools is as high as 70 dBA.[11] It is highest in kindergartens, lower in upper grades, and lowest in high schools. Fluctuation of the noise contributes to annoyance.

The Japanese Ministry of Education has designated a maximum noise level of 55 dBA as the environmental standard for the schools. In the United States, a realistic goal is a background noise level not exceeding 65 dBA.

To achieve low background noise, the interior should have as much sound absorption as possible. The NRC of ceiling materials should be in the range of 0.90 to 0.95.[11] Walls and barriers should also be covered with absorptive materials whenever possible. Such material should be abuse-resistant to prevent damage by students. The large areas of windows should be covered with draperies, which should be hung with 150 percent to 200 percent fullness. Floors should be covered by carpets to reduce surface-generated noise such as footsteps.

In order to avoid an excessive disturbance by intruding speech and background-noise, the floor area should be 4.6 to 5.6 m^2 (50 to 60 ft^2) per student as opposed to 2.3 to 3.2 m^2 (25 to 35 ft^2) per student in conventional classrooms. The increase in total floor-space is not so great, however, because in the open plan there are no corridors.

In addition, the shape of the open-plan space should be carefully designed. For example, a low, dome-shaped ceiling can introduce unwanted reflections, which increase the level of noise.

Furniture should not be considered as acoustical barriers. It can provide some visual separation but will not provide much noise attenuation. As in offices, the effective barriers have to be large.

The acoustics of the open-plan schools are a complicated problem, and proper design requires careful consideration and the assistance of an acoustician. As an alternative, semiopen plans seem to be developing.

REFERENCES

1. U.S. Department of Housing and Urban Development. Transmittal Notice 1390.2. August 4, 1971.

2. Rettinger, Michael. *Acoustic Design and Noise Control.* New York: Chemical Publishing Co., 1973.

3. Berendt, Raymond D., Winzer, George E., and Burroughs, Courtney B. *A Guide to Airborne, Impact, and Structure Borne Noise-Control in Multi-family Dwellings.* U.S. Dept. of Housing and Urban Development, September, 1967.

4. Bishop, Dwight E., and Hirtle, Parker W. Notes on the sound-transmission loss of residential-type windows and doors. *J. Acoust. Soc. Am.* **43**, 880–882 (1968).

5. Sabine, Hale J., Lacher, Myron B., Flynn, Daniel R., and Quindry, Thomas L. *Acoustical and Thermal Performance of Exterior Residential Walls, Doors and Windows.* U.S. Dept. of Commerce, November 1975.

6. Egan, David M. *Concepts in Architectural Acoustics.* New York: McGraw-Hill, 1972.

7. Beranek, Leo L., Blazier, Warren E., and Figwer, J. Jacek. Preferred noise criterion (PNC) curves and their application to rooms. *J. Acoust. Soc. Am.* **50**, 1223–1228 (1971).

8. Performance Data Acoustical Materials. American Board Products Association, 205 West Touhy Ave., Park Ridge, Illinois. 60068. 1975.

9. Acoustics–Architectural Theory & Practice. American Board Products Association, 205 West Touhy Ave., Park Ridge, Illinois 60068. 1975.

10. Sabine, Hale J. Acoustics of open-plan office. *Sound and Vibration*, 26–29 (September 1973).

11. Walsh, David P. Noise levels and annoyance in open plan educational facilities. *J. Architectural Res.* **4**, 5–16 (1975).

7

Procedures in hearing measurement and protection

Lawrence Vassallo, M.S.
and
Joseph Sataloff, M.D.
Hearing Conservation and Noise Control, Inc.
Philadelphia, Pennsylvania

Since the advent of federal regulations to protect the hearing of workers in America, considerable interest has been shown in fostering personnel practices to assure that the work force does not suffer hearing impairment due to noise exposure at work. The term *hearing conservation* is often used to refer to the subject matter of this chapter. However, *hearing conservation* embodies not only personal hearing testing and, if necessary, hearing protection devices, but it includes sound measurement and noise abatement as well.

This chapter will be concerned with two major aspects of hearing conservation programs: monitoring audiometry and personal hearing protection.

MONITORING AUDIOMETRY

The purpose of monitoring audiometry is to check the effectiveness of a hearing conservation program. By comparing serial audiograms, we can determine if the engineering, administrative, and hearing protection programs are actually preventing hearing loss due to noise. Since monitoring audiometry depends on accurate hearing tests, the following factors must be considered if audiometric findings are to be valid and reliable.

Steps in Establishing a Program

Selecting the Test Site. Ambient noise levels in the test chamber must meet the criteria set forth by the American National Standard S 3.1-1960.* It is advisable therefore to obtain an octave-band analysis of the proposed test location prior to purchasing a booth. With this information, the manufacturer can advise a booth suitable for the location. If the area meets ANSI standards and there are no other distracting factors, the room may be suitable for audiometric testing without a booth.

Other factors to consider when selecting a test site are intermittent distracting noises that may come from nearby conversation, ringing telephones, foot traffic, or production noises when doors to the test area are opened. Vision testing should not be conducted in the same area or at the same time as hearing testing.

Equally important in the test site selection is its convenience to the personnel conducting the hearing tests, personnel to be tested, adequate lighting, and ventilation.

The Premanufactured Booth. Ambient noise levels within the booth must meet ANSI standards. When whether to use a premanufactured booth is being considered, other factors in addition to attenuation capabilities must be allowed for. These include:

Room size. The room in which the booth will be placed should be of sufficient horizontal and vertical size to accommodate the booth without it touching walls or ceilings. In some instances, it may be necessary to place the booth further away from the room walls than anticipated in order to reduce or eliminate low frequency sound transmission into the booth.

Position of the booth window and door. Doors can be hung to open from either the right or left side. Whichever mounting is selected, the door should be capable of opening fully and should not interfere with other objects or doors in the room containing the booth. The observation window then should be on the door-opening side, not on the door-hinge side. This places the technician within a step of the door handle.

Position of the audiometer and technician. Most booths have a shelf that attaches directly to the booth under the observation window. The shelf should be large enough to both hold the audiometer and provide desk space for the

*Criteria for Background Noise in Audiometer Rooms.

technician. If a separate table is used, it should not be butted against the booth as this will bypass the shock (vibration) mountings of the booth. In addition, do not use the space between the booth walls and the room walls for storage, nor attach the booth to the room with moldings, steps, or ramps. Ample space should be allowed on the window side of the booth for a chair for the technician. The booth itself should be equipped with a jack panel for the exterior connections of the audiometer to the booth and the earphone cords inside the booth; internal lighting; and a ventilation system. Earphone hangers generally have to be installed by the purchaser. These should be placed above and behind the subject's chair. The subject's chair should have an armrest adjacent to the window for easy observation by the tester of the subject's responses to the test tones. A shelf placed at window level inside the booth serves the same purpose. This places the subject profile-view to the examiner, the most ideal position. Overhead lights or shiny wall surfaces should not cause reflections on the observation window. Such reflections interfere with the ability to observe subject responses.

Booth assembly. Large booths must be assembled onsite. Assembly can be performed by plant personnel, but it is advisable to check with the manufacturer concerning attenuation guarantees if such assembly is not conducted by the manufacturer's representatives.

Rechecking noise levels. After assembly, internal noise level readings should be obtained with the booth door closed and ventilation blower operating. Annually thereafter, noise levels should be obtained and recorded. This is necessary because gaskets may lose their seals, the ventilation blower may become noisy, vibration mounts may harden, or surrounding noise levels may increase.

Testing without a booth. There may be available in the facility a very quiet office space, an unused vault, a sound-treated room, or some similar area in which audiometric tests can be satisfactorily conducted. The area must meet all the prerequisites listed previously in "Selecting the Test Site." The audiometer is placed on a table, and the technician and subject on opposite sides of the table.

This "across the table" arrangement requires that the subject be in a complete profile position, so that the audiometer face cannot be seen. Reflective surfaces must be avoided as these might act as a mirror, thus allowing the subject to see the dial manipulations and hand movements of the technician. In addition, the technician must be especially careful about giving visual cues (head, arm, hand movements) when presenting the tone, or auditory cues that might occur with a mechanically noisy tone presenter (manual audiometry). These cues can influence the subject to indicate perception of the test stimulus when in fact the response is prompted by the visual or extraauditory stimuli. Noisy drive motors of self-recorders can also be distracting to the subject. Discussion during the examination should be avoided.

Security and protection of equipment (especially earphones) lying on a table in a relatively open space is often a problem. Measures should be taken to ensure that audiometric equipment is handled by authorized personnel only.

Selecting Hearing Testing Equipment

Manual Audiometers. Audiometers used in industry must meet ANSI S 3.6-1969 Specifications for Audiometers. For industrial purposes, a limited-range, pure-tone air-conduction audiometer is all that is necessary. However, audiometers capable of levels greater than 70 dB referenced to the standard reference threshold level, and with capability for testing 8000 Hz are recommended by the authors and are essential for otologic diagnoses. Unless a qualified professional is available to perform special tests, equipment with bone conduction, masking, and speech circuits should not be purchased. There are approximately 20 manufacturers of manual audiometers. Selection should be governed by stability of the circuitry, mechanical longevity, ease of operation, availability, speed of repair service, and cost and reliability of service contracts. Average cost of limited range audiometers is approximately $450. Depending on the subject's hearing levels, test time ranges from 2 to 6 minutes. That is, thresholds within the normal range of hearing take less time to determine than do thresholds that exceed the normal range.

Self-Recording Audiometers. Commonly misnamed "automatic" audiometers, self-recording equipment is actually subject-controlled rather than technician-controlled. The subject controls the intensity of the tone as it is presented around his threshold (frequency, ear selection, and threshold recording are built into the test mode of the equipment). There are currently four manufacturers of self-recording audiometers, and selection should be based on the same guidelines listed for manual audiometers. Depending on the type purchased, test time varies from 6 to 12 minutes (30 seconds per test frequency per ear). Hearing levels of the subject have no influence on the programmed time of the equipment. Prices are approximately $2000.

There is no truth to the claims that self-recording audiometry is more valid and reliable than manual audiometry, or that the technicians need not be certified. If the test environment is adequate, equipment properly calibrated, subjects properly instructed, and the technician properly trained and motivated, both techniques should yield valid and reliable results.

Care and Calibration of Audiometers

Currently proposed guidelines to the Occupational Safety and Health Act specifies three types of calibration procedures, which are conducted at intervals of 1 month (biological), 1 year (periodic), and 5 years (exhaustive). Audiometers should be electronically and acoustically checked in accordance with ANSI S 3.6-1969 specifications.

Biological Calibration. The well-cared-for normal human ear is a very stable organ; it does not experience significant fluctuations in day-to-day sensitivity. This factor of biological stability can be used to advantage in checking the stability of the audiometer, in which there may be instability. The use of the human ear to check the output of the audiometer is called *biological calibration*. Calibration subjects should have no thresholds poorer than 25 dB.

It is recommended that a biological calibration of audiometers be conducted at least once a month. However, the testing workload may warrant that the calibration check be conducted more frequently—perhaps semiweekly, weekly, or even daily. It is important to remember that if the audiometer is found to be out of calibration, the last valid audiograms are those that were done up to the time of the *previous* calibration check. All audiograms done *from* the time of the previous calibration check to the detection of the abnormality should be repeated.

In addition to recording the thresholds, also record the date and the time of the calibration check, the serial number and reference standard of the audiometer, and the signature and printed spelling of the tester. Maintain all of these records in a separate "calibration folder." Recording the time of the test is important, especially if a change of 10 dB or more is noted. Subsequent tests on other biological subjects will show, according to the time entered, that immediate follow-up tests did, or did not, substantiate the detected "change."

Each tester should know his or her own thresholds and in an emergency be able to do a "self-audiogram." This is more of an audiometer check rather than a true biological calibration. It can be used as a preliminary procedure if difficulty with an audiometer is suspected, or if biological subjects are not immediately available. In any case, true biological calibrations must be performed routinely.

Periodic Calibration. Federal regulations state that the audiometer must be subjected to an annual (periodic) calibration check or to a periodic calibration within 30 days of a biological check that uncovers: threshold changes greater than 5 dB at any frequency; distorted signals; attenuator or tone-presenter transients; or other severe operating difficulties. We suggest, however, that a careful check be make to ascertain the permanence of the problem before the audiometer is shipped out for repairs. Notify the calibration agency of the specific problem you are having with the audiometer. Delineate between problems and required calibration checks. Ship the audiometer in its original carton if possible. The audiometer *must* be calibrated with its own set of earphones.

When the work is completed, you should obtain from the repair agency a certificate of calibration and a copy of the actual output levels developed by the audiometer as it was subjected to various tests. (The latter is sometimes called a *bench certificate* or a *readout*.)

Exhaustive Calibration. Once every 5 years, an exhaustive calibration check of the audiometer will be performed. This is a deeper check of the audiometer function and must comply with ANSI S 3.6-1969 Specifications for Audiometers. The repair agency should be familiar with the procedures necessary to conduct periodic and exhaustive checks as outlined in the federal regulations.

A certificate of calibration and listing of actual audiometer outputs should be obtained for the plant records upon completion of this work. Keep all certificates and calibration data with the biological calibration logbook.

New audiometers or those returned from repair or calibration service should immediately be subjected to a biological calibration (and a listening check) before being placed into service. See Fig. 7-1 for recording self-listening checks.

If a replacement audiometer is to be used while the plant audiometer is being serviced, it is necessary to conduct biological (and listening) checks before putting it into service. Record all information on temporary equipment as is done with permanent equipment.

Training of Technicians

All personnel conducting industrial tests should have accreditation as occupational hearing conservationists. Accreditation can be obtained by attending a program that meets the guidelines of the Intersociety Committee on Audiometric Technician Training and is recognized by the Council for Accreditation in Occupational Hearing Conservation (CAOHC). To maintain accreditation, the technician should attend a refresher course every 5 years. All training programs should be directed by a professional holding an instructor's certificate issued by CAOHC.

General topics to be covered in an approved program include: Hearing Conservation in Noise; Anatomy and Physiology of the Human Ear; Sound, Psychophysics and Audition; Federal and State Industrial Noise Regulations; The Audiometer; Audiometric Techniques; The Audiogram; Supervised Audiometric Testing; Principles of Noise Analysis; The Hearing Conservationist in the Industrial Setting; Personal Ear Protection: Review and Examination.*

Usually the testing site should be established and audiometric testing equipment installed before the technician is trained (assuming the program is just being started). With the test facilities ready to use prior to the training of the technician, there will thus be no time lag before the newly learned skills are put

*A listing of council-approved programs can be obtained from the council secretary at 1619 Chestnut Avenue, Haddon Heights, New Jersey 08035.

Self-Listening Test of Audiometer Function

Date	Audiometer Serial Number	Make and Model	Jacks Seated	Cords O.K.	Headband Tension O.K.	Earphone Cushions O.K.	Dials Tight	Volume Increases and Decreases	Pitch O.K.	Tone On-Off O.K.	Tone Presenter Inaudible	No Static	No Crosstalk	Tester Signature

Fig. 7-1 Self-listening test of audiometer function.

to use. Given this situation, the new program and the new technician should first test office and managerial personnel. This acquaints them with the program firsthand and gives the newly trained technician additional expertise and confidence in conducting the hearing tests. In spite of excellent training, however, it is advisable that the audiometric technique of the technician be monitored by a plant consultant in hearing conservation.

During the procedures of selecting the test site, purchasing equipment, and getting the technician trained, managerial personnel, labor leaders, and hourly workers should be made aware of the program, its purpose, and conduct. Plans should be made with department managers for scheduling of hearing tests. Every effort should be made to avoid production interruptions, lost time, or overtime. Changes in production, breakdowns, personnel shift changes, and so on may disrupt the best intentioned plans, but it is still important to try to preplan the conduct of the program. These meetings also serve as a time for the development of understanding and cooperation among all concerned and responsible personnel. Equally important is enlisting the cooperation of the personnel department for informing the medical department of changes in work assignments of hourly personnel.

Preemployment Hearing Tests

All employment applicants should receive a hearing test. Included with this should be the taking of a brief otological history and ear examinations.

Preplacement Hearing Tests

There may be situations in some industries where it is impractical or impossible to conduct preemployment hearing tests. In these situations, employees are hired and the hearing tests are delayed until they have been placed in a particular job. In all situations where the workplace noise in which the employees are placed is 85 dBA or more *throughout* the day, employees should be tested as soon as possible.

Annual Hearing Tests

All personnel exposed to 85 dBA or more should be tested annually. Comparison is made of the employee's most recent levels of hearing (thresholds) against his first recorded reading. All tests must be baseline test–i.e., at least 14 hours must have elapsed since the employee's last exposure to noise. Employees can be tested directly after work only if they have worn hearing protectors that effectively reduce their exposure to 80 dBA or less. This obviates the need to test the exposed worker's hearing prior to his workday and reduces scheduling problems.

Additional Hearing Tests

It is recommended that efforts be made to conduct hearing tests whenever employees are terminated, return from layoff, and return from sick leave or after injuries in or out of the workplace.

Plantwide Hearing Testing Program

Employees not in critical noise areas should be retested every 2 to 3 years to illustrate management's concern for all personnel. Hearing can change for reasons other than noise exposure; such changes should be identified, and proper follow-up referrals should be initiated.

Audiogram Interpretation

The hearing conservation program should be conducted with medical supervision. All hearing tests should be evaluated by competent personnel who have received specific training in industrial hearing conservation programs, such as otologists, audiologists, and staffs of speech and hearing centers. The same proficiency should be expected of consultants hired to interpret audiograms and make a medical diagnosis. This can be done only by a physician, especially an otologist.

All employees or prospective employees referred for otological care should receive proper diagnostic testing and follow-up. In addition, communication between the medical department and the otological center should be maintained.

Employees or prospective employees should be referred for otological attention only on the recommendation of the physician or audiologist reading the audiograms for the plant. The retesting recommendations made by the physician or audiologist should be strictly followed, as they will be based on their professional skill combined with their knowledge of current federal regulations governing this area.

Audiogram Records

Figure 7-2 shows the typical serial-type audiogram form recommended for use with the manual audiometer. Each audiometric examination is recorded and kept in sequence. A brief but adequate otological history is taken and recorded with this initial audiogram. Subsequent observations can be recorded in the "Comments" column adjacent to the appropriate entry or in the space at the bottom of the sheet. Theoretically, a record of this type is capable of storing hearing levels covering a 10-year span. Graph audiograms (Fig. 7-3) are not satisfactory for industrial purposes because each audiogram taken must be

HEARING CONSERVATION NOISE CONTROL

(Use Ink Only)

FULL NAME V (LAST) L (FIRST) A (MIDDLE) BIRTH DATE 12-28-29 S. S. #

SERVICE DATE 1-9-63 DEPT. Weave SHIFT 2

Date 1-8-75	RIGHT Yes	RIGHT No	LEFT Yes	LEFT No
Is Eardrum Visible	X		X	
Is Perforation Present		X		X
Is Drum Normal	X		X	
Is Pathology Present		X		X

DATE AND TIME	RIGHT EAR 500	1000	2000	3000	4000	6000	8000	LEFT EAR 500	1000	2000	3000	4000	6000	8000	SERIAL NUMBER / STANDARD	JOB TITLE	Years On Present Job	NOISE EXPOSURE dBA	* Type	Hrs. Per Day	** Time Lapse	Ear Protection Used	TESTER Sign & Print Name	COMMENTS
1/8/75	0	0	10	25	50	35	10	0	5	10	25	55	40	15	11484 ANSI	weaver	12	97	c	6	14	plug		

HISTORY: Anyone in your Family have hearing loss? NO Is your hearing—Good? Fair? X Poor?

Ever had previous measurement? yes Where? here & x mills Been in military service? yes Army 2.5 yrs.

Exposed to any gunfire or loud noises? yes, target shooting, hunting Ever had infections (running ear)? no

Ever had surgery on either ear? no Explain

What antibiotics or other drugs have you taken? penicillin What contagious diseases have you had? measles, chicken pox

Ever worked at a very noisy job? yes If yes, where, type job, length of time weaver, x mills, 5 years

Ever had dizziness? no What type, when does it occur?

Have you ever had noises in your ears? yes What does it sound like? ringing-constant

Do you have any noisy hobbies? yes - gunning Do you have a second job? no

DATE	Comments following periodic or special audiograms (use reverse side for any comments, if necessary)

* C = Continuous (steady) Noise; I = Interrupted Noise; Im = Impact Noise. ** Time interval since most recent exposure to noise at job.

Fig. 7-2 Serial audiogram form.

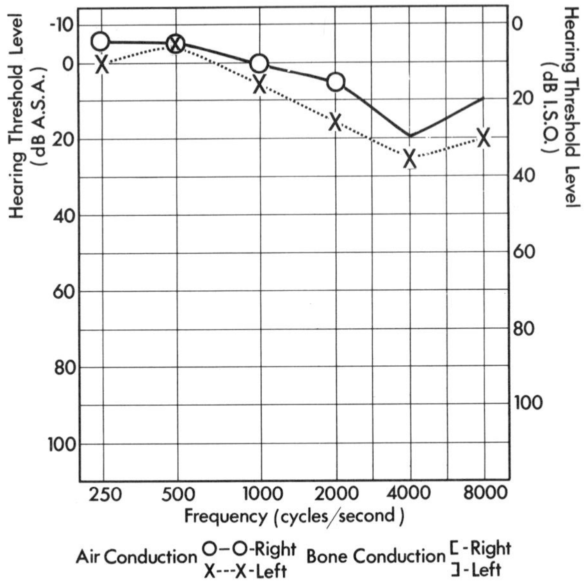

Fig. 7-3 Graph audiogram form.

recorded on a separate card. In addition to creating bulky files, they are more difficult to record and read.

Figure 7-4 shows two typical audiogram cards used with self-recording audiometers. Thresholds are read by taking a midpoint between the peaks and valleys and carrying that line out to the hearing level at the side of the card. In order to be a valid audiogram, there must be at least six crossings per frequency, and the pen tracing must not be wider than 2 dB. The tracing midpoints must be on a horizontal plane. As with the graph type in Fig. 7-3, each test is recorded on a separate card. For ease of interpretation, it is recommended that thresholds be transferred to a serial audiogram. All original recordings are maintained. All audiograms should be maintained for 5 years after employment termination.

Basic information needed on the audiogram includes:

Name
Date of test
Audiometer serial number
Calibration date
Signature and certification of tester
Social Security number or employee identification number
Time of test
Make, model of audiometer
Noise exposure level of employee
Otologic history

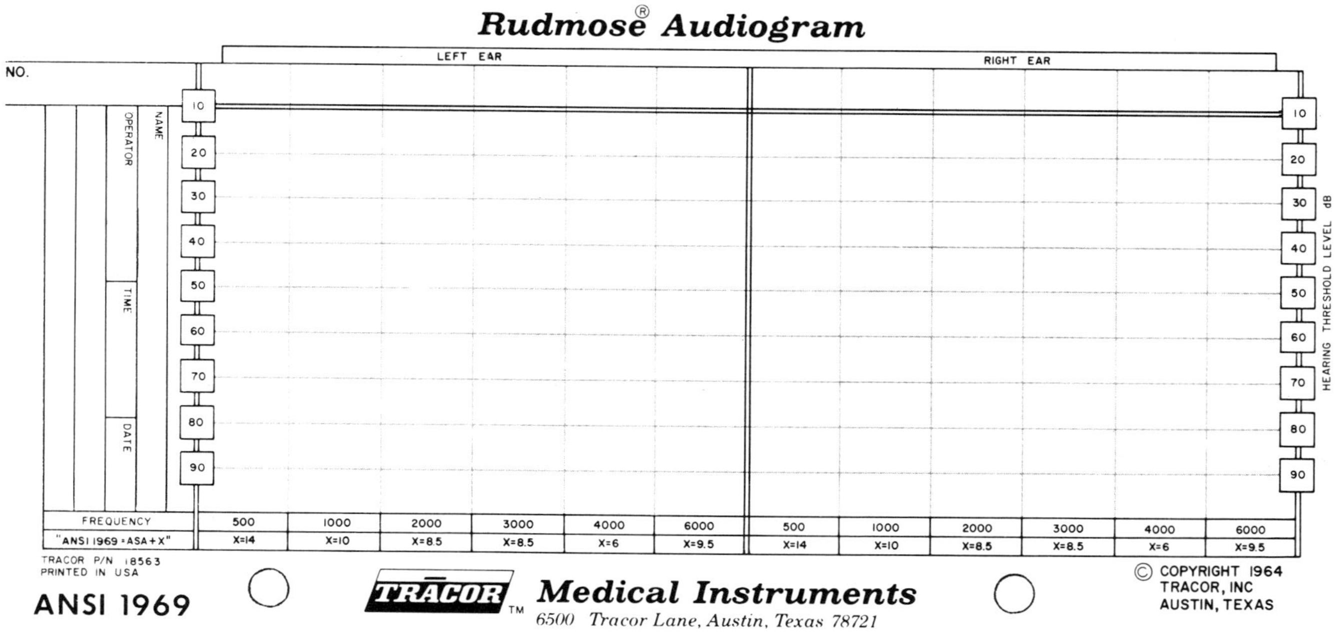

Fig. 7-4 Two types of self-recorder audiogram forms. *(Courtesy of GenRad and Tractor, Inc.)*

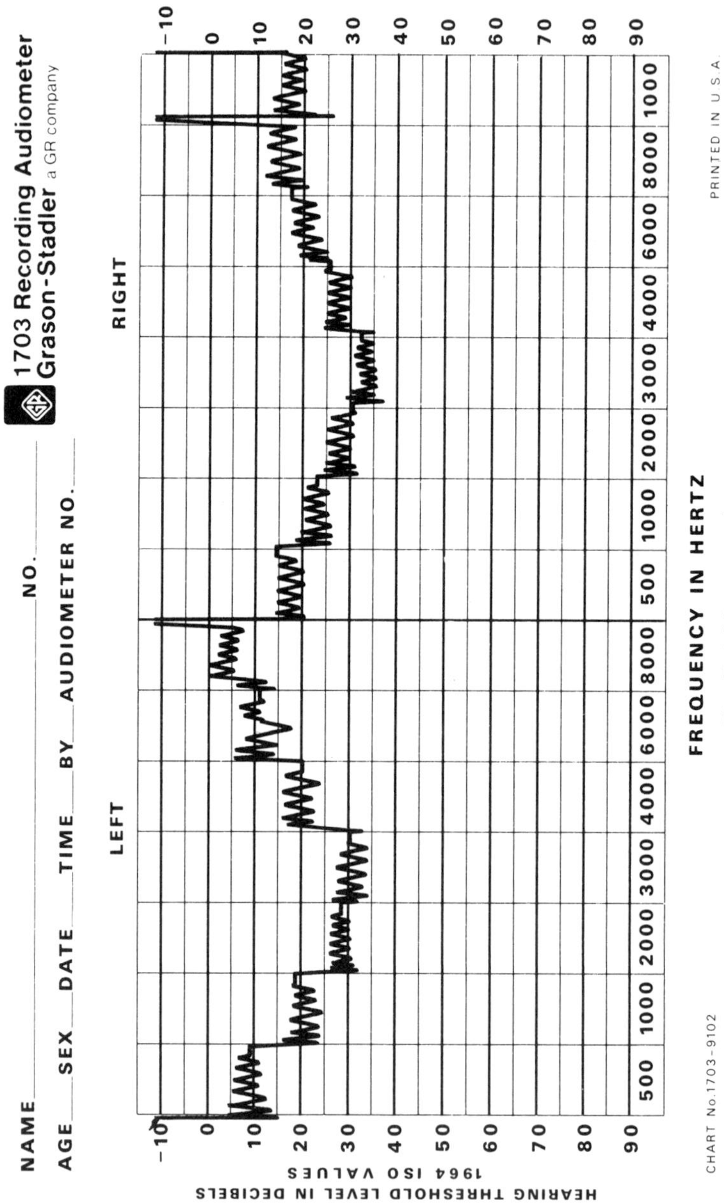

Fig. 7-4 *(Continued)*

Use of Outside Test Facilities

An employer is not absolved of the responsibility of having valid and reliable audiograms just by hiring outside testing services. At the very least, the employer should ascertain the qualifications of contracted personnel as well as their testing equipment and environment. With proper training of plant personnel and the purchase of adequate equipment, the hearing testing program can be very effectively and economically conducted in-house. In determining the feasibility of outside consultants to do the hearing testing, one must consider the peraudiogram cost, the disruption of work, the availability to do retesting at the time intervals required (within 30 days in some cases), and their availability to do preemployment or preplacement testing in the face of high employee turnover. In-house programs mature into ongoing, smoothly revolving programs without great disruption of production.

Summary of Monitoring Audiometry

Audiometry is a system of monitoring the effectiveness of the hearing conservation program. Factors contributing to poor audiometric results are audiometers out of calibration, poor testing environment, poor or inconsistent audiometric technique, and inaccurate recordkeeping. Control of these variables should yield valid and reliable audiometric tests.

HEARING PROTECTION

Industry is responsible for protecting employee hearing from harmful noise. The preferred method is to reduce the levels at the source (engineering controls). Another method is to reduce employee daily noise dose to acceptable levels (administrative controls). Still another procedure is to reduce the noise at the ear of the worker (personal hearing protection). Some government agencies consider personal protective devices an interim step while noise levels are being reduced through engineering techniques.

Characteristics and Effectiveness of Hearing Protectors

Paths of Sound and Attenuation Limitations. The primary area of damage to the hearing mechanism from intense noise is in the cochlea (inner ear). The purpose of hearing protection devices is to reduce (attenuate) the level of noise entering the outer and middle ears before it reaches the inner ear.

Noise can be transmitted through a protected outer ear directly through the protecting device, through a device altered by the wearer, or by the device itself set into vibration by the sound pressure waves impinging on it. The effect then is that transmission of sound to the middle and inner ear is only partially attenuated at the low frequencies, or even at all frequencies.

In addition, vibrations of the skull from impinging sound waves are transmitted to the inner ear by way of the outer and middle ears, or directly to the inner ear. This limits the amount of attenuation attainable with hearing protectors. Maximum attenuation to be expected is approximately 55 dB.

Other factors governing attenuation are the type of protector used (muffs or inserts), the compliance of the material used in the device, the design, and the frequency components of the noise in which the devices are worn.

Measuring Attenuation. The standard method* for measuring attenuation characteristics has recently been revised.† Most manufacturers publish the attenuation curves of their products but the published test data must be carefully examined to determine if a device will be suitable for a given application.

Other Factors to Consider. In addition to providing adequate protection to the auditory system from excessive noise, factors to consider are cost; durability; comfort; ease of use, cleaning, and storing; appearance; and easy replacement of damaged parts (as in muff types). They should be nontoxic, nonflammable, and nonirritating, and have minimum shrinking, hardening, or cracking characteristics.

Types of Hearing Protectors

Circumaural Earmuffs. *Earmuffs* (Fig. 7-5) are plastic domes that cover the ears and are connected with a spring band. The band is usually adjustable to compensate for varying head sizes and ear positions and also provides the tension to seal the domes against the head. Bands may be further adjustable to be worn over the top of the head, to the rear of the head (at the nape of the neck), or under the chin.

With the proper hardware, domes can also be attached to hardhats. In all cases, proper tension must be maintained to effect a proper seal.

Cushions are attached to the domes where they contact the head. These may be filled with foam, air, or liquid and should be compliant to conform to the irregularities of the surfaces they contact. Congenital deformities of the face or skull, mastoid surgery, or certain facial configurations may negate the use of muffs if a seal cannot be obtained.

Generally, the larger domes fitted with compliant cushions give the greater attenuation of low frequency sounds. The domes are also lined with open-cell foam for further sound absorption and damping. The dome opening and inside

*ANSI specification Z 24.22–1957.

†ANSI 3.19–1974 Method for the Measurement of Real-Ear Protection of Hearing Protectors and Physical Attenuation of Earmuffs.

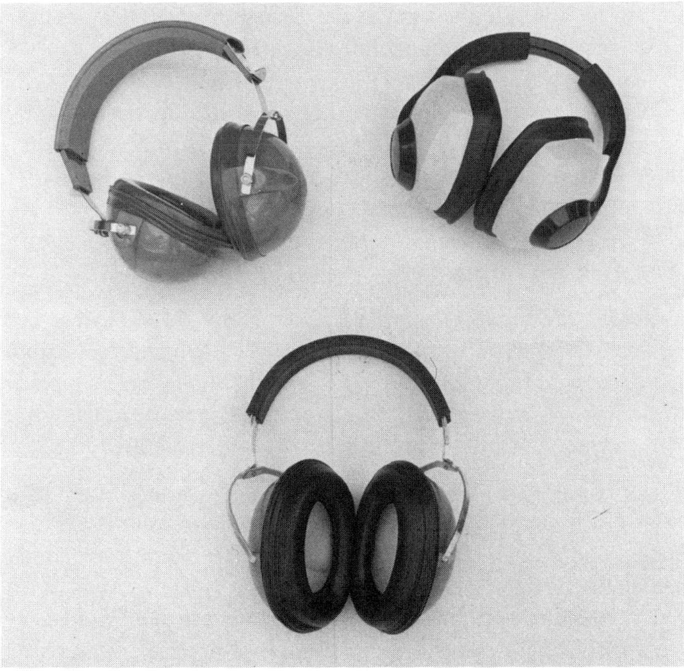

Fig. 7-5 Several types of muff protectors. (*Courtesy of American Optical Corp., Mine Safety Appliances Company,* and *Willson Products Division.*)

dimension should be large enough to encompass the outer ear. For this reason and those mentioned previously, earmuffs must be "fitted."

Earmuffs generally afford the greatest amount of attenuation, are easier to fit, are easier for employees with deformed or missing digits to handle, are more easily monitored by safety personnel, are usually longer lasting if parts can be replaced, and can be reissued when properly rebuilt and sterilized. In addition, they can be used on collapsed or infected ears, are best for short exposure (for instance, for personnel making frequent trips in and out of noise during the day, or personnel or visitors passing through a critical noise area), and in general afford more assurance that attenuation is being provided.

Some disadvantages of earmuffs are that initially they are more expensive to purchase; they require a tight seal against the head, which requires pressure and therefore may be uncomfortable; are usually bulky, heavy, and hot; may not be effective when worn with eyeglasses or hardhats; may cause dermatitis where the cushion contacts the skin (especially if they are not cleaned properly), and are more difficult to store. They may lose attenuation capabilities if the cushions become hard or cracked, or lose the fluid or air filling, or if the headband loses its tension. In addition, if the worker alters them by drilling holes in the domes (usually for "ventilation"), attenuation is lessened or nullified.

Aural (Inserts). Inserts or plug-type protectors fit directly into the ear canal. They come in many configurations and are made of rubber, plastic, or wax-impregnated cotton. Proper fit depends on the ability to make contact along the entire circumference of the ear-canal walls. This generally requires an outward pressure to be exerted by the insert to form a proper seal. Ear inserts are supplied in four general configurations.

Premolded (Sized). Two general types of premolded sized inserts are available: one with the V51-R configuration (Fig. 7-6[A]) and one with a smooth bullet-shaped design (Fig. 7-6[B]). The V51-R ordinarily is supplied in five sizes: extra small, small, medium, large, and extra large. The bullet-shaped design is supplied in small, medium, large, or in numbered sizes from 6 to 12.

Premolded (Universal). The premolded universal inserts are manufactured with two or more flanges on the same stem of progressively larger diameters to accomodate practically any size ear canal, Fig. 7-6(C). Some are manufactured in small, medium, and large sizes.

The V51-R is suitable for most canal configurations except those that are very straight or round. The bullet-shaped and universal designs are most suitable for straight or round canals.

Premolded plugs are less expensive than earmuffs, custom-molded inserts, or canal-cap protectors. They are also less cumbersome to wear, carry, or store, and they do not interfere with nor are they hampered by the wearing of hardhats or eyeglasses. Such plugs are better accepted for use in hot humid workspaces. They are available in a wide range of sizes and shapes, and they are easy to keep clean.

Some disadvantages are that these earplugs require a tight seal of the ear canal to be effective and therefore may cause discomfort. Plugs are easily lost and are harder to monitor by safety personnel. They require sizing of each ear, become hard or shrink if not replaced at necessary intervals, may elicit a cough response on insertion by the wearer, and need to be reseated periodically. In an effort to make the inserts more comfortable, the user will sometimes alter the plugs by drilling holes through them or by cutting off the bungs. This will alter or destroy their attenuation capabilities. Improperly fitted (usually undersized) plugs irritate the ear canals. Some degree of manual dexterity is required for insertion. Inserts of any type should not be used in infected ear canals.

Custom-Molded. Custom-molded inserts are generally supplied in kits (Fig. 7-7) containing silicone material and a fixative agent that are mixed together before insertion into the ear canal and outer ear. Upon curing, a permanent custom fit for the ear is obtained. Another type of custom-molded plug is molded from the ear-canal impression for which special molding materials are used. The fin-

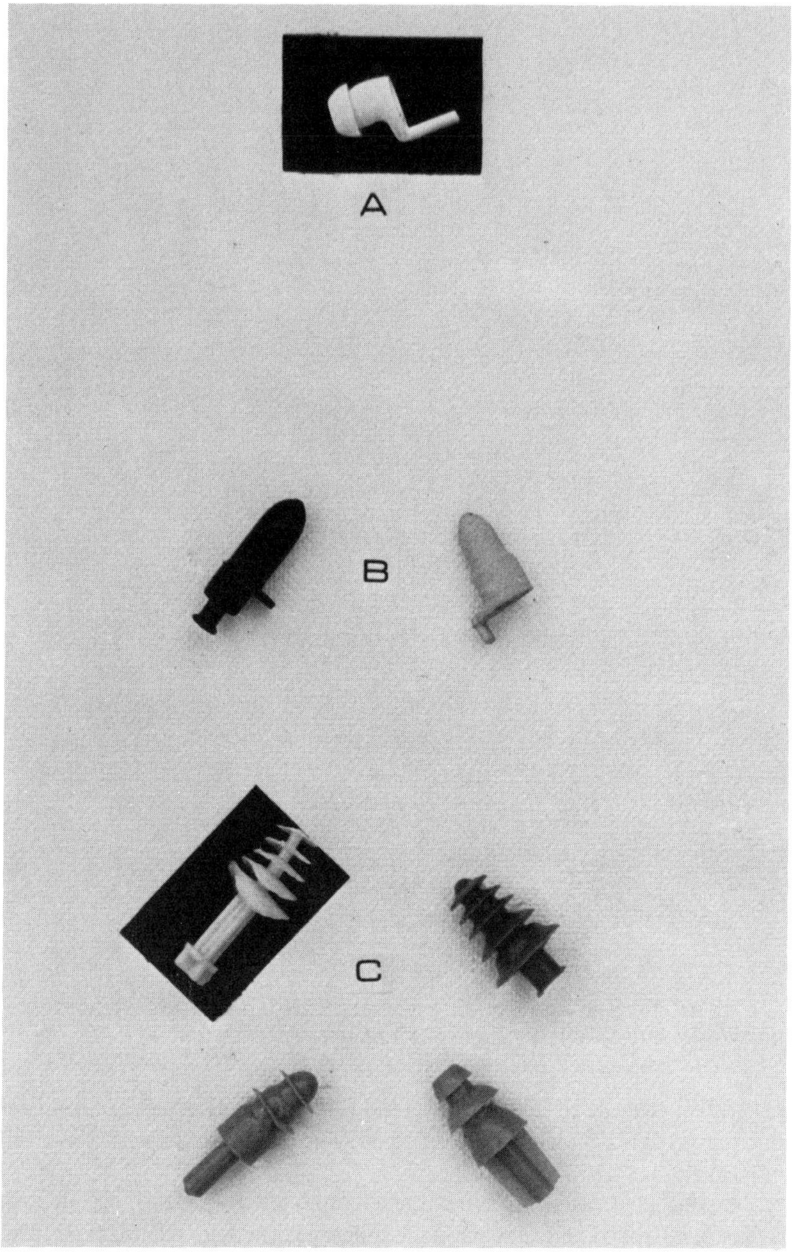

Fig. 7-6 Insert-type protectors. (*Courtesy of Mine Safety Appliances Company, Surgical Mechanical Research, Inc., 3M Company, Flents Products Company,* and *Willson Products Division*.)

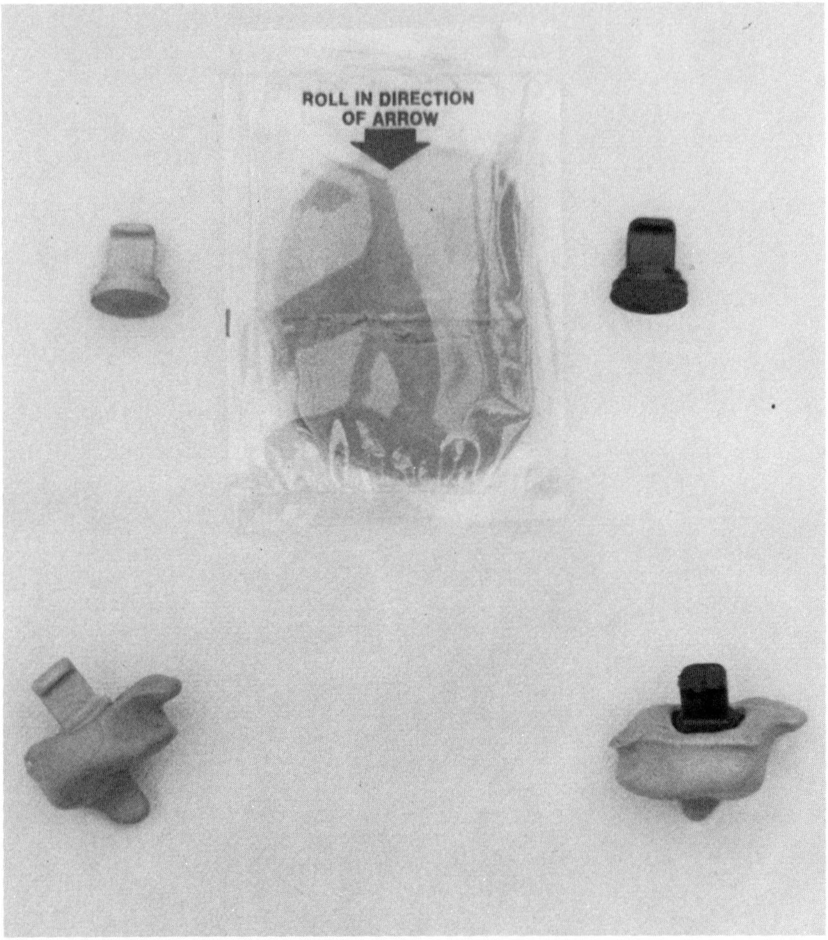

(*a*)

Fig. 7-7 Kit (*a*) and finished (*b*) custom-molded protectors. (*Courtesy of Marion Health and Safety, Inc., and Soundmaster Products, Inc.*)

ished product is made of a hard plastic and is usually expensive when compared to the first type of molded plug described. Both are more expensive than the premolded protectors.

The obvious advantages of custom-molded plugs are the personal nature of the protectors and the time spent with the employee in their manufacture. They generally have a longer service life and are easily inserted by the user after proper instruction.

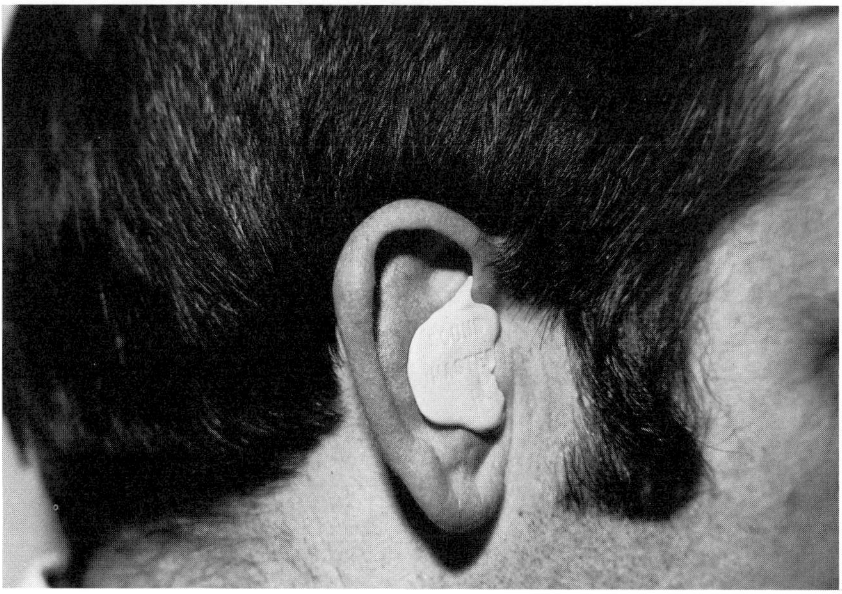

(*b*)

Fig. 7-7 (*Continued*)

The time and money spent in the manufacture of custom-molded plugs is an initial disadvantage, but may be amortized over their relatively long life (lossage not included). Some shrinking may take place, which requires redipping in special fluids to restore the outside dimensions. Shelf life of some of the kits has been a problem. Some complaints can be expected concerning perspiration forming behind the material in the outer ear, and the weight of the material in the outer ear.

Formable, Moldable, or Malleable. Materials used in these types are wax-impregnated cotton, sponge rubber, spun glass, and moldable silicones (sometimes enclosed in a highly resilient rubber jacket) (Fig. 7-8). Size or shape of the canal is not a fitting problem provided that the quantity of the material is adequate to fill the ear-canal entrance. Discomfort is not a frequent complaint although some materials have a tendency to remain in the canal after the plug is removed. The sticky or tacky feeling of the material while in the ear may also be objectionable. This class of protectors is generally the least expensive on initial fitting. However, they are considered a "throwaway" item and over a period of time could become expensive to maintain. They usually provide excellent attenuation and are less variable in this characteristic than the other types of inserts. The materials must be premolded by hand before insertion into the ears, thus

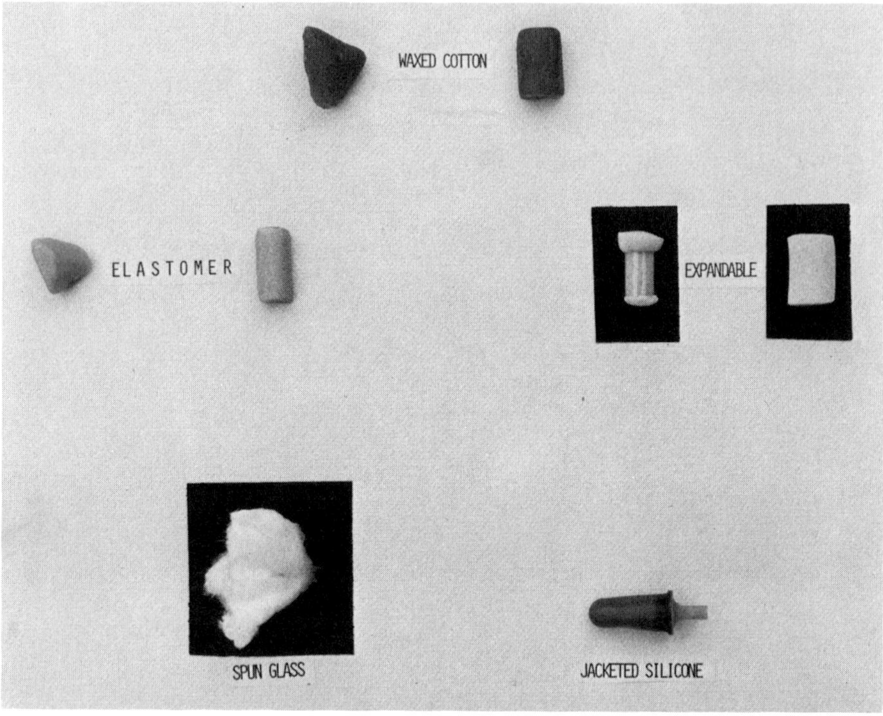

Fig. 7-8 Moldable types of protectors. (*Courtesy of Flents Products Company, Marion Health and Safety, Inc.,* and *Bilsom International, Inc.*)

presenting the problem of transferring any oils, acids, metals, or other foreign substance from the hands to the material being handled. Hands should be clean before handling moldable plugs but the worker on the job may not always be able to follow this suggestion.

Superaural (Canal Caps). Sound attenuation with canal caps is achieved by sealing the external opening of the ear canal (Fig. 7-9). The caps, which are made of soft rubber, are held in place by a spring headband. Although sizing is not a problem, the wearer must be taught how to position the caps properly at the canal entrances, and also how to maintain proper headband tension. Compared to other devices, it is more difficult for inspectors to judge whether canal caps are being properly worn. Also, there appears to be no middle ground in the acceptance of canal cap devices: either they are "too uncomfortable" or they are accepted totally.

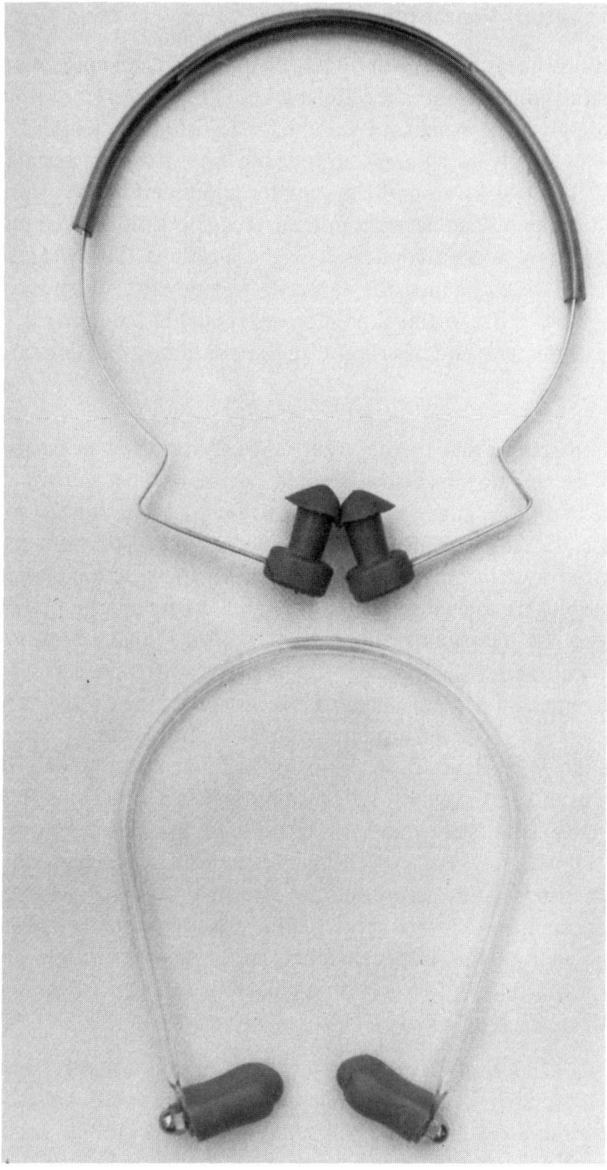

Fig. 7-9 Canal-cap protectors. (*Courtesy of H. E. Douglass Engineering Sales Company* and *Marion Health and Safety, Inc.*)

Fitting and Care of Protectors

Muffs. Openings in earmuffs must be large enough to accommodate the earflap. In addition, the domes frequently are marked "top" or "front" and must be placed accordingly. The headband should be adjusted to the head and should hold the domes snugly to all areas around the ears. If eyeglasses are worn, the user must check for leaks around the temples and ascertain that glasses with bifocal and trifocal lenses can be worn in their proper positions. Wigs and soft caps should not interfere with a good seal of the cushions. The wearer should be shown how to place and adjust the earmuffs and practice the procedure several times for the issuer. The wearer should be instructed in the proper cleansing and rinsing of the dome seals and inserts and to have damaged parts or hardened seals replaced.

Premolded Inserts. Prior to the issuing of inserts, the ear canals should be checked for discharge or excessive wax. At the same time, the size, shape, and configuration of the canal are evaluated in order to determine the size and type protector most suitable for proper fit. In making this inspection, the earflap is pulled up and away from the head, which reveals the true configuration of the canal. The earplug is then inserted and rotated to its proper position. At this time, the wearer is questioned about comfort and sound reduction. The same procedure is followed with the opposite ear. If a different size or style is required in the opposite ear, the wearer is so instructed. The insert is removed by pulling it to one side to avoid a suction and then gently easing it from the canal. The user should be instructed to perform insertion by using one hand to pull the earflap up and out (usually by reaching across the back of the head) and using the other hand to insert and rotate the plug to its proper position. Proper removal procedures should also be explained as well as proper washing, rinsing, and storing methods. Emphasis should be given to rinsing all soap residue from the protectors and to carrying them in the box or pouch supplied. The user should also be instructed to inspect the devices frequently for signs of hardening, shrinking, or fracturing and to seek immediate replacement. Resizing should be conducted before each reissuance of inserts.

Custom-molded. Generally, a representative of the company supplying the ear-mold kits will instruct all medical and safety personnel responsible for making the custom-molded plugs; thereafter, those instructions should be strictly followed. Instructions to the wearer for inserting and caring for the protectors are similar to those given for the wearers of premolded inserts.

Moldable. Most moldable plugs first have to be shaped into a cone, the small end of which is then pressed into the ear canal. Further molding with the fingertip is generally required until a proper seal has been obtained. Instructions in hand cleanliness are obviously important. If the material is to be used again, a suitable carrying case must be provided and used. Dispensing machines or stations should be conveniently located for easy access for resupplying the users. Some of the moldable materials can be washed, and this procedure should be thoroughly explained.

Canal Caps. The angular placement of the caps on the headband requires that the caps be placed in the proper ear. The caps will be marked *R* and *L*, unless they can be rotated to assume the proper angle, in which case they need no marking. The headband should then be adjusted to fit snugly against the head. As with the earmuff type, headband tension is important for obtaining a proper seal. Cleansing procedures are similar to those recommended for insert-type protectors.

Starting the Program

Ordering Hearing Protection. The more successful hearing protection programs allow employees to choose from suggested categories of hearing protection. If inserts are satisfactory for a particular work area, the employee should be allowed a choice of from three to five different types. If muffs are necessary for exposure levels, a choice of two styles should be available. Generally, 10 percent will select earmuffs if given a choice between inserts and muffs. The remaining population requiring protection will use inserts. If sized inserts are purchased, the distribution shown in Table 7-1 should serve as a guide for ordering. Also to be considered in obtaining the initial supply are factors concerning the personnel turnover rate and employees who lose or want to replace their protectors. The latter factor could run as high as 25% or more.

A good selection of inserts to have on hand would consist of 50% of the V51-R type and 50% of the flanged type in several styles.

Once the moldable or custom-molded protectors are introduced, nearly all employees will want them. An administrative decision should be made at the outset as to which types will satisfy the particular needs of the employer and employee. Thereafter, only those protectors supplied by the employer should be permitted.

TABLE 7-1 Distribution Guide of Sized Inserts to Order.

Size	Percent to Order	Size	Percent to order
Extra small	5		
Small	10	or Small	15
Medium	25	or Medium	65
Large	40	or Large	20
Extra large	20		

Education. Management and supervisory personnel should first be instructed in the importance of hearing protection and in their responsibility in enforcement. In fact, enforcement policies should be prepared as part of joint meetings with management and labor representatives. As part of the educational program, all management and supervisory personnel are fitted with hearing protection. These should be worn whenever they enter a noisy area, even if they are just passing through. Earmuffs are especially effective due to their high visibility.

Hourly employees should then have educational sessions concerning hearing and the wearing of protective devices. These can be run by the medical and safety personnel. (Excellent films, booklets, and posters are available.) The sessions should also be used to describe the program's function, and its enforcement and replacement policies; to teach the care and replacement of protectors; and to answer questions concerning the wearing of hearing protectors. The hearing testing program should also be explained at this time.

Schedules for Fitting. The results of the noise survey will serve as a guide for posting proper signs in those departments having critical noise levels. This information will also be used to determine the types of protectors suitable for each department. Usually, one department is fitted at a time, starting with that having the higher noise levels. Approximately 15 minutes per person should be set aside for the initial fitting. To save time in a large facility, the fitting team could go to the workers, using area office spaces as temporary headquarters. In fact, fitting "in the field" is sometimes preferable because there is immediate realization and appreciation of the noise reduction. Time schedules should be worked out in advance so that department managers can plan for expected interruptions. The fitting team should carry or display only those protectors previously determined as suitable for the department being fitted. Fishing-tackle boxes are excellent for transporting necessary supplies.

Setting up for Fittings. The protectors should be attractively displayed and labeled. Sizing tools or kits are available, or several sets of different sizes should be used for the fitting trials. Bowls of washing and rinsing solutions should be set up for sanitizing the devices after each trial. Use throwaway speculums for ear examinations with an otoscope, and cotton-tipped applicators for removing loose wax from the canals. Employees with infected, reddened, or impacted canals should be referred to a physician before being fitted with protectors.

Recordkeeping. In addition to verbal instructions concerning the fitting and hygiene of the protectors, each employee should receive handouts with the same information. Employee records of the type and size of protectors issued should be kept and perhaps should have a place for the employee to sign. Master records should also be kept for inventory and informational purposes. If fitting has been deferred for medical evaluation, record the diagnosis and prognosis of the physician, including when fitting will be suitable. Notes of "excuse" should be handled immediately by telephoning the physician to inquire when the patient can participate in the hearing conservation program. Additional recordkeeping entails the issuance of earmuff-type protectors and the signed acknowledgement by the employee that these are to be returned on termination of employment. Also, records should be kept of earmuffs loaned to employees. All infractions of rules governing use of the protectors as noted by safety or supervisory personnel should also be noted on the employee hearing-testing records.

Generally, after the initial issuance, all replacements due to loss or damage should be charged to the employee (moldable plugs excepted). All reissuances of sized protectors should be preceded by a sizing procedure if several months have elapsed since the last fitting.

Follow-up. A well-planned program will collapse quickly if initial efforts are not followed up. This entails continuing-education programs, strict enforcement policies, frequent inspection of the devices by medical and safety personnel, treating complaints as problems and working individually with employees with problems, following up on medical excuses, maintaining accurate records, and using audiometric results as a guide to the effectiveness of the program.

SUMMARY

Successful hearing protection programs have the following characteristics:

1. They are mandatory and have complete management support.
2. They provide more than one type of protector so that some free selection is available.
3. They allow for trial periods with free exchange of protectors.
4. They allow a break-in time for employees with problems.
5. Supervisory personnel wear protection, preferably muffs.
6. They have the cooperation of labor leaders and representatives.
7. They have a person designated with the responsibility for running the program.
8. Education, promotion, and encouragement are constant ingredients.
9. Hearing protection is properly fitted–and worn!
10. Monitoring audiometry is kept up-to-date, and the results are used for constant evaluation of the hearing conservation program.

REFERENCES

American Industrial Hygiene Association. Industrial Noise Manual. Third Ed. American Industrial Hygiene Association, Akron, 1975.

Botsford, J. Ear protectors—Their characteristics and uses. *Sound and Vibration* **6**, 24 (1972).

Flugrath, J., and Turberville, J. The effectiveness of earplugs. *Sound and Vibration* **6**, 32 (1972).

Guild, E. Ears can be protected. *Noise Control* **4**, 33 (1958).

Maas, R. Hearing protection programs can succeed. *National Safety News* **101**, 77 (1970).

Murphy, A. With teamwork and dedication hearing conservation programs can be successful. *ASSE Journal,* June (1971), p. 21.

Olishifski, J. B., and Harford, E. R. (eds). *Industrial Noise and Hearing Conservation*. National Safety Council, Chicago, 1975.

Sataloff, J. *Hearing Loss.* Philadelphia: J.B. Lippincott Co., 1966.

Sataloff, J., and Michael, P. *Hearing Conservation.* Springfield: Charles C Thomas, 1973.

Vassallo, L., and Glorig, A. Recommended techniques for accurate audiometric tests. *Sound and Vibration* **10**, 12 (1975).

Recommended Sources of Additional Information

Films

Ear Protection and Noise (motion picture, sound, 10 minutes).
American Academy of Opthalmology and Otolaryngology
1966 Inwood Road
Dallas, Texas 75235

The Sound of Sound (motion picture, sound, 16 minutes).
American Optical Corp.
Safety Products Division
Southbridge, Massachusetts 01550

Protect Your Hearing (motion picture, sound, 16 minutes).
David Clark Company
360 Franklin Street
Worcester, Massachusetts 01604

It Takes Two (motion picture, sound, 18 minutes).
Price Filmakers
3491 Cahuenga Blvd.
Hollywood, California 90028

Listen While You Can (motion picture, sound, 15 minutes).
I. C. Webb Company
301 North Madison Street
Rockford, Illinois 61110

For Good Sound Reasons (motion picture, sound, 17 minutes).
Willson Products Division
Second and Washington Streets
Reading, Pennsylvania 19603

Published Material

American National Standards Institute
1430 Broadway
New York, New York 10018

Council for Accreditation in Occupational Hearing Conservation
1619 Chestnut Avenue
Haddon Heights, New Jersey 08035

Directory of Training and Education Resources in Occupational Safety and Health
Department of H.E.W.
National Institute for Occupational Safety and Health
Division of Training
Cincinnati, Ohio 45202

Directory of Noise Control Products: Personal Protection, Audiometric Rooms, Audiometers
National Safety News
425 N. Michigan Avenue
Chicago, Illinois 60611

List of Personal Hearing Protectors and Atteunuation Data (HEW Publication No. NIOSH 76-120)
Superintendent of Documents
U.S. Government Printing Office
Washington, D.C. 20402

8

Industrial and construction noise

George M. Diehl*
Consulting Engineer
Machinery Acoustics
Phillipsburg, New Jersey

Many people must live and work in areas where noise exceeds acceptable levels. Industrial and construction noise crossing area boundaries frequently provokes community annoyance complaints, and many types of industrial and construction machinery operate at noise levels high enough to damage workers' hearing. Thus, industrial and construction machinery are targets of noise control legislation on the federal, state, and local levels, and various noise control techniques are being applied to bring them into compliance.

Some machines respond fairly easily to noise reduction procedures, while others do not. In such cases, at the present time, machine operators and those working nearby must wear personal hearing protection equipment.

*Formerly Manager, Sound and Vibration Section, Ingersoll Rand Research Inc., Phillipsburg, NJ.

TYPES OF MACHINERY INVOLVED

Industrial and construction machinery cover a very wide range of equipment. In general, however, the noise crossing industrial plant and construction site boundaries, is dominated by relatively few types of machines. Similarly, noise control work in industrial plants is usually done on a priority basis, depending upon the noise level of a machine and the number of machines of that type. Machine A may produce a higher sound level than machine B, but there may be only one machine A, and 20 B machines. This would expose many more workers to machine-B noise than to machine-A noise. It therefore becomes a matter of sound level and the number of machines in use.

Investigations have shown that in refineries, chemical plants, and similar industrial operations, noise reduction priority ratings could be assigned in the following order:

Machine Type	Priority Rating
Centrifugal compressors	1
Turbines	2
Centrifugal pumps	3
Reciprocating compressors	3
Gears	3
Rotary compressors	4
Gas turbines	5

This would not necessarily be the order in other projects. For example, in a gas-pipeline compressor station, the turbine may be the major noise source, dominating the noise in nearby residential areas and requiring extensive noise control measures before anything else can be treated.

A similar procedure is used to establish priorities in transportation and construction noise. For example, the U.S. Environmental Protection Agency has determined that medium- and heavy-duty trucks contribute the most sound energy to the environment of any highway vehicle (see Chapter 10). In construction equipment, pile drivers produce much higher sound levels than anything else, yet they do not contribute as much sound energy to the environment as dump trucks, portable air compressors, and concrete mixers, which are the most widely used items of construction equipment.

The EPA list of noise sources, in descending order of sound energy contributed to the environment, follows:

Dump truck
Portable air compressor

Concrete mixer
Jackhammer
Scraper
Dozer
Paver
Generator
Pile driver
Rock drill
Pump
Pneumatic tools
Backhoe

NOISE CONTROL LEGISLATION

Noise immission standards have only an indirect control over the noise radiated by a machine. They state maximum permissible sound levels in workplaces; acceptable daytime and nighttime levels in residential, commercial, and industrial areas; and maximum permissible noise crossing industrial and construction site boundaries. The measured noise in such cases may be produced by a single machine or by a combination of many kinds of machinery.

The Occupational Safety and Health Standard for Occupational Noise Exposure is an example of a noise immission standard. The maximum permissible sound level at a worker's ears, and the time of exposure, are not related directly to the noise produced by any one machine, but depend upon the total noise in the area, where the worker is located with respect to the machine, and other factors. For this reason, noise immission standards or their intent, must not be confused with product noise emission regulations.

Many state, city, and local governments have written noise control codes listing maximum permissible sound levels crossing industrial and construction site boundaries into residential and commercial zones. These levels may be stated either in terms of overall A-weighted sound pressure levels, or in terms of octave-band levels.

If machinery noise reduction at the source is not feasible, for engineering or economic reasons, construction contractors must resort to other techniques to be in compliance with the code. For example, they can group machines strategically, or space them, or locate them properly with respect to boundaries and/or the surrounding terrain.

Product Noise Emission Standards

Product noise emission standards, as their name indicates, have a direct control over the noise produced by a machine. They state that the maximum noise radiated by a machine must not exceed a certain level, when the noise is measured according to a prescribed test code. Permissible noise may be stated in terms of overall A-weighted sound pressure levels, octave-band sound pressure levels, one-third octave-band sound pressure levels, or a combination of these—all measured at a fixed distance from the machine, under known or measurable room conditions. Alternatively, maximum noise emission may be stated in terms of sound power, or sound power level—independent of distance from the machine and independent of room environment.

For regulatory purposes, a popular rating, one easily checked by enforcement officers, is the overall A-weighted sound pressure level, measured at a fixed distance from the machines.

The Noise Control Act of 1972 (Public Law 92-574) provides that ". . . the Administrator of the Environmental Protection Agency shall publish a report identifying products, or classes of products, which in his judgment are major sources of noise, and shall then publish noise emission regulations for those products." The act does not establish the exact noise measurement procedure, but gives the administrator of the EPA the authority to do so. The procedure can be based on either sound pressure level or sound-power level, but in either case, the noise emission will be related to the machine and not to the environment or boundaries.

ACOUSTIC DESIGN CRITERIA

One of the first decisions that should be made before starting a noise reduction program on a machine is to establish an acoustic design goal. This should be slightly lower than the final desired level, but not too much lower, because the costs of noise reduction increase rapidly as design goals are lowered.

They must be set somewhat below what is actually needed, however, to allow for errors in measurement, unexpected acoustic leaks, differences in quality of acoustic materials, and, if a group of machines is involved, to allow for differences between machines.

Even though the objective may be to meet a certain noise immission standard, such as the Occupational Safety and Health Act, or to comply with some community noise ordinance, the design criterion must be either the sound power level of the machine or the sound pressure level at a fixed distance from the machine. That is, the noise must be related directly to the machine and not to anything else.

SOUND TEST CODES

Most large industrial compressed-air machinery, such as axial flow compressors, centrifugal, positive displacement, and reciprocating compressors, are tested according to American National Standard S5.1-1971.* This same code is also used for sound tests on small pneumatic tools such as grinders, hoists, chipping hammers, and impact wrenches. Hydraulic pumps of all sizes are sound-tested according to the same principles set forth in ANSI S5.1.

In the case of large machinery, the code requires the measurement of sound-pressure levels at a distance of 1 meter from the nearest major surface of the machine under test. First, a preliminary survey is made all around the machine, at a distance of 1 meter from the nearest major surface, at a height of 1.5 meters above the floor or walkway, to determine the location of maximum A-weighted sound-pressure level. This is called the *primary microphone location*. Additional microphone locations are established at each end of the equipment train, and at the center of the sides of each casing in the train. All microphone locations are at a horizontal distance of 1 meter from the nearest major surface of the machine, and at a height of 1.5 meters above the floor or walk level.

At each microphone location, overall A-weighted sound pressure level and octave-band sound pressure levels are measured. The background sound levels are measured in the same way, but at only one of the microphone locations.

According to the code, approximate free-field conditions exist if a 6-dB drop-off can be measured in each octave band of interest as the microphone is moved away from the machine in all directions. The code states that if this 6-dB drop-off cannot be obtained, then suitable corrections should be made to the readings.

Sound Power and Sound Power Level

Making "suitable corrections" is not as easy as it may seem. It is obvious that measured sound pressure levels are affected by the environment where the machine is installed, and that this effect must be either calculated or measured in order to adjust measured sound pressure levels to some standard condition, such as free-field.

The sound power level of a machine is independent of measurement distance from the machine, and, although radiated sound power depends upon the acoustic impedance into which it is radiated, it is practically independent of environment for most machinery installations. These features make sound power an excellent way to express the noise rating of a machine.

Unfortunately, most industrial machinery is installed in locations where it is difficult to determine sound power. One of the best techniques now being used is called the *two-surface method*.[1] In this procedure, octave-band sound pressure levels are measured on the surfaces of two hypothetical parallelepipeds over the machine being tested. The second imaginary "box" is farther away from the

*Test Code for the Measurement of Sound from Pneumatic Equipment.

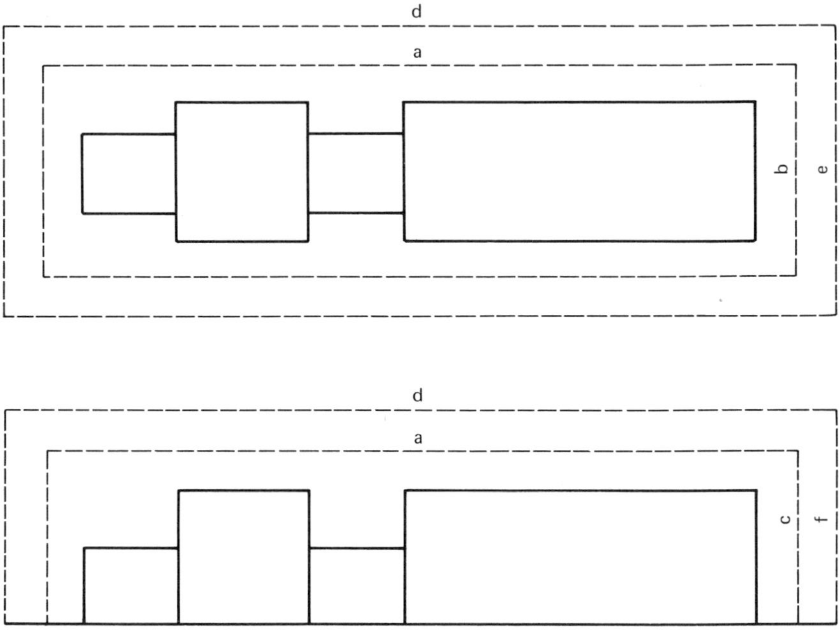

Fig. 8-1 Two-surface method for determination of sound power level.[1]

machine than the first, as shown in Fig. 8-1, and therefore its area, S_2, is greater than that of the first box, S_1.

Also, the sound pressure levels measured on the surface of S_2 will be less than those measured on surface S_1 because they are farther away from the machine. From the difference in average sound pressure levels, and the ratio of the areas of the two "boxes," a correction can be determined (Fig. 8-2) so that sound power level can be calculated by the equation

$$PWL = SPL_1 + 10 \log S_1 - C$$

where

PWL = the sound power level re 10^{-12} watt
SPL_1 = the average sound pressure level measured on parallelepiped S_1, in decibels re 20 μ Pa
S_1 = the area of parallelepiped S_1, in m^2
C = the correction, from Fig. 8-2.

Both sound pressure level and sound power level are expressed in terms of decibels, but the numbers are entirely different. This can cause confusion when communicating with the public concerning machinery noise and could result in

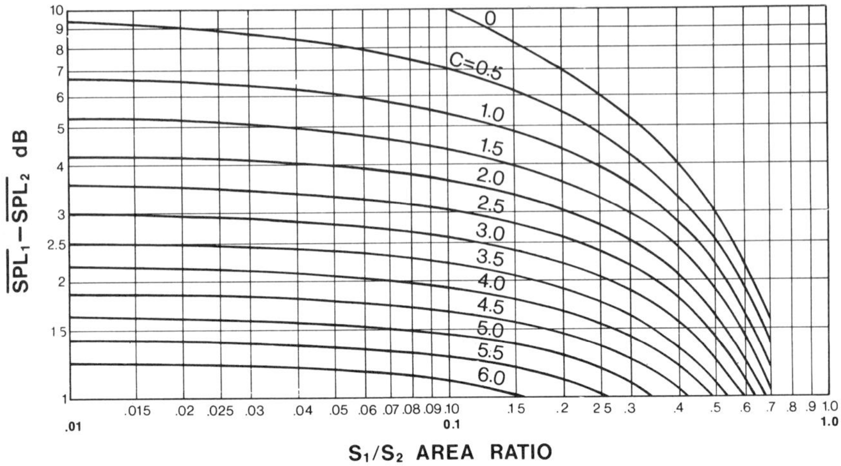

Fig. 8-2 Sound power level correction, two-surface method.[1]

confusion in regulatory action also. The Noise Control Act of 1972 requires labeling on products designated as major noise sources, to inform prospective users of the level of the noise the product emits. The use of decibels for both sound pressure level and sound power level would require a considerable amount of explanation and education.

One way to avoid this is to express sound power level in bels. One bel is equal to 10 decibels. Thus, a sound power level of 105 dB would be indicated as 10.5 bels. Another way is to express the sound power output in terms of acoustic milliwatts. That is, express it in terms of sound power instead on in terms of sound power level. A number of different proposals are under consideration, and the matter has not been settled.

Construction Machinery

Portable air compressors, rock drills, paving breakers, and large drill rigs are sound tested according to ANSI S5.1. In the case of construction machinery of this type, sound pressure level measurements are made at a distance of both 1 and 7 meters.

The EPA has published noise emission standards for portable air compressors, as required by the Noise Control Act of 1972. The standard includes a test procedure that is a modified version of ANSI S5.1, but requires the measurement of A-weighted sound pressure levels at a distance of 7 meters only. Octave-band data are not required.

Recently the International Organization for Standardization (ISO) has produced draft standards for paving breakers, pneumatic tools, and portable air compressors, based on sound power level instead of sound pressure level, but they have not yet been approved. Although sound power and sound power level are excellent ways to express the acoustic output of machines, there is a strong tendency to use A-weighted sound pressure levels for regulatory purposes.

EFFECT OF ENVIRONMENT

The noise measured around any industrial or construction machine is not entirely due to the machine itself, but also to the environment in which the machine is installed.

If the machine is in a free field, there is no reflected noise component, but if it is in a hard-walled room where there are many reflections, the noise is almost uniform all over the room, and there is no decrease in sound pressure level with distance from the machine. In the usual industrial location, the environment is semireverberant, and the measured sound pressure level depends upon the "room constant," expressed in terms of square meters (see Chapter 6). If the room constant is low, the environment is reverberant. If the room constant is high, the environment is nearer to a free field. In a true free field, the room constant is infinity, but generally it may be something less than 100 m^2, up to several thousand square meters, depending upon the materials used in the room construction, and the octave band under consideration. For example, a factory area 30 meters long, 15 meters wide, and 8 meters high, with concrete floors, walls, and ceiling, would have a room constant of approximately 85 m^2.

Whenever sound power level is to be determined, the room constant must be known, or it must be calculated or measured. There are several ways to do this.[1] Conversely, when the sound power level of a machine is known, the sound pressure level that it will produce in a particular location depends upon the room constant of that location. It is important to keep this in mind when writing or interpreting purchase specifications or sound guarantees.

PURCHASE SPECIFICATIONS

Purchasers of industrial and construction machinery are not particularly interested in how much noise a machine makes on the equipment manufacturer's test stand; they want to know how much noise it will make on their own property. On the other hand, machinery manufacturers cannot guarantee, or even predict, the noise a machine will make on the customer's property unless they know what the room constant is. The problem is not as difficult as it may seem at first; nevertheless, it must be understood that the accuracy of the prediction depends upon how accurately the acoustic properties of the machine can be determined, and how accurately the acoustic properties of the customer's environment can be established.

Sound specifications for purchased machinery should set maximum permissible, or acceptable, limits on the noise produced by the machine, and should be referred to the machine. They should not be based on noise at workers' ears, nor on how much noise the machine will emit at construction site boundaries—unless those locations can be stated as exact distances from the machine.

Sound specifications that say, for instance, "The noise produced by the machine must not exceed the levels stated in the Occupational Safety and Health Act," are meaningless. If the specification says, "The sound level at the operator's position must not exceed the 8-hour exposure limit stated in the Occupational Safety and Health Act," and if the operator location is fixed, then the specification does have a definite meaning.

Sound specifications or guarantees for machines that have no fixed operator location, such as centrifugal compressors, reciprocating compressors, and pumps, generally are written in terms of sound-pressure levels at a fixed distance from the machine, when tested under essentially free field conditions. Alternatively, the guarantee, or estimate, may be in terms of sound power level.

Not very long ago, sound power level was always calculated from measurements of sound pressure levels taken with a sound level meter set on "flat" response, or with C-weighting. Today, "overall A-weighted sound power level" is gaining wide popularity (see Chapter 2). This number is calculated by applying the A-network corrections to the octave-band sound pressure levels, calculating A-weighted octave-band sound power levels, combining the weighted octave-band sound powers, and converting to overall A-weighted sound power levels.

A typical sound guarantee may be similar to the following:

> The XYZ Company guarantees that on the date of delivery the average A-weighted sound-pressure level of the [machine] will not exceed [85] dBA re 20 μ Pa at a distance of 1 meter from the nearest major surface of the machine, when sound measurements are made in accordance with American National Standards Institute Test Code for the Measurement of Sound from Pneumatic Equipment (ANSI S5.1-1971), and when the machine is operated under the load and/or other operating conditions specified in the purchase order. The A-weighted sound pressure levels are based on measurements made in a free field above a reflecting plane, or corrected to this condition.

If octave-band sound pressure levels are wanted instead of overall A-weighted sound pressure levels, then these levels would be listed instead of the overall level shown.

Alternatively, the guarantee may state that "on the date of delivery, the A-weighted sound power level will not exceed [93] dBA re 10^{-12} watt."

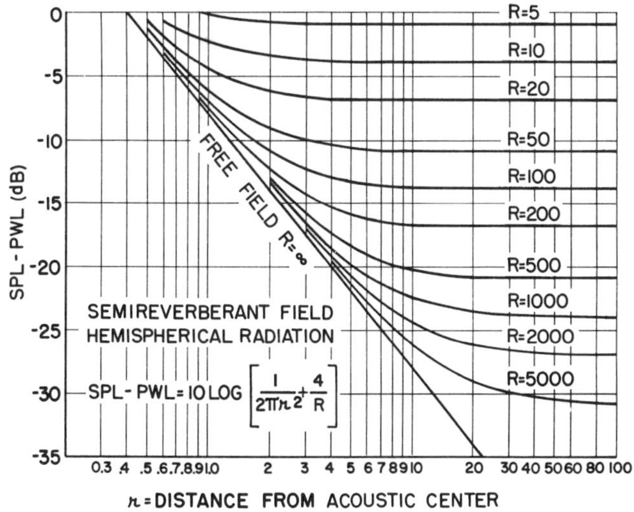

Fig. 8-3 Effect of room constant on sound-pressure level.

It is not necessary to qualify this with measurement distance, or room constant, or environmental conditions because sound power is practically independent of those things.

Even if the purchasers know the sound power level of the machine previously determined by the manufacturer, they still must calculate the sound pressure level the machine will produce when it is installed on their own premises.

This can be done by means of Fig. 8-3. First the room constant, R, is either estimated, calculated, or measured by one of the well-known procedures[1] (see Chapter 6). Figure 8-3 then gives the value in decibels to be subtracted from the sound-power level to obtain the expected sound pressure level.

EXAMPLE: The sound power level of a machine has been found to be 105 dB re 10^{-12} watt. Calculate the expected sound pressure level at a distance of 10 meters, in an area where the room constant is 200 m^2.

SOLUTION: 1. From Fig. 8-3, when $r = 10$ meters, and $R = 200\ m^2$, SPL-PWL = -17.

2. $SPL = PWL - 17 = 105 - 17 = 88$ dB.

Construction Machinery

The preceding type of calculation is usually not necessary in the case of construction equipment, which is almost always used outdoors.

Sound pressure levels, generally measured at distances of 1 and 7 meters, provide the information needed to determine operator noise exposure, and to predict community annoyance or compliance with zoning regulations.

Some cities have developed product noise emission standards for construction equipment that set maximum permissible A-weighted sound pressure levels at distances of 50 feet (15.24 meters) and 100 feet (30.48 meters). These standards are often used by contractors in their purchase specifications for construction machinery to be used in those cities. The Noise Control Act of 1972 provides that, in the future, no state or political subdivision may adopt or enforce regulations for new products that set limits on noise emissions from those products, that are not identical to the federal regulation.

NOISE REDUCTION PROCEDURES

The noise from industrial and construction machinery may be reduced by internal design changes and by external design procedures.

Certainly, the best way to reduce noise is to prevent its generation in the first place—by internal design changes alone. But if economic feasibility must be considered along with technical feasibility, then the best way often includes additional noise control after the machine has been built. In fact, the best way may be to apply external noise reduction alone, and not do anything to the original internal design. There are several reasons for this:

1. Noise reduction design goals are being set lower and lower by federal, state, and local governments.
2. It is often impossible to reduce the noise of high-horsepower, high-energy machines to these low levels by internal design changes alone.
3. When large, high-horsepower machines have been designed, developed, and tuned to maximum efficiency, machinery manufacturers—and purchasers as well—are reluctant to do anything internally that may reduce that efficiency.
4. External sound control usually can reduce the noise to acceptable levels without disturbing the internal design, in the case of large industrial machinery.
5. There are some examples of construction machinery where neither internal design nor external methods can reduce noise to permissible levels and still keep the machines operational.

Therefore, it is important that external design techniques must not be considered as "second class" noise reduction, but as a perfectly acceptable approach.

Internal sound control includes such things as selection of the best combination of centrifugal compressor impeller blades and diffuser vanes; proper design of internal passages in compressors and pumps; controlling internal clearances and tolerances; selection of speeds, number of compressor stages, and other operating parameters; controlling dynamic balance; avoiding mechanical resonances; and many other factors that are now well known to machinery designers.

External sound control includes sound absorption, sound isolation, vibration damping, vibration isolation, and muffling. In general, a combination of several of these techniques or principles must be used to obtain the necessary noise reduction, depending upon the particular class of machinery and the acoustic criterion to be met. Examples of internal and external noise control techniques and of their effectiveness will illustrate this.

EFFECTIVENESS OF NOISE REDUCTION TECHNIQUES

Centrifugal compressor noise can be controlled to a limited extent by internal design changes, but large, high-speed high-horsepower machines require additional external sound control when they must meet low noise criteria. Figure 8-4 shows a typical centrifugal compressor. In these machines, certain combinations of impeller blades and diffuser vanes are noisier than others, and noise may be produced at a frequency that can be calculated quite accurately. Noise reduction can be accomplished by selecting the proper combination of blades, and by controlling clearances between rotating and stationary vanes.

Even though this procedure can reduce noise appreciably, it is not enough to reduce the noise of high-horsepower machines much below 85 dBA at 1 meter,

Fig. 8-4 Centrifugal compressor. (*Courtesy of Ingersoll-Rand Company.*)

because turbulence noise then becomes the dominant factor, and this can be reduced by an acoustic enclosure.

Acoustic Enclosures

A typical acoustic enclosure (Fig. 8-5) is used to reduce the noise of a gas turbine driving a centrifugal compressor. In this installation, the heavy-walled compressor did not have to be enclosed to meet the noise criterion.

To be most effective, acoustic enclosures must be tight. Cracks and openings should be kept to a minimum, since relatively small openings can destroy the effectiveness of an otherwise excellent—and expensive—enclosure.

Figure 8-6 shows how the transmission loss of an enclosure can be decreased by a small opening. *Transmission loss* (*TL*) is a measure of sound-isolating ability and is equal to the number of decibels by which sound energy is reduced in transmission through a partition or enclosure (see Chapter 7). From Fig. 8-6, it can be seen that an opening ($TL = 0$), with an area of only 1 percent in a wall whose transmission loss is 50 dB, reduces the overall transmission loss to only 20 dB.

Fig. 8-5 Gas-turbine-driven centrifugal compressor.

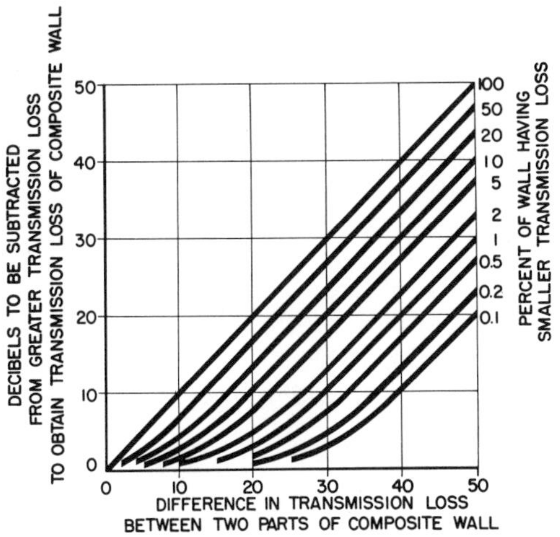

Fig. 8-6 Transmission loss of composite wall.

Fig. 8-7 Unsilenced portable air compressors on city construction project.

Fig. 8-8 Quieted portable air compressor.

Portable Air Compressors

Portable air compressors, shown on Fig. 8-7, have been designated as a major noise source by the EPA. Their noise can be reduced by external design techniques. It would be a waste of time and money to attempt reduction by internal design changes in the compressor, because the driving internal-combustion engine is the major noise source. The situation will probably remain this way until quieter engines can be obtained.

Small compressors can be reduced to acceptable sound levels without much trouble, but the large units require complete enclosures (Fig. 8-8), including sound absorption and sound isolation. Vibration damping must be applied to the large, flat panels, and the engine must be mounted on vibration isolators. In addition, air must be brought in through acoustically treated ducts, and there must be an adequate exhaust silencer on the engine. This treatment reduces the noise from about 110 dBA to 85 dBA at a distance of 1 meter.

Paving Breakers and Rock Drills

Paving breaker noise, another major noise source, can be reduced from about 108 dBA at 1 meter to about 102 dBA by a strap-on muffler (Fig. 8-9). Exhaust noise is the largest component of the total noise, and it is fairly wide-band. It

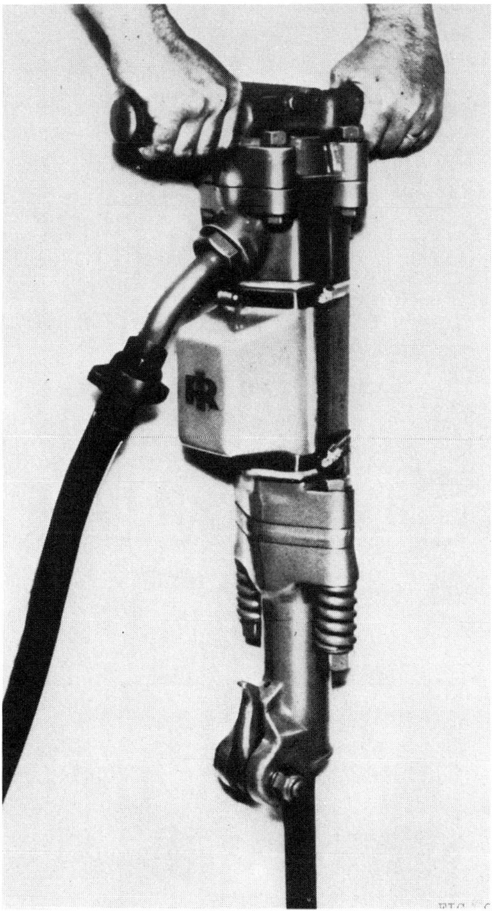

Fig. 8-9 Paving breaker with strap-on muffler.

must be reduced before any measurable improvement can be made in the overall level. These exhaust mufflers rely on gradual expansion, direction changes, and multiple-hole discharge openings as shown on Fig. 8-10. They contain practically no absorption, because absorptive-type mufflers almost always "freeze up." Moisture in the air turns to ice when the exhaust air expands to atmospheric pressure. This stops the breaker.

Breakers with integral mufflers are slightly more effective, and reduce the noise to about 100 dBA at 1 meter (Fig. 8-11).

Ringing of the steel, or moil point, caused by a combination of transverse and longitudinal vibration, is the next largest component of paving breaker noise, which accounts for most of the noise at the high frequency end of the

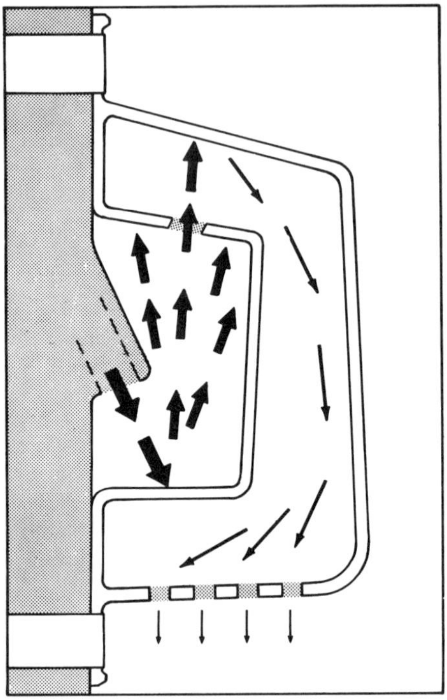

Fig. 8-10 Paving breaker exhaust muffler.

Fig. 8-11 Paving breaker with built-in muffler.

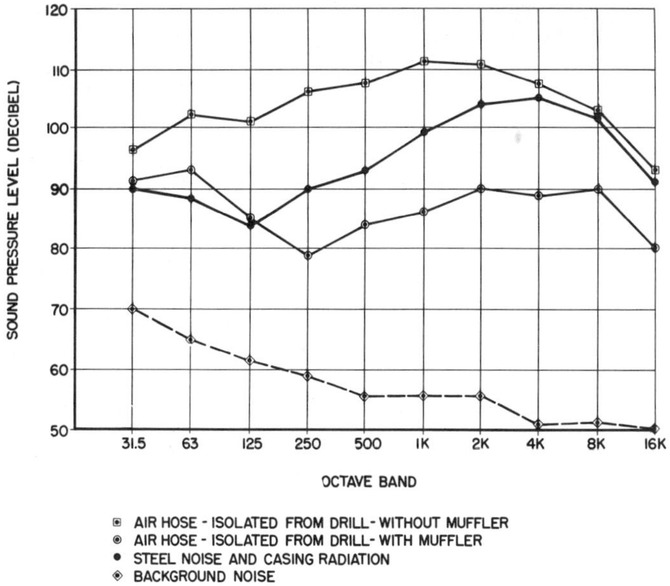

Fig. 8-12 Paving breaker exhaust noise, plus casing and moil-point noise.

spectrum. Various damping techniques have been tried to reduce this high-frequency ringing, but so far damping reduces the overall noise only about 2 dBA.

Figure 8-12 shows the exhaust noise and the casing and moil-point noise. The top curve shows the octave-band sound pressure levels at the end of a hose, 20 meters long, attached to the paving-breaker exhaust. This is almost entirely due to the exhaust.

The third curve from the top shows the same thing except with a large muffler at the end of the hose.

The second curve from the top shows the octave-band levels 1 meter from the paving-breaker casing when the exhaust air is piped away and muffled. This then represents the casing plus moil-point vibration component of the total noise.

Large Drills

Exhaust mufflers are also used to reduce the noise of large crawler drills. The exhaust air is carried up to the top of the drill tower to remove it as far as possible from the operator, and then it is passed through the muffler at the top of the tower before exhausting to atmosphere (Fig. 8-13).

For locations that must comply with strict noise control codes, large drills could be totally enclosed (Fig. 8-14). The acoustic enclosure provides high TL

Fig. 8-13 Exhaust muffler on crawler drill.

Fig. 8-14 Enclosed crawler drill.

for the drill body and drill steel-radiated noise, and in addition, acts as a large absorptive-type muffler for the exhaust air, which is directed out of the top of the tower. This treatment could reduce the noise at 7 meters from 104 dBA to 82 dBA. More development work is needed, however, to make these large drill enclosures practicable.

COST OF NOISE REDUCTION

Noise reduction projects should always have an acoustic design goal. It is a mistake to design for more noise reduction than is really needed because the costs increase rapidly as permissible levels get lower and lower.

It is also a mistake to set the acoustic design goal too high because current noise-control legislation seems to force acceptable levels lower and lower. This places the acoustical engineer in the position of shooting at a moving target. However, he can make the best estimate by setting the acoustic goal several decibels below the present target to allow for errors in measurement, differences between machines due to manufacturing tolerances, unavoidable acoustic leaks, and differences in acoustical materials.

REFERENCE

1. Diehl, George M. *Machinery Acoustics*. New York: Wiley-Interscience, 1973.

9

Aircraft and airport noise

William C. Sperry

Office of Noise Abatement and Control
U.S. Environmental Protection Agency
Washington, D.C.

NOISE SOURCES AND CONTROL

Aircraft noise originates from both the propulsion system and the airframe. The propulsion system is the major noise source of the two to the extent that substantial noise control must be applied to it before the airframe noise becomes a significant factor. However, the state-of-the-art of source-noise control has nearly reached that status. Exceptionally good progress has been made, and more is expected in the near future, in the development of quieter aircraft engines with the result that airframe-noise control must now be given high priority.

The propulsion systems of aircraft take several forms. For example, piston or gas-turbine engines drive propellers (or, in the case of helicopters, rotors), and gas-turbine engines produce thrust as a result of the rate of momentum of high-velocity gas discharging from nozzles. Whatever form the propulsion system may have, it contains its own unique noise-generating mechanisms. Figure 9-1 represents the two most common types of

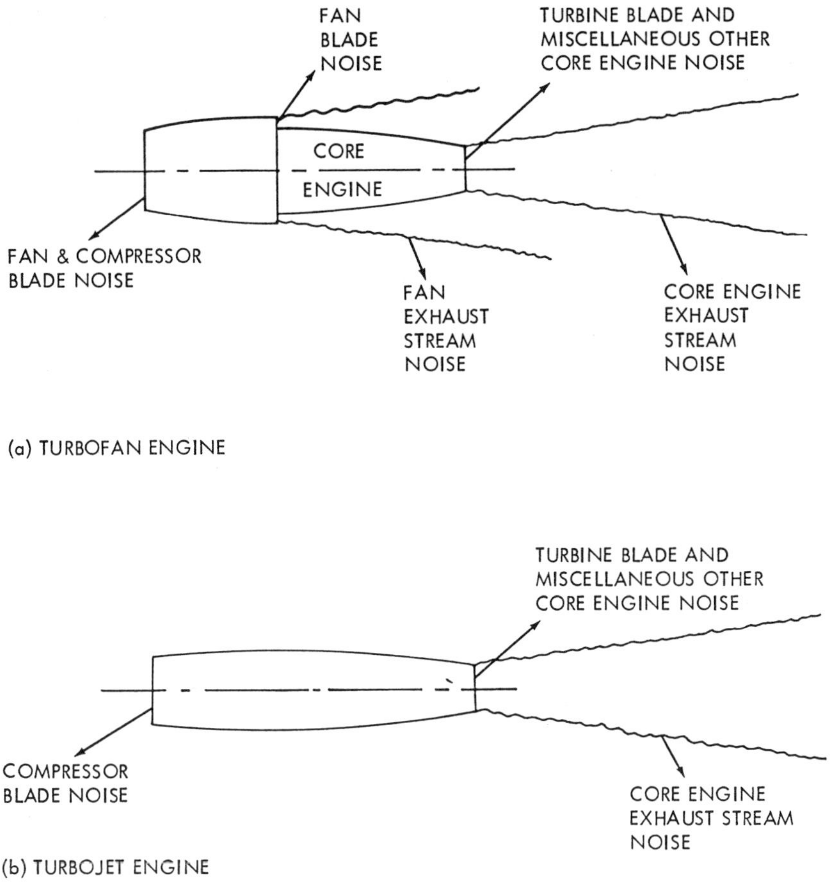

Fig. 9-1 Major sources of noise from aircraft turbofan and turbojet engines. *a*, Turbofan engine. *b*, Turbojet engine.

gas-turbine engines in use. The first type is a turbofan composed of a core engine component consisting of a compressor, combustor, and turbine, and a front-fan component. The second type is a turbojet consisting only of the core engine component. It should be noted that the term *turbojet engines* is used in Federal Aviation Administration regulations to denote all gas-turbine engines (whether turbojet or turbofan) that derive their propulsion by means of gas discharge. The major noise sources illustrated in Fig. 9-1 are: core engine and fan exhaust streams; compressor and fan blades, whose noise radiates from the inlet, and fan exhaust nozzles; and turbine blades, combustor, and miscellaneous sources whose noise radiates from the core engine exhaust nozzle.

A more definitive description of one of the major sources of noise is presented in Fig. 9-2, which shows two typical cases of jet exhaust flow such as might discharge from the core engine.[2] The first case represents the flow pattern that would result from subsonic flow from a converging nozzle. The pattern, however, is very similar to that which would result from a converging-diverging nozzle operating at design conditions; that is, for ideally expanded flow. For overexpanded and underexpanded flow, the pattern would be different. The second case represents the flow pattern that would result from a subsonic underexpanded (overchoked) nozzle. The noise fields would be different for each case.

Sound-power reduction of the sources of noise can be a very difficult method of noise control because of the highly sensitive capability of the human ear. The magnitude of the problem can be illustrated by using the exhaust-stream noise

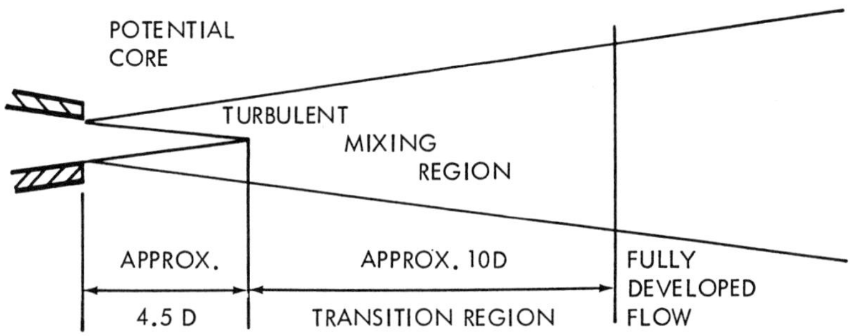

(a) SUBSONIC JET FLOW

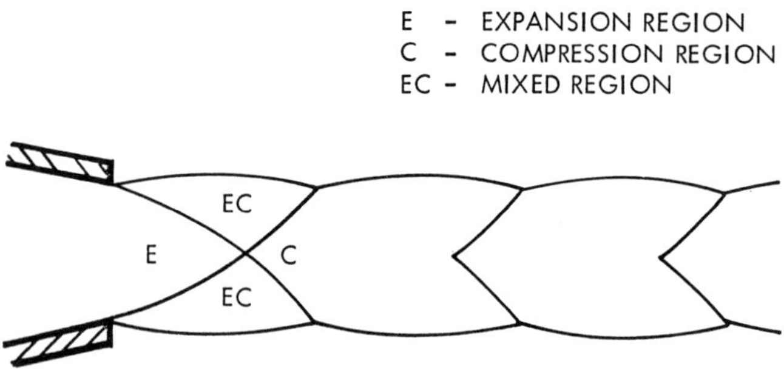

(b) OVERCHOKED JET FLOW

Fig. 9-2 Jet flow converging nozzles. *a*, Subsonic flow. *b*, Overchocked jet flow.

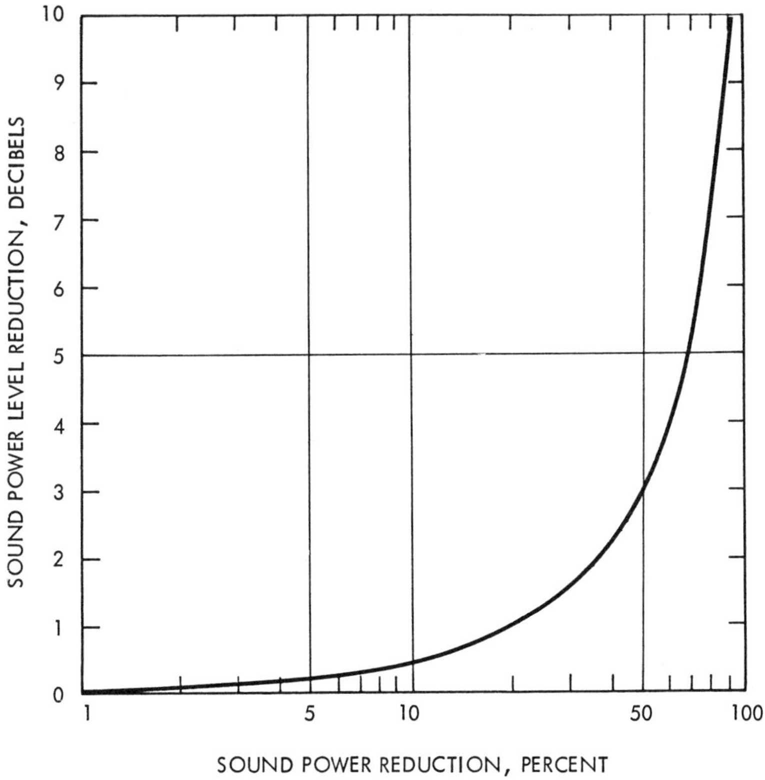

Fig. 9.3 Relationship between sound power and sound power level.

source as an example and referring to Fig. 9-3. The energy converted to noise from a jet stream is only a very small fraction of the total energy available in the high-velocity jet stream.[2] Hence, suppression techniques are handicapped by the need to provide a radical change to a very small portion of jet-stream energy without significantly affecting the majority. At high noise levels, a reduction in acoustic power of 5 dB may not seem to be particularly noteworthy insofar as the human ear is concerned, although it actually represents a 70 percent reduction in acoustic power, as shown in Fig. 9-3. Therefore, a suppression device that accomplishes a 5-dB reduction may indeed be providing a radical change to the acoustical energy.

Figure 9-4 further illustrates the problem discussed above by showing approximate relationships among mechanical, turbulent, and acoustical power in a jet stream under conditions of subsonic and ideally expanded flow.[2] The mechanical power is assumed to vary as the third power of velocity, which is equivalent to 30 dB of mechanical power level for each decade change in Mach number. The turbulent power is assumed to be a maximum of 1 percent of the mechanical

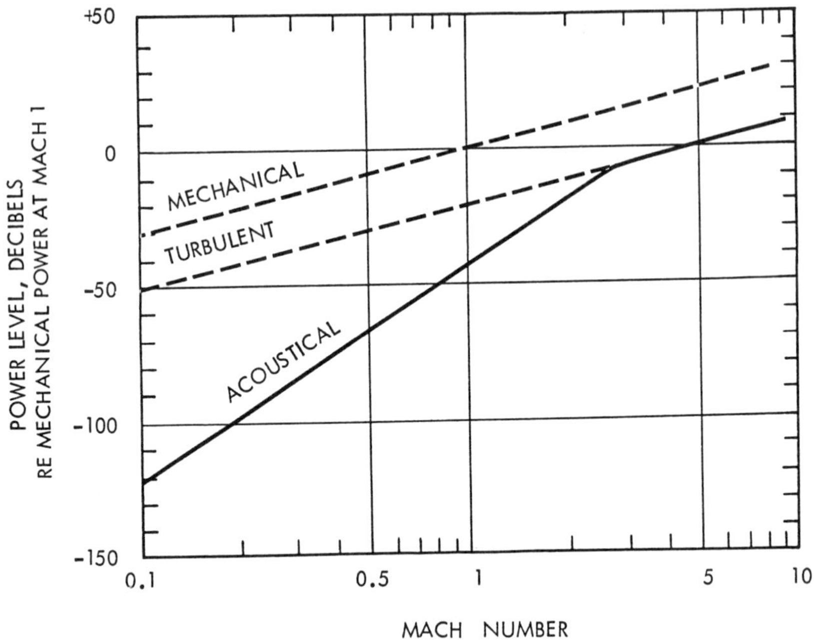

Fig. 9-4 Jet-stream power relationships.

power at all velocities, which translates to a 20-dB difference. And the acoustical power is assumed to vary as the eighth power of the velocity, which is equivalent to 80 dB of acoustical power level for each decade change in Mach number. The preceding assumptions, while only idealized approximations, are reasonable enough for the purpose of this example, including the relationship of acoustical to mechanical power shown at unity Mach number. It can be seen from Fig. 9-4 that the acoustical power varies from an infinitesimal percentage of the mechanical power at Mach 0.1, to about 0.01 of 1 percent at Mach 1.0, to 1 percent at about Mach 2.8 where the turbulent power forms the limit of noise for all higher velocities.

Figure 9-5 illustrates the relative noise characteristics of high- and low-bypass-ratio turbofan engines, which is discussed further in the next section.[3] The ordinate represents the maximum noise that would occur on a line parallel to the engine centerline, and the abscissa represents the engine thrust. It is seen that the fan noise decreases substantially less with thrust reduction than the jet-stream noise. Fortunately, however, fan noise can be attenuated effectively by remedial sound absorption treatment (SAM) applied judiciously to the surfaces of flow passages. Jet-stream noise, on the other hand, is more difficult to reduce in a feasible manner. The principal result is that noise abatement effectiveness is

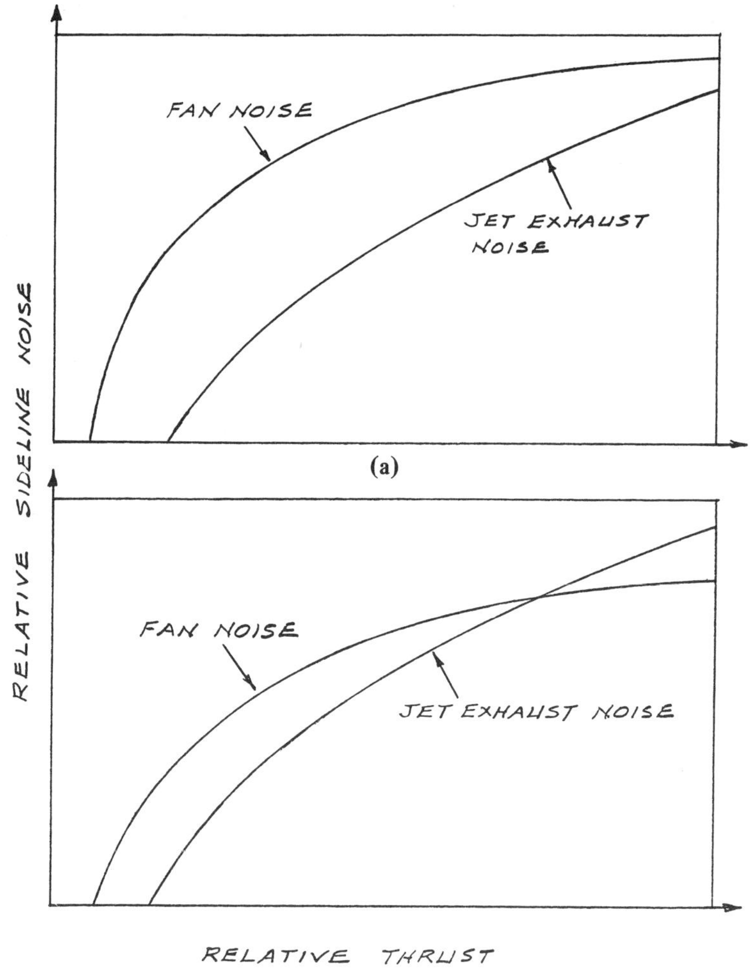

(b)

Fig. 9-5 Noise characteristics of turbofan engines. *a*, High bypass ratio engine. *b*, Low bypass ratio engine.

limited at the higher thrust settings by the difficulty in lowering the jet-stream noise floor. The noise characteristics of the turbojet engine would not include the fan noise component.

An important aspect in effecting noise reduction for turbofan and turbojet engines is illustrated in Fig. 9-6. The jet exhaust curve is shown to vary as the eighth power of the velocity and is identical to the acoustical power curve of Fig. 9-4. However, jet exhaust noise, which is generated downstream of the core engine nozzle, is merely one component of the composite core engine noise. The remain-

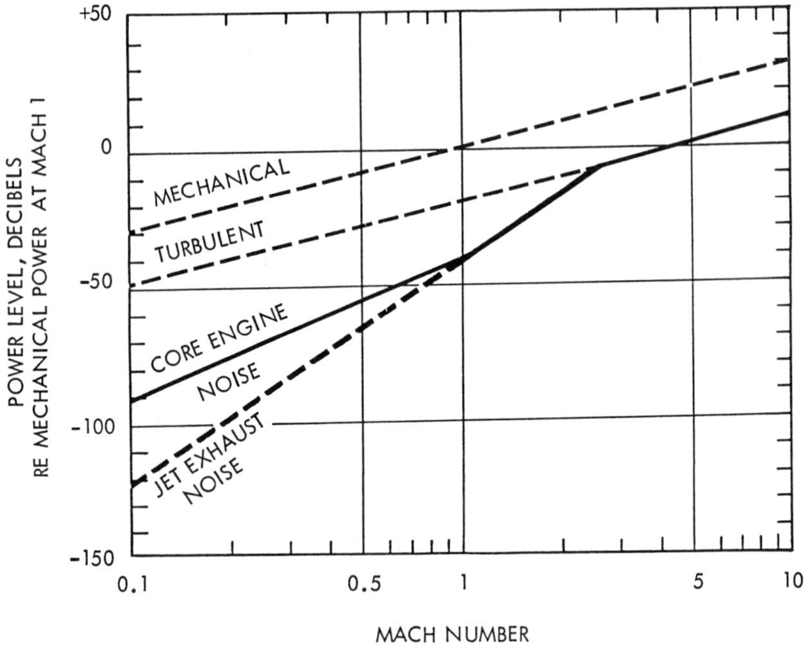

Fig. 9-6 Core engine power relationships.

ing components are generated uspstream of the engine nozzle and dominate at velocities below Mach 1, while above Mach 1 the jet exhaust noise dominates.[3] Possible contributing sources of the subsonic core engine noise include the turbine and the interaction effects of combustion and hot flow with the turbine, combustor, and miscellaneous discontinuities in the flow passages. Noise control of these sources is difficult but not impossible.

Sound power reduction of the source is difficult to accomplish without serious performance degradation. The ideal solution, simply stated, is to devise some method of decreasing the efficiency of conversion of mechanical to acoustical energy. This is particularly difficult with jet-stream noise because the mechanism of noise generation is fundamentally inefficient, and few methods to further decrease the efficiency are obvious. Fortunately, the other noise sources (e.g., fan, compressor, and turbine) result from more efficient conversion mechanisms, and more techniques are available for reducing the efficiency. Other methods of noise control, such as directivity change and frequency shift, offer opportunity for effective exploitation. Sound absorption treatment of the surfaces of inlet and fan exhaust ducts, guide vanes, struts, and splitters presents additional opportunities for controlling fan and compressor noise, especially since strong and durable materials are now available for engine application. The same treatment techniques can be used for turbine noise as well, but the materials must be capable of functioning in a high temperature environment. Variable nozzles and guide vanes have potential for appreciably reducing jet exhaust noise and fan and compressor inlet noise by virtue of providing more efficient mechanical performance.

Engine noise control is difficult, particularly by the method of sound power reduction, but it is basic and must be exploited to the maximum extent feasible. Power-plant builders must utilize an integrated team of engineers and scientists who are expert in acoustics, aerodynamics, thermodynamics, turbomachinery, structures, and materials in order to explore all technical aspects of the problem. Engine designers must consider second-order effects in their designs, not merely to accomplish a fraction of a percent gain in performance, but to reduce the efficiency of conversion of mechanical to acoustical energy.[3] Designers must innovate and exploit all potential possibilities. Compressor discharge bleed might be employed for filling guide-vane wakes and overexpanded nozzles, to accelerate the mixing of hot and cold flows, and for various methods of boundary layer control. Variable control systems should be studies for optimizing engine performance at all operating conditions pertinent to noise abatement, and not just the cruise mode. Rotor-blade design should be reexamined for maximum performance at lowest tip speeds and relative velocities using variable geometry or other techniques. Methods for changing noise tonal quality, reducing tone-to-noise ratios, and replacing discordant with concordant sounds should be considered as possible methods of improving subjective response.

Airframe noise (also referred to as *nonpropulsive, aerodynamic*, or *self* noise) is defined as the noise generated in flight from sources other than the engine, auxiliary power units, and other mechanical accessories. Airframe noise, as illustrated in Fig. 9-7, is produced by aerodynamic boundary layers or the turbulence created by air passing over and around the wings, fuselage, nacelles, flaps, land-

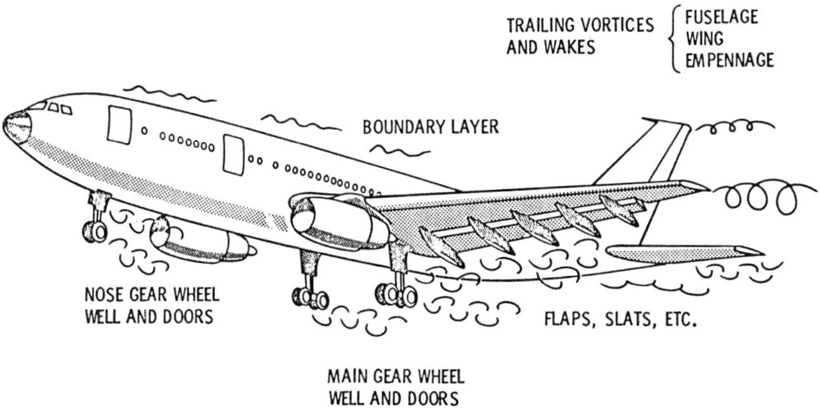

Fig. 9-7 Airframe noise sources.

ing gear, wheel wells, doors, windows, and miscellaneous cavities and protuberances that tend to disrupt smooth flow.[19,52] The most controlling performance parameters in airframe noise generation are the flap settings and the speed of the aircraft. Turbulence—and therefore airframe noise—increases with flap angle but depends to a large extent on the flap design.

It is becoming more apparent that airframe noise must be considered in the design of future aircraft if significant further noise reduction is to result. Decreasing engine noise to levels near or below those due to the airframe will have little effect on further aircraft noise reduction, unless the airframe noise itself is reduced.

In summary, it is well known that noise represents wasted energy, and properly designed noise control can redirect some of that energy to performance. But, since noise is a small percent of the total energy, until recently it has been considered by the engine and airframe designers to be a second-order effect and thus neglected. That is, the wasted noise energy was not considered to be of sufficient magnitude to be worth recovering. However, when noise and its control became important, the designers were forced to consider second-order effects related to noise and found that there were benefits beyond those that could be attributed to the energy transfer of noise to performance. In other words, there was a fallout of performance improvement resulting from the increased knowledge of engine and aircraft design, which can be attributed to the requirements for noise control. This phenomenon, a recent development, should become more effective with time.

NOISE CONTROL AS A SYSTEM CONCEPT

Protection of the public health and welfare from aircraft noise is accomplished most effectively by exercising four noise-control options taken together as a system:

1. *Source control,* consisting of the application of basic design principles or special hardware to the engine/airframe combination, which will minimize the generation and radiation of noise.
2. *Path control*, consisting of the applicaton of flight procedures that will minimize the generation and propagation of noise.
3. *Receiver control*, consisting of the application of procedures such as restrictions on the type and use of aircraft at the airport, which will minimize community noise exposure.
4. *Land-use control*, consisting of the development or modification of airport surroundings for maximum noise-compatible usage.

In general, the primary approach for noise abatement is to attempt to control noise at the source to the extent that aircraft would be acceptable for operations at all airports and en route. And in principle, aircraft noise can be controlled extensively at the source by massive implementation of technology. In practice, however, the technological capability for complete control without exorbitant penalties is not yet available and may never be. Symbolic representations of massive noise control would be a "cork" in the exhaust nozzle with 100 percent performance loss or various "porous corks" with lesser performance losses. Although the noise would be eliminated or substantially abated, such unreasonable treatment would obviously not be tolerated. Insistence that full protection of the public health and welfare be accomplished solely by source control, therefore, could have the effect of preventing the development of most new aircraft and grounding the existing fleet.

The source-noise-reduction accomplishments to date have been substantial, and the extensive programs in progress sponsored by the National Aeronautics and Space Administration, the U.S. Department of Transportation, and the U.S. Department of Defense, as well as the in-house programs of the aircraft and engine manufacturers, hold promise for further sizable gains. Nevertheless, there has to be a limit, or floor, beyond which it is technologically impractical or even impossible to proceed. The following sources of noise have been identified as potential noise floors that may be relatively near at hand: jet stream, engine core, and flow-surface interaction.

Noise from the jet-engine exhaust stream mainly results from the mixing of the high-velocity gas discharge with the ambient air. The noise source is usually defined as a volume distribution of convected quadrupoles* whose strength is proportional to the relative jet-stream velocity to the eighth power. The absolute noise level for any given velocity is dependent upon various factors such as exhaust-nozzle size and shape and various influences upstream of the nozzle, such as geometry, roughness and turbulence scale. Current methods of jet-noise reduction involve the use of exhaust-noise suppressors, which break up the main jet and, in effect, change the manner in which it mixes with the ambient air. Such suppressors have been most effective at the higher jet velocities, where the noise is greatest, but are accompanied with significant penalties in thrust, drag, fuel consumption, and airplane empty weight.

The most effective procedure to control jet-stream noise without excessive penalties is to reduce the jet velocity but maintain thrust by increasing the mass flow. The technique used for turbofan engines is to increase the bypass ratio. Incidentally, high-bypass-ratio turbofan engines, which are efficient for subsonic airplanes, were developed originally for performance, not for noise control. The noise levels at the lower jet-stream velocities are higher than would be expected based on jets mixing noise only. There is evidence that other sources of noise are important in this velocity range. For example, sources generated inside the engine, commonly referred to as *core engine noise, may dominate.*

Core engine noise is defined as the noise produced by the gas-generator portion

*Elementary sound sources are classed as monopoles, dipoles, or quadrupoles representing periodically changing volume, force, and shear, respectively.

of the gas-turbine engine, either solely, or as influenced or amplified by the fan discharge, tail pipe, and any other portion of the exhaust system. Core engine noise is assumed to radiate only in the aft-engine quadrant, and its sources are generated upstream of the tail-pipe exit plane. Core engine noise does not include compressor-generated noise radiating from the engine inlet nor fan-generated noise radiating from either the engine inlet or exhaust ducts. It may, however, include compressor-generated noise transmitted downstream through the engine flow passages or fan-generated noise enhanced by interaction with the core engine noise or with the gas stream.

Flow surface interaction noise is produced by the interaction of flows with solid surfaces of the aircraft, and can result from propulsive and nonpropulsive sources. An example of a propulsive source is a powered-lift aircraft where the interaction of the jet engine exhaust with the wing and flap surfaces can be significant noise sources. Nonpropulsive noise is produced by aerodynamic boundary layers or the turbulence produced by air passing over and around the airframe and its various components and is commonly referred to as *airframe noise.*

It appears at this time that airframe noise is the limiting source in the sense that there is no technology conceivable that will permit noise levels lower than the self-noise generated by a reasonably clean airframe. Although it is possible that the noise of the jet exhaust stream or core engine could "bottom out" at levels higher than those for the airframe, the technological capability for substantially lowering the levels of the jet and core sources appears to be more promising than for the airframe source.

Path control can be an effective option for substantial reduction of aircraft noise. Furthermore, it has the advantage that the results are additive to those obtained by source control. However, specialized flight procedures are limited because of the need to maintain the highest degree of safety. Therefore, insistence that full protection of the public health and welfare be accomplished solely by flight procedures is not feasible at this time and probably never will be. Nevertheless, all aircraft can be flown safely in various modes that produce a wide range of noise levels. And, at the least, those safe modes, which will minimize the generation and propagation of noise, should be identified and standardized.

The major problem with aircraft noise in terms of numbers of people exposed, occurs in the vicinity of airports. This problem could be relieved by the application of various operating restrictions at the airport. Extensive use of airport restrictions, however, is cost-effective only if all feasible source and path control options have been implemented. Otherwise, airport restrictions may result in unnecessary damage to the local and national economy.

A concept that merits consideration is that the airport authorities in some cases, and the FAA in other cases, would impose restrictions on the aircraft operators as needed (curfews, quotas, weight and type limitations, preferential runway use, noise abatement takeoff and approach procedures, landing fees, etc.) to ensure that the airport neighborhood communities are noise-compatible consistent with the requirements of health and welfare. The restrictions available to the airport operator would be those approved by the FAA, the Civil Aeronautics Board, and the U.S. Environmental Protection Agency. The highest degree of safety must be maintained, and interstate and foreign commerce requirements must be considered. Restrictions involving flight safety and air-traffic control would be the sole responsibility of the FAA.

After all feasible noise-control measures have been applied to the aircraft—by design, treatment, or modification of the source; by flight and air-traffic control procedures; and by design, location, and use of airports—the noise may still be a problem at some locations. In this event, noise-compatible land use is probably the only remaining solution. The land-use control option is more easily exercised in the siting and development of new airports than as a remedial measure for existing noise-impacted communities. For the latter case, the costs of land-use control may be so high that maximum effort must be devoted to implementing the source, path, and receiver control options taken together as a system.

MEASURES OR RATING OF NOISE

The environmental effects of aircraft noise are difficult to quantify except for the relatively small number of cases of intense sound that occur in close proximity to high-energy noise sources. The detrimental effects of intense sound, such as pain and disorientation in humans, are so well known and guarded against that it really is not a major source of environmental degradation. On the other hand, less severe sound, ranging from barely perceptible levels to levels capable of impairing the hearing of human beings after many years of exposure, may have some influence, probably negative, on almost every living organism. It is for this range of noise levels that judgments on environmental quality are most difficult and to which measures and ratings of aircraft noise are directed.

One significant difference between noise and most other environmental pollutants is that noise is dissipated by the atmosphere in relatively short distances and in a relatively short time after the source has ceased to function. Noise does not linger in the environment and its dissipation is a transformation of mechanical energy into heat in such infinitesimal quantities as to be difficult to measure accurately except under well-controlled or defined test conditions. Consequently, associated with aircraft noise (where the propagation paths are long and the atmospheric media are variable) are problems involving accuracy in measuring its physical properties as well as rating its effects on the environment.

The most important noise receivers are human beings, and noise descriptors must be related to human response to the maximum extent feasible. The effects of aircraft noise on human beings that have been identified are known by such psychoacoustic and physioacoustic terminology as *loudness, noisiness, annoyance, intrusiveness, speech interference, sleep interference, startle, auditory fatigue, temporary threshold shift*, and *permanent threshold shift.* There is no single descriptor of noise, however, that is responsive to all of these human response indicators although effort has been and is continuing to be devoted to that purpose.

The primary element in any criterion for determining human response to aircraft noise is the evaluation measure for describing the single-event noise of an aircraft flyover. Three basic physical properties of sound pressure can be measured: amplitude, frequency distribution, and the time variation of both. For ranges of these properties meaningful to human beings, the combinations required would be infinite. It is necessary, therefore, to develop a simpler set of scales than the basic physical properties to evaluate the noise, preferably a single-number descriptor. In addition, another measure is required that will rate the effects of the cumulative noise exposure of many aircraft flyovers during a specified time period.

Aircraft-noise signatures, which involve interrelated spectral, temporal, and spatial functions of sound pressure, are among the most intricate of the common noise sources. Accurate descriptions of such sound referred to only a single point in space would include time histories over many seconds of many values of the rate of pressure fluctuation. The voluminous documentation required for precise definition of aircraft noise over many points in space is practical only for research, development, and design purposes.

The search for a single-number descriptor of aircraft noise has been extensive. The end result to date, used widely by the aviation community, is *effective perceived noise level (EPNL)* in units of EPNdB. This measure evolved from a simpler concept, called *perceived noise level (PNL)*, in units of PNdB, which in turn evolved from the still simpler *A-weighted level (AL)*, in units of AdB.* EPNL evaluates four factors of the aircraft-noise signature: absolute level, broadband frequency distribution (spectrum), maximum tone, and flyover duration.[28]

Simply stated, EPNL consists of instantaneous PNL corrected for the presence of the maximum tone and flyover duration. Both PNL and AL are methods for weighting the noise spectrum by deemphasizing the low frequency and emphasizing the high frequency contributions. However, AL provides more suppression for the low frequencies and less amplification for the high frequencies than does PNL. Neither AL nor PNL recognize tonal content, nor does either provide a penalty for long duration noise, whether due to flight velocity or directivity character of the noise-radiation pattern. Consequently, AL is less stringent in rating noise than is PNL, which in turn is less stringent than EPNL.

There are many noise exposure measures used by various nations for rating the

*The author believes that the use of these symbols is consistent with the nomenclature and with national and international aviation regulations, but realizes that such use is not conventional everywhere.

cumulative effects of aircraft flyover noise over specified periods of time. In general, the noise is cumulated as either the logarithmic summation or average of weighted single-event noise levels. This logarithmic cumulation process is commonly referred to as summation or averaging "on an energy basis" in order to clearly indicate that the process is not arithmetic. Identification, national origin, and brief descriptions of the most common noise exposure measures follow.[28] All measures relate to a 24-hour period and have decibellike units unless otherwise indicated. Some of the following also appear in Chapters 2 and 3.

1. Composite Noise Rating (CNR), United States. CNR uses PNL with discrete or discontinuous corrections for number of flyover events, for nighttime noise exposure (2200 to 0700 hours), and for seasons of the year.

2. Noise Exposure Forecast (NEF), United States. NEF is the summation on an energy basis of EPNL and was developed as a more accurate substitute for CNR with continuous as opposed to discrete correction for number of flyover events and for nighttime noise exposure (2200 to 0700 hours).

3. Equivalent Level (L_{eq} or EQL), United States. L_{eq} is the average on any energy basis of AL over a specified period of time (not necessarily 24 hours).

4. Day-Night Level (L_{dn} or DNL), United States. L_{dn} is L_{eq} for a 24-hour period with a correction of 10 dB added to the single-event noise levels (AL) occurring in the nighttime (2200 to 0700 hours).

5. Community Noise Equivalent Level (CNEL), United States. CNEL is the average on an energy basis of time-integrated AL with a correction factor of 3 dB added to the single-event noise levels occurring in the evening (1900 to 2200 hours) and a correction factor of 10 dB added to the single-event noise levels occurring in the nighttime (2200 to 0700 hours).

6. Aircraft Sound Description System (ASDS), United States. ASDS, in units of minutes, is the arithmetic summation of the time that aircraft noisc cxcccds a specified level (usually 85 AdB).

7. Weighted Equivalent Continuous Perceived Noise Level (WECPNL), International. WECPNL is the average, on an energy basis, of EPNL with appropriate weightings for evening (1900 to 2200 hours) and/or nighttime (2200 to 0700 hours) noise exposure.

8. Noise and Number Index (NNI), United Kingdom. NNI is based upon the average on an energy basis of maximum PNL over a specified time period (not necessarily 24 hours) taking into account the number of flyover events.

9. Noise Pollution Level (NPL or Lnp), United Kingdom. NPL takes into account L_{eq} and the effect of the magnitude of the time variation of noise level.

10. Isopsophic Index (N), France. N is based upon the average on an energy basis of maximum PNL, taking into account the number of flyover events and appropriate weightings for evening and nighttime noise exposure.

11. Mean Annoyance (Q), Germany. Q is based upon the average on a modified energy basis of AL over a specified time period (not necessarily 24 hours).

12. Total Noise Load (B), Netherlands. B is based upon the average on a modified energy basis of AL taking into account the number of flyover events and time of day.

13. Noisiness Index (NI), South Africa. NI is the average, on an energy basis, of tone-corrected AL modified for duration, time of day, and season of year.

A representative chart of the various cumulative noise exposure measures employed in assessing aircraft noise exposure throughout the world is presented in Fig. 9-8. The designated symbol, name, sponsor, and noise evaluation measure utilized for each individual measure is displayed. The noise measures enclosed by rectangles in Fig. 9-8 refer to those measures used primarily outside the United States. WECPNL is represented as the only internationally sponsored

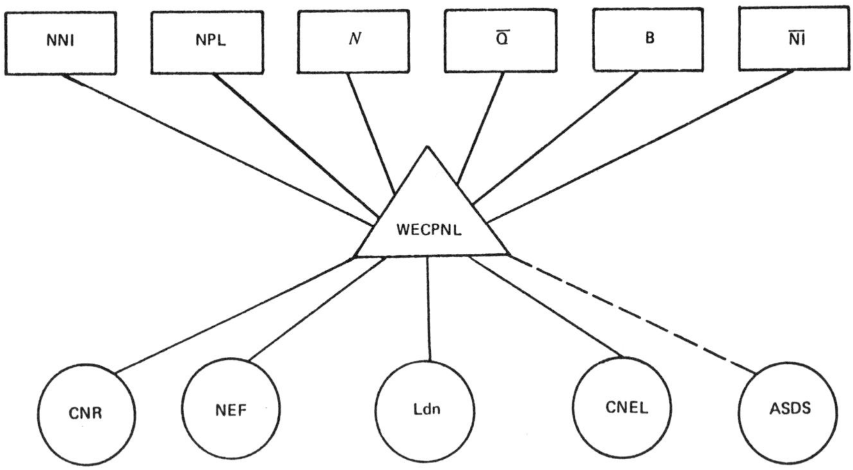

CUMULATIVE NOISE EXPOSURE MEASURES

SYMBOL	NAME	SPONSOR	NOISE EVALUATION MEASURE
WECPNL	WEIGHTED EQUIVALENT CONTINUOUS PERCEIVED NOISE LEVEL	INTERNATIONAL (ICAO)	EPNL
CNR	COMPOSITE NOISE RATING	USA (VARIOUS)	PNL
NEF	NOISE EXPOSURE FORECAST	USA (VARIOUS)	EPNL
Ldn	DAY-NIGHT LEVEL	USA (EPA)	AL
CNEL	COMMUNITY NOISE EQUIVALENT LEVEL	USA (CALIFORNIA)	AL
ASDS	AIRCRAFT SOUND DESCRIPTION SYSTEM	USA (FAA)	AL
NNI	NOISE AND NUMBER INDEX	UNITED KINGDOM	PNL
NPL	NOISE POLLUTION LEVEL	UNITED KINGDOM	AL
N	ISOPSOPHIC INDEX	FRANCE	PNL
$\overline{Q}$	MEAN ANNOYANCE LEVEL	GERMANY	AL/PNL
B	TOTAL NOISE LOAD	NETHERLANDS	AL
$\overline{NI}$	NOISINESS INDEX	SOUTH AFRICA	AL

Fig. 9-8 Cumulative noise exposure ratings used for aircraft.

measure (triangle). The measures used in the United States are enclosed by circles and located in the lower section of the figure.

The solid lines indicate that, because of the basic similarities in methodology (time-averaging on an energy basis of weighted noise levels), any one measure can be relatively easily compared to another by means of the international standard. The broken line connecting ASDS with WECPNL in Fig. 9-8 indicates that there is no obvious way to translate the values of one to another, not even by a "rule of thumb."

The cumulative noise exposure measures used most in the United States for aircraft noise are CNR, NEF, CNEL, and L_{dn}. When formulating empirical relationships for converting L_{dn} to the other measures just mentioned, care must be taken regarding the assumptions for the conversion factors. For example, several "rules of thumb" have been established regarding the relationships of L_{dn} to CNR and to NEF. L_{dn} has been expressed as being approximately equivalent to CNR minus a constant 35, with a degree of error in the range of ± 3. In regard to NEF, L_{dn} has been considered to be approximately equal to NEF plus a constant 35, with a range of error of ± 3. At best, these relationships are fair approximations for they assume that AL used in L_{dn} is a constant 13 dB lower than PNL or EPNL. The accuracy of that assumption is heavily dependent on the spectral distribution of the measured noise, which is in turn dependent on the distance to the source, the atmospheric conditions, and the presence or absence of tonal components in the source spectrum. In reference to the relationship of L_{dn} to CNEL, since CNEL is obtained using virtually the same methods as L_{dn}, with a small difference in the correction factors used for other than daytime noise levels, the results of CNEL are within fractions of a decibel of L_{dn}.

Of the four cumulative noise exposure measures for aircraft most common to the United States, NEF has the most background material associated with it in the sense that it has been most extensively used by the aviation community. The trend for the future, however, appears to be in the direction of L_{dn}. Both NEF and L_{dn} are methodologies for predicting single-number ratings of the cumulative noise that will intrude into airport communities during a 24-hour period. Figure 9-9 shows the factors accounted for by the measure and included in the computations for PNL, AL, EPNL, SEL, NEF, and L_{dn}. The first two items on the chart represent the amplitude and frequency distribution of sound pressure, which, when weighted in specified manners, will give PNL or AL. If, in addition, the effects of the maximum tone and the noise duration are included, the results are EPNL, which is the noise evaluation measure used by the FAA to certify aircraft. On the other hand, if the duration is combined with AL (without tone correction), the result is SEL. If the remainder of the items indicated are included into specified mathematical models, the results will be NEF or L_{dn}. The results of the computations are most useful when the values at individual positions on the ground are combined into equal NEF or L_{dn} contours and plotted on maps of the airport and its neighborhoods as indicated in Fig. 9-10.

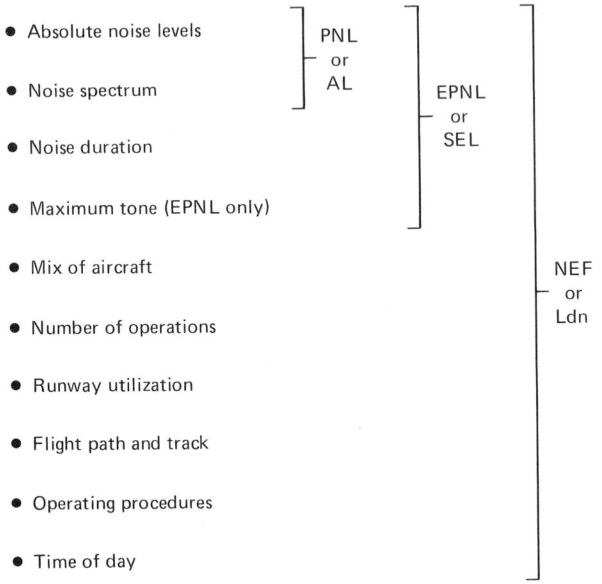

Fig. 9-9 Factors included in computations for NEF and L_{dn}.

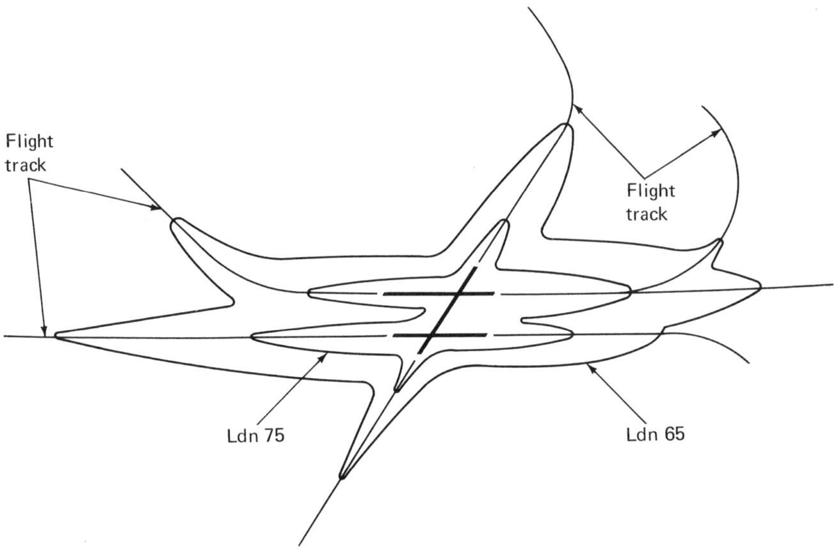

Fig. 9-10 Example of cumulative noise exposure contours around an airport.

LEGISLATIVE AND REGULATORY ACTIVITIES

In 1968, Public Law 90-411 amended Section 611 of the Federal Aviation Act of 1958 to require that, in order to afford present and future relief and protection to the public from *unnecessary* aircraft noise and sonic boom, the FAA prescribe and amend such regulations as it may find necessary to provide for the control and abatement of aircraft noise and sonic boom. In addition, PL 90-411 provided specifications that must be considered by the FAA in prescribing and amending aircraft-noise and sonic-boom regulations.

The Noise Control Act of 1972 (Public Law 92-574) supersedes Public Law 90-411 and further amends Section 611 of the Federal Aviation Act of 1958 to include the concept of "health and welfare" and to define the responsibilities of, and interrelationships between, the FAA and the EPA in the control and abatement of aircraft noise and sonic boom. Specifically, the Noise Control Act requires that, to afford present and future relief and protection to the public *health and welfare* from aircraft noise and sonic boom, the FAA, *after consultation with the EPA*, prescribe and amend such regulations as the FAA finds necessary to provide for the control and abatement of aircraft noise and sonic boom.

The Noise Control Act also requires that the EPA submit to the FAA proposed regulations to provide such control and abatement of aircraft noise and sonic boom–including control and abatement through the exercise of any of the FAA's regulatory authority over air commerce or transportation or over aircraft or airport operations–as the EPA determines is necessary to protect the public health and welfare.

In prescribing and amending standards and regulations, Section 611 of the Federal Aviation Act as amended requires that the FAA consider whether any proposed standard or regulation is:

1. consistent with the highest degree of safety in air commerce or air transportation in the public interest
2. economically reasonable
3. technologically practicable
4. appropriate for the particular type of aircraft, aircraft engine, appliance, or certificate to which it will apply.

These considerations of safety, economics, technology, and suitability are constraints on the noise regulatory actions, that must be made compatible with the requirement of protection to the public health and welfare. To achieve compatibility, the regulations must be carefully constructed, comprehensive, and definitive instruments for exploiting the most effective and feasible technology, flight procedures, and operating controls available.

The regulations proposed by the EPA are based upon a comprehensive study undertaken by the EPA and reported to Congress.[23] The "Aircraft/Airport Noise Study," which was completed in August 1973, was required to investigate the:

1. adequacy of FAA flight and operational noise controls
2. adequacy of noise-emission standards on new and existing aircraft, together with recommendations on the retrofitting and phaseout of existing aircraft
3. implication of identifying and achieving levels of cumulative noise exposure around airports
4. additional measures available to airport operators and local governments to control aircraft noise.

The study was implemented by a task force composed of six task groups whose product consisted of a report to Congress and six volumes of supporting data (one volume for each task group).[16-21]

As a result of the study, the EPA determined that an effective program to protect the public health and welfare with respect to aircraft noise would require the development and proposal to the FAA of regulations in three complementary areas:

1. Flight procedures regulations standardizing various modes of noise-abatement operations
2. Type and airworthiness certification regulations controlling noise-source emissions in the design of new aircraft and by modification or phaseout of certain portions of the existing fleet
3. An airport noise regulation that would limit aircraft-noise impact in communities near airports by requiring development of mutual compatibility between airport operations and land-use patterns.

Currently, there are five significant noise control regulations, which have been prescribed by the FAA and which have a substantial influence on the design, development, and operation of aircraft. Titles, part numbers of the Federal Aviation Regulations (FAR), and brief descriptions follow.

1. Noise Standards: Aircraft Type Certification, FAR 36.[10,69] FAR 36 prescribes noise standards for the issue of type certificates, and changes to those certificates, for subsonic transport category airplanes and for subsonic turbojet-powered airplanes regardless of category. This rule initiated the noise-abatement regulatory program of the FAA under the statutory authority of Public Law 90-411.

2. Civil Aircraft Sonic Boom, FAR 91.55.[15] The control of sonic boom within the United States and its territories is accomplished by means of an operating rule that prohibits supersonic flights of civil aircraft except under terms of a special authorization to exceed the speed of sound (Mach 1.0). Authorization to operate at a true Mach number greater than unity over a designated test

area may be obtained for special test purposes. A flight outside of a designated test area at supersonic speeds may be authorized if the applicant can show conservatively that the flight will not cause a measurable sonic boom overpressure to reach the surface.

3. Noise Standards for Newly Produced Airplanes of Older Type Design, FAR 21.183(e) and FAR 36 Amended.[24] This rule requires new production turbojet and transport category subsonic airplanes to comply with FAR 36, irrespective of type certification date.

4. Acoustical Change Approvals, FAR 36 Amended.[36,53] Under FAR 36, an "acoustical change" is any voluntary change in the type design of a transport category or turbojet engine-powered airplane that may increase the noise levels created by the airplane. The purpose of these amendments is to tighten the conditions under which an applicant for an acoustical change approval must show that modifications of certain turbojet or transport category airplanes will not increase the takeoff or sideline noise levels of those airplanes.

5. Noise Standards for Propeller Driven Small Airplanes, FAR 21 and FAR 36 Amended.[38,67] This rule prescribes noise standards for:

a. The issue of normal, utility, acrobatic, transport, and restricted-category type certificates for propeller-driven small airplanes (12,500 lb or 5670 kg maximum weight).
b. The issue of standard airworthiness certificates and restricted-category airworthiness certificates for newly produced propeller-driven small airplanes of older type designs.
c. The prohibition of "acoustical changes" in the type design of subsonic-transport-category large airplanes and turbojet-powered airplanes that would increase their noise levels beyond specified limits.

FAR 36 was the first type-certification regulation for aircraft noise prescribed by any nation. A comprehensive, highly technical rule, it is appropriate to the sophisticated sound source (aircraft) that it is designed to regulate. The development of the basic concepts inherent in FAR 36 was, for the most part, the result of the experience of government and industry not only of the United States but of France and the United Kingdom as well. Government representatives of these nations formed a "tripartite" working group, which provided most of the initial ideas and coordinated the technology upon which FAR 36 is based. Subsequently, representatives of Germany, Japan, the Netherlands, and Sweden participated in the working group and made valuable contributions. The seven nations involved represented the major aircraft manufacturing nations of the world (except for the USSR) and were able to pool their specialized knowledge and experience to provide a substantial technology base for the development of aircraft-noise regulations.[11]

It must be emphasized that various national and international standard organizations* provided invaluable background data in such areas as noise measurement procedures, electronic equipment characteristics, atmospheric attenuation of sound, and noise evaluation measures. Where these standards organizations had issued approved citable documents that were relevant, they were included as references in FAR 36. In many cases, however, the areas were so new that the only documents available were draft working papers. Where appropriate, such material was included (but not referenced) in FAR 36.

The United States was the first nation to take advantage of this wealth of information and include it in a noise control regulation published in November 1969 as FAR 36. Shortly thereafter, in December 1969, the International Civil Aviation Organization (ICAO) recommended standards that were subsequently, adopted by the Council of ICAO in April 1971, as Annex 16 to the Convention of International Civil Aviation.[13] FAR 36 and Annex 16 are essentially the same rule; FAR 36, however, was slightly more stringent. Most nations have adopted Annex 16 as their type-certification rule for aircraft noise, and, because the rules are so similar, no major problems have arisen in regard to reciprocity between the United States rule FAR 36 and Annex 16 for the rest of the world.

Notwithstanding the actions taken by the FAA for the control of aircraft noise, the prevailing opinion in the United States among environmental organizations, Congress, and the EPA is that much more should and can be done to reduce the community noise impact caused by aircraft. The EPA, as discussed previously, is committed to the development and proposal to the FAA of noise-control regulations pertaining to three complementary areas. Nine projects have been identified, three for flight procedures, five for type certification, and one for airport, which will have one or more regulations associated with each.

A report is prepared for each project which provides definitive information on such matters as health and welfare, technology, cost-effectiveness, and criteria for noise levels, measurements, and analyses. Each project report provides the background for the preparation of one or more notices of proposed rulemaking (NPRM), which is the format of the regulations proposed by the EPA for promulgation by the FAA under Section 611 of the Federal Aviation Act as amended by the Noise Control Act of 1972.

The status of the EPA activities relating to the nine aircraft/airport noise-control projects follows.

Flight Procedures

All aircraft can be operated in many safe modes that generate a variety of noise signatures. And these signatures, when propagated to the ground, translate to

*American National Standards Institute (ANSI), International Electrotechnical Commission (IEC), International Standards Organization (ISO), and Society of Automotive Engineers (SAE).

different noise levels. The path-control option is represented by flight procedures, which minimize the generation and propagation of aircraft noise. The EPA is considering three separate sets of proposed regulations within this option, which would standardize a few flight procedures that are safely and routinely conducted at present or that have been extensively tested for safe effective results.

Takeoff. Aircraft are capable of utilizing a wide variety of departure procedures, and the objective of this project is to investigate a number of the ones commonly used for their contribution to the reduction of community noise impact. For example, the five takeoff procedures that follow are used by aircraft operators for various purposes.

1. *FAA/ATA.* The FAA and the Air Transport Association (ATA) developed a procedure, which the FAA described in an advisory circular and which has been used by some airlines for several years. However, because of the dissatisfaction of other airlines with that procedure, the ATA now recommends that a different one be followed. In addition, the FAA no longer believes that its advisory circular presents the best takeoff procedure from the standpoint of minimal community noise. Consequently, the FAA as well as the EPA is examining the subject and intends to propose a regulation.
2. *ALPA/NWA.* The Airline Pilots Association recommends a procedure that is similar, except for minor details, to the one routinely used by Northwest Orient Airlines (NWA). Both organizations claim benefits for community noise impact, fuel consumption, wear and tear on engines, and safety.
3. *FAR 36.* The test conditions for aircraft noise certification permit compliance with noise levels to be met at the takeoff measuring point using a power or thrust reduction to specified limits. The FAR 36 procedure, however, is not used in routine operations at airports. It should be investigated, therefore, for its suitability as a standard operation procedure. If it is determined to have no advantage compared to the others, it should be eliminated from the FAR 36 takeoff test conditions and a more appropriate one substituted.
4. *Reduced Thrust.* Some airlines use less than full takeoff thrust in order to reduce wear and tear on the engines. The reduced-thrust procedure should be investigated for its contributions to noise—that is, whether the overall noise impact is increased or decreased.
5. *DCA.* Noise abatement procedures are prescribed by the FAA for departures at Washington National Airport (DCA). The noise impact benefits of the DCA procedures should be compared with those of the others to determine whether the DCA procedures should be used at other airports or whether they should be supplanted at Washington by more effective ones.

These takeoff procedures can be placed into one of three classifications regarding noise abatement, depending upon the location of noise-sensitive areas relative to the takeoff runway; that is, sideline, near downrange, and far downrange. Each classification would permit less noise at the designated location than would result from the other two. All five procedures involve reduction in thrust or power, the differences being the position of the airplane when cutback occurs, the extent of the cutback, the position of the airplane for cleanup, and the flap-retraction schedule.

The EPA is considering proposing to the FAA a regulation requiring the use of at most two approved noise abatement procedures for takeoff except where safety requirements dictate otherwise.

Approach and Landing. On August 29, 1975, the EPA officially submitted to the FAA the following three proposed regulations on approach and landing procedures for noise control:[47–49]

1. "Reduced Flap Setting Noise Abatement Approach for Turbojet Engine-Powered Airplanes."
2. "Visual Two-Segment Noise Abatement Approach for Turbojet Engine-Powered Airplanes."
3. "Two-Segment ILS Noise Abatement Approach for Turbojet Engine-Powered Airplanes."

The first rule would require that, when approaching for a landing, each civil turbojet engine-powered airplane use the minimum certificated flap setting for that airplane that is appropriate to each phase of that approach and landing. An approach made with less than full landing flaps would reduce aircraft noise approximately 2 to 3 dB as compared to a full-flap approach, since the airframe drag with the reduced settings would be less, thereby requiring less thrust.

The second rule would require that each civil turbojet engine-powered airplane, when making an approach for a landing under visual weather conditions on a runway having appropriate electronic guidance equipment, use an approved, visual two-segment approach procedure. The approach would consist of a two-part descent with the upper, or first-segment, glide angle flown at about 5° to 6° depending upon the airplane, and with the second segment being flown using the existing Instrument Landing System (ILS) glide angle. It is predicted that noise reduction would amount to as much as 10 to 14 dB over land areas located 3 to 7 nautical miles (5556 to 12,964 m) from the approach runway. This could mean a substantial decrease in noise impact for many airport communities. The noise reduction due to steeper approaches results from two factors: increased height above the ground and reduced thrust setting.

The third rule when implemented would supersede the second rule and would require that (1) each civil turbojet engine-powered airplane have special landing instrumentation and (2) when making an approach for a landing on a runway having appropriate electronic guidance equipment, use an approved all-weather two-segment approach procedure. The flight profile, operating procedures, and noise reductions would be practically the same as under the second rule. The difference is that the second rule is capable of being implemented only under special limited conditions (visual), while the third rule is capable of being implemented under Instrument Flight Rules (IFR) conditions. The second rule could be effective immediately but would be an interim measure to be supplanted at a later date by the third rule, which could be effective only when appropriate cockpit instrumentation (e.g., glide-slope computers) are developed and installed.

Public hearings were held on the three proposed regulations on November 6, 1975, in Washington, D.C. On November 29, 1976, the FAA published in the *Federal Register* an amendment to FAR 91 requiring the use of minimum certificated flap settings, a proposed rule on delayed landing flaps, and a notice of decision not to prescribe a two-segment approach under either visual or instrument (ILS) conditions.[62,64] The reason given by the FAA for the non-prescription actions was "... that the two-segment approaches under visual and instrument conditions are not consistent with the highest degree of safety in air commerce and are not in the public interest."

Minimum Altitudes. On December 6, 1974, the EPA officially submitted to the FAA the following proposed regulation on minimum altitude procedures for noise control: "Noise Abatement Minimum Altitudes within Terminal Areas: Turbojet Powered Airplanes."[33,40] The proposed rule would require that all turbojet-powered civil airplanes must enter the airport terminal area and descend to land in a prescribed manner. The proposed rule would make mandatory one set of procedures that are now FAA "advisory" (meaning voluntary) and amend another set of procedures that are now regulatory. As proposed, civil turbojet engine-powered airplanes approaching an airport for landing would be required to:

1. enter the terminal area of that airport at a height of 10,000 ft (3048 m) above the airport and remain at that height until further descent is required for a safe landing
2. descend below a height of 5000 ft (1524 m) after entering the descent area established by FAA for the direction of the landing runway
3. intercept the glide-slope at a height of not less than 3000 ft (914 m).

Items (1) and (2) are currently advisory, and item (3) would amend a current regulation that establishes a height of 1500 ft (457 m) as the minimum.

Public hearings were held on this proposed regulation on March 5, 1975, in Washington D.C. The FAA published in the *Federal Register*, on November 29, 1976, a notice of decision not to prescribe this proposal,[63] because the requirements would not be practical to implement and would interfere with air traffic control (ATC). As an alternative, the FAA has issued an internal directive

(Order) entitled "Local Flow Traffic Management," which states that "... the FAA believes that direct management of the ATC Function using a total systems approach is far preferable to regulations of the kind proposed by EPA, which seriously limit ATC flexibility." According to the FAA, the new Order will provide "... an enhanced effort in noise abatement, substantial fuel savings, and a more efficient ATC system."

Type and Airworthiness Certification

A type-certificate signifies that an aircraft type design has been demonstrated to conform to FAA standards on airworthiness and noise. Each aircraft requires an airworthiness certificate that signifies that the specific aircraft has been manufactured in accordance with its FAA certified type design and has subsequently been maintained according to regulations. The EPA is considering five sets of proposed noise control regulations within the source noise control option, which would be implemented by means of type and/or airworthiness certification.

Retrofit and Fleet Noise Level. On January 28, 1975, the EPA officially submitted to the FAA the following two proposed regulations concerning the control and reporting of noise of existing turbojet airplanes:[35,42,43]

1. "Civil Subsonic Turbojet Engine-Powered Airplanes Noise Retrofit Requirements."
2. "Fleet Noise Level Requirements."

The first rule would require that existing subsonic turbojet engine-powered airplanes, regardless of weight, meet the requirements of FAR 36. The rule would be applicable to United States air carrier and commercial operators and to certain foreign air carriers. The rule would not be applicable to airplanes having experimental, provisional, or restricted airworthiness certificates. The EPA NPRM is based upon the requirements of an earlier FAA NPRM for which there had been no further public action. However, the EPA requirements would be more stringent than those of the FAA as airplanes less than 75,000 pounds (34,020 kg) maximum weight (e.g., business jets) would not be exempted.

The project report details the options available to the owners of airplanes that must comply, the cost-effectiveness of those options, and the relief to the noise-impacted population that would result. The following five options are available to the owners of existing noisy airplanes that cannot comply with FAR 36:

1. Replace noisy airplanes with quieter airplanes.
2. Retrofit noisy airplanes with new engines.
3. Retrofit noisy engines with noise suppressors.

4. Retrofit noisy engines with "Quiet Nacelles."
5. Retrofit noisy airplanes with "Refanned" engines.

The Quiet Nacelle technology was initiated originally by NASA and developed into "flight worthy, flight weight, capable of being certificated" hardware by the FAA through contracts with Boeing and Douglas. Noise reduction for Quiet Nacelles results from the application of sound absorption material (SAM) to the walls of the engine flow passages; engine modifications are not involved.

The Refan technology was developed by NASA through contracts with Pratt and Whitney, Boeing, and Douglas, and is applicable only to JT8D engines, which propel the B-737, DC-9, and B-727 airplanes. Noise reduction for Refanned engines results from extensive engine modifications whereby more energy is extracted from the turbine and transferred to a new and larger fan component. The modified engine has a higher bypass ratio, which reduces the noise from the engine core component and increases the noise from the engine fan component. Therefore, Quiet Nacelles with SAM treatment are required as part of the Refan technology. Refan retrofit is almost nine times more expensive than Quiet Nacelle retrofit but is significantly more effective.

The second EPA-proposed rule would require that each airline operator submit annually to the FAA the airline's Fleet Noise Levels (FNL) and all data necessary for FNL computations. The FNL is a single-number evaluator of the noise generated by a fleet of airplanes referred to each of the three noise measuring points of FAR 36 (sideline, takeoff, and approach). FNL is expressed by the logarithmic average of the product of the noise level of each type of airplane and the number of operations performed by that type of airplane over the previous year. The FNL rule would require that the FAA publish annually in the *Federal Register* the FNLs for each fleet operator and the relative differences between the operator's actual FNLs and the FNLs that would be achieved if the noise levels of the operator's airplanes were equal to those required by FAR 36.

The EPA-proposed rule on FNL is only an annual reporting requirement, which would establish a noise "figure of merit" for each airline. It would permit the public to judge on a quantitative basis the relative differences in noisiness between those airlines that were exploiting noise-control technology and those that were not. However, the establishment of an FNL would provide a regulatory mechanism, if necessary, for lowering noise levels whenever future technology becomes current. Should it be necessary to establish a mandatory FNL, it would, of course, be accomplished by separate rulemaking action.

Public hearings were held on the Retrofit and Fleet Noise Level proposed regulations on March 18, 1975, and April 17, 1975, respectively, in Washington, D.C. The FAA published in the *Federal Register*, on November 29, 1976, a notice of decision not to prescribe the EPA's proposed Fleet Noise Level regulation.[66] The reason given by the FAA for the nonprescription action is that

". . . the FAA believes that no constructive purpose would be served by generating FNL values, under the proposed regulation, and furnishing these values to airport or air carrier management." Although the EPA did not object to the FAA's nonprescription action, they did inform the FAA that the EPA may elect to publish air-carrier FNL data from time to time in order that the public may be informed of the general status of the air-carrier fleet.

The FAA published in the *Federal Register*, on December 23, 1976, an amendment to FAR 91 that would require that certain airplanes not presently complying with FAR 36 must now comply in accordance with a phased time schedule.[66] The options available to the airplane operators are essentially some form of retrofit or replacement of noisy airplanes by quieter ones. This new rule, commonly referred to as the "retrofit/replacement" rule, is similar to the retrofit rule proposed by the EPA except that airplanes weighing 75,000 pounds or less, and airplanes engaged in foreign commerce, would not be required to comply. The FAA, therefore, used the amendment as a notice of decision not to prescribe those two aspects of the EPA retrofit proposal.

The reason given by the FAA for the nonprescription action pertaining to lower weight range of jet aircraft is ". . . the FAA believes that the noise reduction benefits, technological practicability, cost effectiveness, and economic impact of applying the proposal requirements, across the board, to all turbojets weighing 75,000 pounds or less have not yet been adequately determined." The reason given by the FAA for the nonprescription action pertaining to foreign commerce airplanes is based upon ". . . the potentially disruptive effect that action by the United States at this time could have on the orderly development of international noise abatement standards . . . ."

Supersonic Transport Aircraft. On February 27, 1975, the EPA officially submitted to the FAA the following proposed regulation on the control of noise of existing and future types of aircraft capable of exceeding the speed of sound: "Aircraft Noise Requirements: Civil Supersonic Airplanes,"[41,45] The EPA recommendations were that:

1. Initial production versions of current SST types to be subject to one of eight regulatory options proposed by the EPA, the particular option to be determined after public hearings.
2. Later production versions of current SST types to meet the FAR 36 standards now in effect.
3. Future SST types to meet FAR 36 standards in effect on the date of application of the type certificate.

"Initial production versions" refers to airplanes upon which substantive productive effort was commenced prior to the date the regulation is proposed. "Substantive productive effort commenced" as defined in the proposed rule means that parts have been fabricated or delivered or are on order (in a legally binding financial commitment) for the airplane in question equivalent in total value to 5 percent or more of the selling price of the airplane.

There are only two types of current SSTs: the English-French Concorde and the Russian TU-144. It is estimated that there are 16 Concorde and up to 14 TU-144 airplanes that can be classed as initial production versions of current SSTs. Thus, it is estimated that a maximum of 30 SSTs would be subject to one of the eight options, which range from outright ban to no regulation whatsoever. In other words, the EPA suggested (no rule was proposed) that the current SSTs committed to production should be considered separately, but, if so, only on an airport-to-airport basis for possible operations in specific cases where the noise impact would be minimal. All other SSTs would be required to comply with FAR 36.

Public hearings were held on the supersonic transport proposed regulation on May 16, 1975, in Los Angeles, California, and on May 22, 1975, in Washington, D.C. Further action is the responsibility of the FAA in accordance with the sequence of actions specified by the Noise Control Act.[70]

Modifications to FAR 36. FAR 36 is a type certification regulation applicable to certain kinds of airplanes designated as types. Only one airplane of each type need be tested, and the results of the tests are assumed valid for all individual airplanes produced of that type.

FAR 36 has several purposes; its main purpose is to provide requirements that will influence the design of aircraft to include implementation of source noise-control technology to the maximum extent feasible. Source control consists of the application of basic design principles or special hardware to the engine/airframe combination that will minimize the generation and radiation of noise. And, as used here, feasibility means that the noise control technology will be compatible with the regulatory constraints of safety, economics, technology, and suitability.

Another purpose of FAR 36 is to provide meaningful noise levels for specific types of aircraft, which will be useful in predicting the noise impact in airport neighborhood communities. It must be understood that since FAR 36 is a type-certification regulation, the data resulting from the certification process will be limited in their extent insofar as community noise-impact studies are concerned. Nevertheless, FAR 36 should require the acquisition and reporting of data that can be utilized in such studies.

Still another purpose of FAR 36, which is fundamental to the other purposes, is the setting of standards for the acquisition and reduction of aircraft noise and flight performance data. Without standards, noise measurements of aircraft have no real credibility, and, without good standards, noise requirements may not be effective. Imprecise meanings, careless terminology, or inadequate testing procedures can lead to circumvention of the intent of the rule. In other words, loopholes may exist that can be exploited by those manufacturers with airplanes that have problems meeting the noise requirements.

The objective of the amendments of FAR 36 is to strengthen its original purposes. Since the promulgation of FAR 36 in 1969, noise-control technology has advanced substantially. The noise compliance requirements should reflect this technology growth and be structured with respect to time to encourage the application of future technology whenever it is determined to be feasible.

The significance of community noise impact is better understood now than it was in 1969. Consequently, an amendment to FAR 36 should include requirements for the acquisition and/or reporting, to a reasonable extent, of data useful for predicting the impact of aircraft noise.

The procedures for type certification of aircraft noise also are better understood now. The considerable experience gained in the United States and abroad should be utilized to simplify techniques, improve accuracy, and close loopholes. Advancements in the electronic data acquisition and reduction systems should be reflected in the standards in order to improve efficiency and reduce costs.

On October 1, 1976, the EPA officially submitted to the FAA the following proposed regulation on modifications to FAR 36: "Noise Levels for Turbojet Powered Airplanes and Large Propeller Driven Airplanes."[52,56] The proposed rule would require nine amendments to FAR 36, most of which pertain to test procedures. The most significant amendment, however, relates to noise levels and is based upon the assumption that the technology of source noise control is time-dependent in the sense that it is based upon the results of past, present, and future programs of research, development, and demonstration (RD&D), which are defined as follows:

1. *Current technology* includes "shelf item" hardware and commonly known (state-of-the-art) techniques and procedures that have been used effectively by most manufacturers for many applications.

2. *Available technology* includes "shelf item" hardware and commonly known (state-of-the-art) techniques and procedures that have been used effectively by some manufacturers for some applications. Also included are the results of RD&D that have not been put into practice but are available for implementation. Some performance testing may still be necessary but this technology has been certified for airworthiness or, by adequate ground and/or flight testing, determined to be capable of being certified.

3. *Future technology* represents the outcome of RD&D programs now in progress, which have not been verified, but of which the results to date indicate high potential to a reasonable degree of confidence. Included are present RD&D programs that are being conducted with sufficient personnel, funding, and time to carry the programs to conclusion. Definitive results are expected

in the relatively near future for acoustical and operational performance, economics, and flight safety. The nature of the expectations is positive because predictions of nonviable results would have been cause for earlier termination of the RD&D programs.

The current technology requirements proposed by the EPA to be effective immediately would be the same as those proposed by the FAA (either NPRM 75-37 or NPRM 75-37C).[50, 57, 69] The available and future technology requirements proposed by the EPA would be progressively more stringent and would be effective on January 1, 1980, and January 1, 1985, respectively. Public hearings were held on this proposed regulation on December 15, 1976, in Washington, D.C. Further action is the responsibility of the FAA in accordance with the sequence of actions specified by the Noise Control Act.

Propeller-Driven Small Airplanes. On December 6, 1974, the EPA officially submitted to the FAA the following proposed regulation on small propeller-driven airplanes (12,500 lb or 5670 kg maximum): "Noise Standards for Propeller Driven Small Airplanes."[32,39] The proposed rule would require that specified noise standards be complied with for airplanes in three time-dependent categories of noise-control technology. The categories and range of compliance noise levels dependent upon weight are:

1. Current: for type certificate applications made on or after October 10, 1973 (79 to 93 EPNdB).
2. Available: for type certificate applications made after January 1, 1975, and newly produced airplanes of older type design manufactured after January 1, 1977 (79 to 91 EPNdB).
3. Future: for type certificate applications made after January 1, 1980, computed by a formula, which would range from 74 EPNdB for a 1,250-lb (567-kg) airplane to 89 EPNdB for a 12,500-lb (5670-kg) airplane.

One month after the EPA submittal to the FAA of the proposed regulation, the FAA prescribed a rule on the same subject.[38] Although the EPA NPRM and the FAA rule are similar, there are three significant differences where the EPA NPRM is more stringent.

1. *Noise Measure.* The compliance noise levels for the NPRM are given in terms of Effective Perceived Noise Level (EPNL) and for the FAA rule are given in terms of A-weighted Level (AL). The EPA position is that EPNL is a more reliable and more stringent measure for assessing human response to aircraft noise. The use of EPNL would be consistent with FAR 36 where EPNL is required for noise measurements of both turbojet and large propeller-driven airplanes. EPNL imposes a penalty on noise containing a strong discrete frequency component and on noise of long duration; both features are inherent in

many propeller sounds. In addition, both EPNL and AL deemphasize the importance of low frequency relative to high frequency noise, but EPNL to a lesser extent.

2. *Compliance Noise Levels.* The EPA-proposed compliance noise levels for current and available technology are about 3 dB lower than those required by the FAA rule taking into account the fact that EPNL averages approximately 14 dB higher than AL for aircraft noise. Additionally, the EPA proposal includes compliance noise levels for future technology, which are not considered in the FAA rule. The future technology noise level requirements are as much as 13 dB lower than the FAA requirements for available technology.

3. *Performance Correction.* The EPA proposes a more stringent performance correction than that required by the FAA rule. The performance correction is intended to ensure that airplanes with better takeoff and climbout capabilities receive credit; conversely, poor-performance airplanes are penalized. High-performance airplanes climb higher and faster than those with lesser capabilities and, as a consequence, produce less noise on the ground beyond the airport boundaries. The EPA performance correction, in addition to being more accurate and realistic than that of the FAA, includes a velocity correction that allows adjustment for the noise exposure time caused by airplane speed.

Public hearings were held on the EPA proposed regulation for propeller-driven small airplanes on March 3, 1975. The FAA published in the *Federal Register*, on December 23, 1976, a notice of decision not to prescribe the noise measure and the compliance noise levels proposed by the EPA.[67] The reason given by the FAA for the nonprescription action pertaining to the noise measure is "... the FAA has determined that noise emission characteristics of propeller-driven small airplanes do not need regulatory penalties to account for tone or duration...." The reason given by the FAA for the nonprescription action pertaining to the noise levels is that "... existing regulations provide adequate noise control and abatement by achieving a reduction in noise level that was imposed after consideration of economic and technological impact."

Short-Haul Aircraft. A potential direction of growth in air transportation is in high-payload short-range aircraft, capable of vertical, short, or reduced takeoff or landing (VTOL, STOL, or RTOL) operations from urban-center airports. Because of the high cost of land in cities, such airports would necessarily be constrained to small areas within a high-density population. For example, runway lengths associated with those "short-haul" aircraft are being considered as 1000 ft (305 m) for VTOL, 2000 ft (610 m) for STOL, and 4000 ft (1219 m) for RTOL. Although most general-aviation-category aircraft have adequate range and are capable of operating from runway lengths less than 4000 ft (1219 m), they are not satisfactory for short haul because of their relatively small payload capacity. Consequently, ideal short-haul aircraft need some form of augmented lift to permit high-payload operations from relatively short runways.

Unfortunately, augmented-lift devices are potentially noisy, and augmented-lift (short-haul) aircraft may have a unique noise problem compared to conventional (nonaugmented-lift) aircraft. The FAR 36 rules were developed for conventional airplanes—that is, airplanes that have a comparatively narrow range of performance parameters such as lift/drag and thrust/weight ratios. In the absence of any other rule, short-haul aircraft (except rotorcraft) would have to comply with FAR 36, which means that, for comparable weights, they would be no noisier than conventional aircraft. However, because of significant differences in their ranges of performance parameters, short-haul aircraft may be noisier than conventional for comparable runway-length operating capability.

The EPA is considering proposing to the FAA one or more regulations for the control of noise from short-haul aircraft. More rigid noise controls than are provided by FAR 36 are necessary for short-haul aircraft to insure that the urban population is adequately protected from the close proximity of those aircraft. In other words, noise control for short-haul aircraft can be effected by implementing the source, path, and receiver control options but the land-use control option for abating residual noise may not be feasible because of the high cost of land in urban areas.

Airports

The impact of aircraft noise on airport neighborhood communities can, in principle, be controlled to any extent by limitations on type of aircraft, requirements for their operating procedures, and restrictions on number and time of day of their movements. The EPA is considering proposing several regulations within the receiver-control option that would reduce noise impact by imposing controls on aircraft at airports.

Airports. On January 13, 1976, the EPA officially submitted to the FAA the following proposed regulation relating to airports: "Airplane Noise Requirements for Operation to or from U.S. Airports."[51] The proposed rule would require that no civil-transport-category airplane may operate from an airport in the United States unless (1) that airplane complies with the noise-level requirements for subsonic-transport-category airplanes of FAR 36 or (2) that airplane had flight time before December 31, 1974. Although the proposed rule is directed, primarily, to preventing any substantial numbers of Concorde and TU-144 supersonic airplanes from operating from United States airports, it is applicable to subsonic airplanes as well. In fact, the proposed rule probably would be restrictive to many more subsonic than supersonic airplanes and in that sense is not discriminatory. In effect, this EPA proposal is the first component of a comprehensive airport noise control package.

Public hearings were held on this proposed regulation on April 5, 1976, in Washington, D.C. Further action is the responsibility of the FAA in accordance with the sequence of actions specified by the Noise Control Act.

On October 22, 1976, the EPA officially submitted to the FAA the following proposed regulation on the control of noise at airports: "Airport Noise Regulatory Process."[61] The proposed rule would require that every airport in the United States, serving air carriers holding a certificate of public convenience and necessity from the CAB, must develop and implement a noise abatement plan. Under the proposed rule, a plan must be developed and implemented for both 2 years and 5 years from promulgation and thereafter on a continuing basis every 5 years.

The key elements of the EPA's airport noise regulatory process are:

1. determination of boundary-line noise levels at the airport in order to determine the extent of planning needed at the facility
2. development of a noise-abatement plan by the airport proprietor
3. presentation of the plan to all interested parties at a public hearing
4. submission of the plan, revised pursuant to public hearing as appropriate, to the FAA administrator for review
5. inclusion of a plan as part of amended Airport Operating Certificate
6. implementation of the plan by the proprietor.

In order for the airport proprietor to carry out the noise abatement planning and implementation effectively, the EPA has designed and developed a methodology called the Airport Noise Evaluation Process (ANEP). The ANEP methodology is based upon the use of the Day-Night Average Sound Level (L_{dn}), which the EPA has recommended for use in all community noise studies and planning efforts.

Public hearings were held on this proposed regulation on January 17, 1977, in Washington, D.C. Further action is the responsibility of the FAA in accordance with the sequence of actions specified by the Noise Control Act.

EXTENT OF THE PROBLEM: POPULATION IMPACT

Modern society has come to recognize the economic and social benefits of commercial air transportation as essential, rather than merely desirable. The benefits accrue to the metropolitan area served by the airport, to the aviation community, and ultimately to the entire nation. While airports and the provision of air-transport services provide the surrounding community with significant benefits, such services can and have resulted in certain detriments.

During the last two decades, there has developed an increasing number of complaints about noise in the vicinity of airports. This has been brought about because jet aircraft are, in fact, noisy; frequency of operations has steadily increased, and urban encroachment around airports continues. These are not idle complaints; there are sufficient data to support the contention that aircraft noise has created social and economic costs to some persons living and working in the vicinity of airports. On the other hand, the existence of noise will make other persons better off if aircraft can operate at less cost because their design, development, flight procedures, number of movements, or growth are not restricted by requirements for the control of noise. Such has been the case in the past, and it will continue to be so, unless dedicated effort directed to implementing the source, path, and receiver noise control options taken together as a system will lead to acceptable noise levels. It would appear, then, that some economic and social inequity may continue to exist between those incurring costs as a result of aircraft noise and those responsible for its generation.

As a result of current United States judicial decisions, landowners in the vicinity of an airport may recover compensation for interference with the reasonable use of their property caused by aircraft noise. The courts have placed liability for aircraft noise on the airport operator, who, in determining the type of service to be provided, brings the noise to the nearby communities and effectively determines what the noise levels and total noise exposure in those communities will be. Social costs, however, have not been recognized by the courts, and it is not suggested that they should for purposes of monetary compensation, for that would not represent a valid solution to the noise problem.

Social costs are related to the intrusiveness of noise and its effects on the health and welfare of individuals. Annoyance and interference of speech, sleep, and concentration are examples of social costs that are difficult to identify, much less quantify in monetary terms. However, such social costs do exist, and the objective of government is to eliminate (through legislative and regulatory actions) this residual inequity even if its magnitude cannot be precisely measured. Although the social costs are not easily quantified, reasonable estimates can be made for the population exposed to significant levels of aircraft noise. Also, judgments can be made regarding certain aspects of human response such as annoyance, which will provide at least one yardstick for measuring the impact of aircraft noise.

The number of people estimated to be impacted by aircraft noise, with cumulative levels ranging from annoyance to hearing impairment, is a significant portion of the national population. For example, Fig. 9-11 shows estimates for 1972 aircraft operations from two different sources.[23,29] The first, taken from the EPA Aircraft/Airport Noise Report published in 1973, estimates that 16 million people were exposed to L_{dn} levels of 60 dB or greater. The second, based upon extrapolation of data from the DOT 23-Airport Study Report published in 1974,

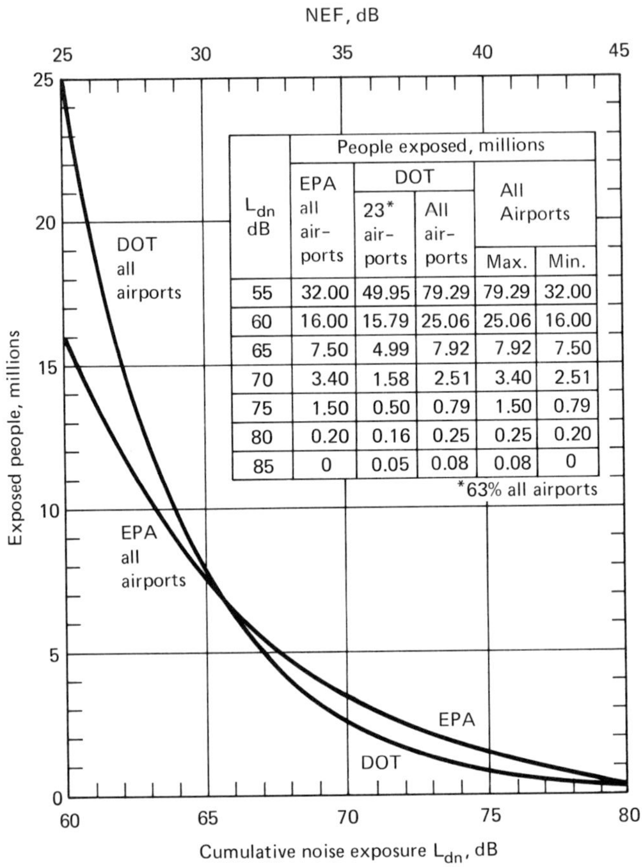

L_{dn} dB	People exposed, millions: EPA all airports	DOT 23* airports	DOT All airports	All Airports Max.	All Airports Min.
55	32.00	49.95	79.29	79.29	32.00
60	16.00	15.79	25.06	25.06	16.00
65	7.50	4.99	7.92	7.92	7.50
70	3.40	1.58	2.51	3.40	2.51
75	1.50	0.50	0.79	1.50	0.79
80	0.20	0.16	0.25	0.25	0.20
85	0	0.05	0.08	0.08	0

*63% all airports

Fig. 9-11 Population exposed to aircraft noise.

estimates that almost 16 million people were exposed to L_{dn} 60 or greater from only 23 of the major airports. Extending the latter estimate to the total national population by assuming that aircraft operations from the 23 airports contribute 63 percent of the national aircraft-noise exposure, the estimated population exposed to $L_{dn}{}^{60}$ or greater would be about 25 million people, or over 10 percent of the total national population.

The EPA "Levels Document" identifies L_{dn} 55 as the noise exposure level that should not be exceeded in order to protect persons against some forms of annoyance with an adequate margin of safety.[27] Although neither the EPA nor the DOT studies predicted numbers of people exposed to levels as low as L_{dn} 55, estimates were made by extrapolation and listed in Fig. 9-11. Therefore, the estimated population exposed to L_{dn} 55 or greater would be about 32 or

79 million people according to extrapolations of the EPA and DOT data, respectively, which represent a very large spread of about 16 to 40 percent of the total national population. Although there is no evidence to support the accuracy of either of the extrapolated values, it is reasonable to assume that the truth would lie somewhere between. Furthermore, it is reasonable to assume that the results of both studies for all L_{dn} values including extrapolations can be combined to form an envelope enclosing realistic values of exposed people. Such an envelope is tabulated as the maximum and minimum values in Fig. 9-11.

The importance of aircraft noise can be demonstrated to be significant by examining the impact of annoyance on the exposed population. Figure 9-12, taken from the EPA "Levels Document," estimates that noise exposure ranging from L_{dn} 55 to 85 would highly annoy 17 to 70 percent of the people, respectively.[27] The quantitative effect of aircraft noise in terms of the number of highly annoyed people can be estimated by relating the number of exposed people to the percent highly annoyed. The results are shown in Fig. 9-13 in the form of an envelope where it is seen that the estimated number of highly annoyed people exposed to L_{dn} 55 or greater ranges widely from about 5.4 to

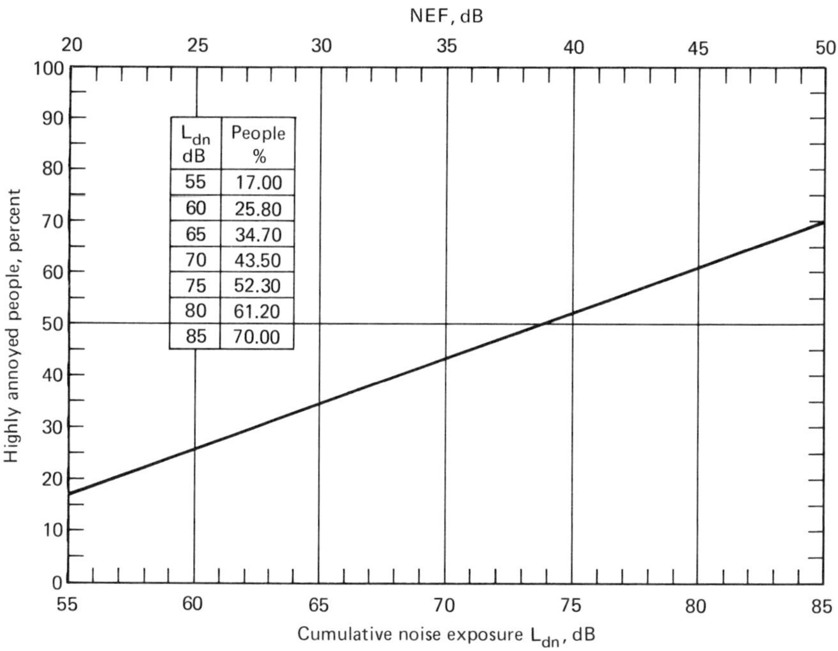

Fig. 9-12 Percent population highly annoyed by noise.

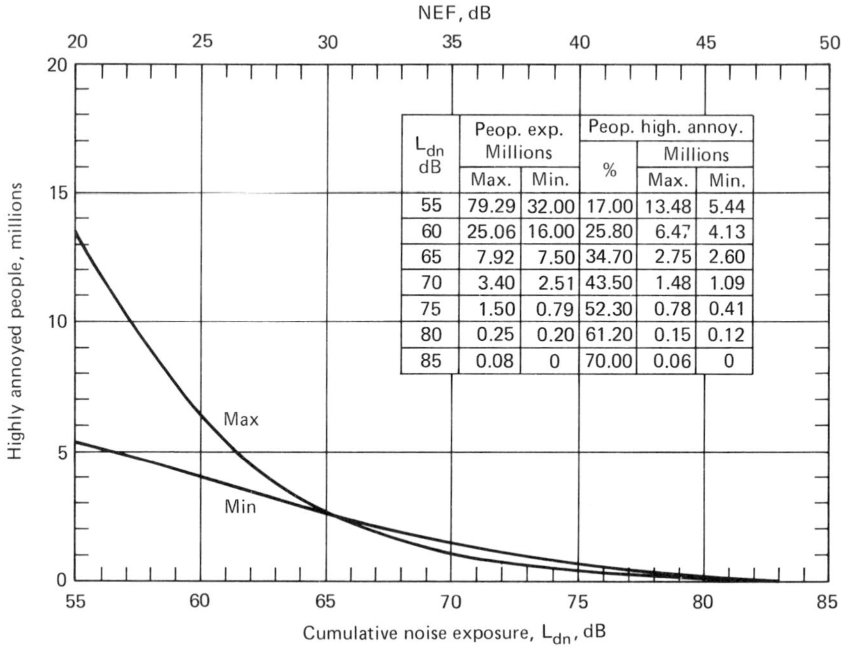

L_{dn} dB	Peop. exp. Millions		Peop. high. annoy.		
			%	Millions	
	Max.	Min.		Max.	Min.
55	79.29	32.00	17.00	13.48	5.44
60	25.06	16.00	25.80	6.47	4.13
65	7.92	7.50	34.70	2.75	2.60
70	3.40	2.51	43.50	1.48	1.09
75	1.50	0.79	52.30	0.78	0.41
80	0.25	0.20	61.20	0.15	0.12
85	0.08	0	70.00	0.06	0

Fig. 9-13 Population highly annoyed by aircraft noise.

13.5 million. The range narrows rapidly to where both studies concur on about 2.3 million highly annoyed people exposed to L_{dn} 66 or greater. And for levels of L_{dn} 80 or greater, the estimated range of highly annoyed people is 120,000 to 150,000. These indeed are significant portions of the national population that suffer at least one social inequity (annoyance) due to aircraft noise.

It would be improper to terminate this discussion without mentioning that public hostility to aircraft noise has grown to such proportions that its potential for inhibiting the continued orderly growth and attendant economic and social benefits of air transportation is very real. So, in the simplest of terms, noise exposure should be reduced to the practical minimum because it both degrades our environment and inhibits our ability to accommodate the rapidly expanding demand for air-transport services.

REFERENCES

The evaluation and control of aircraft and airport noise has been studied extensively since the early 1950s. The effort expended in terms of personnel and money is greater than any other class of noise sources. The published information resulting from those studies is much too voluminous to be included here. Consequently, the following list of references, presented in chronological order, provides only the basic framework of background material for Chapter 9. Most of these references, however, contain their own reference material, which must be studied if more comprehensive background information is sought. Specific references are cited in the text when it is important to do so but not every document listed is cited.

1. Federal Aviation Act of 1958, Public Law 85-726, August 23, 1958.
2. Sperry, W. C. (ed.), *Fundamental Study of Jet Noise Generation and Suppression*, Wright Air Development Division, United States Air Force, WADD Technical Report 61-21, December 1960.
3. Gruschka, H. D. (ed.), *Noise Generation and Suppression in Aircraft*, Proceedings of a Short Course Held at The University of Tennessee Space Institute, Tullahoma, Tennessee, January 29–February 2, 1968.
4. Control and Abatement of Aircraft Noise and Sonic Boom, Public Law 90-411, July 21, 1968.
5. Sperry, W. C., *Aircraft Noise Evaluation*, FAA Report, FAA-NO-68-34, September 1968.
6. National Environmental Policy Act of 1969, Public Law 91-190, January 1, 1970.
7. Airport and Airway Development Act of 1970, Public Law 91-258, May 21, 1970.
8. Clean Air Act, Title IV–Noise Pollution, Public Law 91-604, December 31, 1970.
9. Noise Control Act of 1972, Public Law 92-574, October 27, 1972.
10. *Noise Standards: Aircraft Type Certification*, Federal Aviation Regulations Part 36, 34 FR 18364, November 18, 1969, Amended (36-1), 34 FR 18815, November 25, 1969.
11. *Report of the Special Meeting on Aircraft Noise in the Vicinity of Aerodromes*, International Civil Aviation Organization (ICAO), Doc 8857, NOISE (1969), Montreal, November 25–December 17, 1969.
12. *Civil Aviation Research and Development Policy Study*, Joint DOT-NASA: Report, DOT-TST-10-4 and NASA SP-265; Supporting Papers, DOT-TST-10-5 and NASA SP-226; March 1971.
13. *Aircraft Noise: Annex 16 to the Convention of International Civil Aviation*, International Civil Aviation Organization (ICAO), First Edition August 1971, Second Edition October 1976.
14. *The Long Range Needs of Aviation*, Report of the Aviation Advisory Commission and Technical Annexes, Volumes I and II, January 1973.

15. *Civil Aircraft Sonic Boom*, Federal Aviation Regulations Part 91.55, 38 FR 8051, March 28, 1973.

16. *Legal and Institutional Analysis of Aircraft and Airport Noise and Apportionment of Authority Between Federal, State, and Local Governments*, EPA Aircraft/Airport Noise Study Report (Task Group 1), EPA NTID 73-2, July 27, 1973.

17. *Operations Analysis Including Monitoring, Enforcement, Safety, and Cost*, EPA Aircraft/Airport Noise Study Report (Task Group 2), EPA NTID 73-3, July 27, 1973.

18. *Impact Characterization of Noise Including Implications of Identifying and Achieving Levels of Cumulative Noise Exposure*, EPA Aircraft/Airport Noise Study Report (Task Group 3), EPA NTID 73.4, July 27, 1973.

19. *Noise Source Abatement Technology and Cost Analysis Including Retrofitting*, EPA Aircraft/Airport Noise Study Report (Task Group 4), EPA NTID 73.5, July 27, 1973.

20. *Review and Analysis of Present and Planned FAA Noise Regulatory Actions and their Consequences Regarding Aircraft and Airport Operation*, EPA Aircraft/Airport Noise Study Report (Task Group 5), EPA NTID 73.6, July 27, 1973.

21. *Military Aircraft and Airport Noise and Opportunities for Reduction without Inhibition of Military Missions*, EPA Aircraft/Airport Noise Study Report (Task Group 6), EPA NTID 73.7, July 27, 1973.

22. *Public Health and Welfare Criteria for Noise*, EPA Technical Document 550/9-73-002, July 27, 1973.

23. *Report on Aircraft/Airport Noise*, Report of the Administrator of the EPA in Compliance with Public Law 92-574, Senate Committee on Public Works, Serial No. 93-8, August 1973.

24. *Noise Standards for Newly Produced Airplanes of Older Type Designs*, Federal Aviation Regulations Part 21.183(e) and Part 36 Amended (36-2), 38 FR 29569, October 26, 1973.

25. *Recommended Noise Abatement Takeoff and Departure Procedure for Civil Turbojet Powered Airplanes*, FAA Advisory Circular AC 91-39, January 18, 1974.

26. *Aircraft and Airport Noise Regulations*, EPA Notice of Public Comment Period, 39 FR 6142, February 19, 1974.

27. *Information on Levels of Environmental Noise Requisite to Protect Public Health and Welfare with an Adequate Margin of Safety*, EPA Technical Document 550/9-74-004, March 1974.

28. Pearsons, K. S., and Bennett, R. L., *Handbook of Noise Ratings*, National Aeronautics and Space Administration Contractor Report, NASA CR-2376, April 1974.

29. Bartel, C., Sutherland, L. C., Simpson, L., *Aircraft Noise Reduction Forecast, Volume I–Summary Report for 23 Airports*, Department of Transportation, DOT-TST-75-3, October 1974.

30. Bartel, C., Coughlin, C., Moran, J., and Watkins, L., *Airport Noise Reduction Forecast, Volume II–NEF Computer Programs Description and User's Manual*, Department of Transportation, DOT-TST-75-4, October 1974.

31. Bishop, D. E., *Community Noise Exposure Resulting from Aircraft Operations: Application Guide for Predictive Procedure*, Aerospace Medical Research Laboratory, AMRL-TR-73-105, November 1974.

32. *Noise Certification Rule for Propeller Driven Small Airplanes*, EPA Project Report, November 25, 1974.

33. *Minimum Altitudes for Noise Abatement*, EPA Project Report, November 25, 1974.

34. Modig, C., *Civil Aviation Studies and Interagency Coordinating Organizations*, EPA Report, Volume I (A Background History, with Emphasis on Organizations Dealing with the Aircraft Noise Issue) EPA 550/9-74-019A, Volume II (Appendices), EPA 550/9-74-019B, December 1974.

35. *Civil Subsonic Turbojet Engine-Powered Airplanes (Retrofit and Fleet Noise Level)*, EPA Project Report, December 16, 1974.

36. *Acoustical Change Approvals*, Federal Aviation Regulations Part 36 Amended (36-3), 39 FR 43830, December 19, 1974.

37. Modig, C., *Information of FAA Certification of Aircraft*, EPA Report EPA-550/9-75-022, January 1975.

38. *Noise Standards for Propeller Driven Small Airplanes*, Federal Aviation Regulations Part 21 and Part 36 Amended (36-4), 40 FR 1029, January 6, 1975.

39. *Noise Standards for Propeller Driven Small Airplanes*, EPA NPRM 74-39, 40 FR 1061, January 6, 1975.

40. *Noise Abatement Minimum Altitudes Within Terminal Areas: Turbojet Powered Airplanes*, EPA NPRM 74-40, 40 FR 1072, January 6, 1975.

41. *Aircraft Noise Certification Rule for Supersonic Civil Aircraft*, EPA Project Report, January 24, 1975.

42. *Civil Subsonic Turbojet Engine-Powered Airplanes Noise Retrofit Requirements*, EPA NPRM 75-5, 40 FR 8218, February 26, 1975.

43. *Fleet Noise Level Requirements*, EPA NPRM 75-6, 40 FR 8222, February 26, 1975.

44. *Federal Aircraft Noise Research, Development, and Demonstration Programs: FY73-FY75*, EPA Report, EPA 600/2-75-003, March 1975.

45. *Aircraft Noise Requirements: Civil Supersonic Airplanes*, EPA NPRM 75-15, 40 FR 14093, March 28, 1975.

46. *First Report on Status and Progress of Noise Research and Control Programs in the Federal Government*, EPA Reports, Volume I (Report on Status and Progress), Volume II (Research Panel Reports), EPA 550/9-75-023, June 1975.

47. *Reduced Flap Setting Noise Abatement Approach for Turbojet Engine-Powered Airplanes*, EPA NPRM 75-35(I), 40 FR 44256, September 25, 1975.

48. *Visual Two-Segment Noise Abatement Approach for Turbojet Engine-Powered Airplanes*, EPA NPRM 75-35(II), 40 FR 44259, September 25, 1975.

49. *Two-Segment ILS Noise Abatement Approach for Turbojet Engine-Powered Airplanes*, EPA NPRM 75-35(III), 40 FR 44256, September 25, 1975.

50. *Subsonic Transport Category Large Airplanes and Subsonic Turbojet Powered Airplanes: Proposed Noise Reduction Stages and Acoustical Change Requirements*, FAA NPRM 75-37, 40 FR 51476, November 5, 1975.

51. *Airplane Noise Requirements for Operation to or from U.S. Airports*, EPA NPRM 76-1, 41 FR 6270, February 12, 1976.

52. Sperry, W. C., and Gray, D. C., *Noise Standards for Aircraft Type Certification (Modifications to FAR Part 36)*, EPA Project Report, August 1976.

53. *Noise Type Certification and Acoustical Change Approvals*, Federal Aviation Regulations Part 36 Amended (36-5), 41 FR 35053, August 19, 1976.

54. *A Proposed Unified National Strategy for Noise Abatement and Control*, U.S. Environmental Protection Agency, Draft, October 1976.

55. *Aircraft Noise Measurement and Evaluation Specifications*, FAA NPRM 76-21, 41 FR 47342, October 28, 1976.

56. *Noise Levels for Turbojet Powered Airplanes and Large Propeller Driven Airplanes*, EPA NPRM 76-22, 41 FR 47358, October 28, 1976.

57. *Noise Standards; Aircraft Type and Airworthiness Certification: Proposed Alternate Noise Reduction Stages and Acoustical Change Requirements for Subsonic Transport Category Large Airplanes, and for Single Engine Transport Category Airplanes*, FAA NPRM 75-37C, 41 FR 47375, October 28, 1976.

58. *FAR Part 36 Compliance Regulation, Final Environmental Impact Statement*, Department of Transportation, Federal Aviation Administration, November 10, 1976.

59. *Report of Committee on Aircraft Noise, Fifth Meeting*, International Civil Aviation Organization (ICAO), Montreal, November 15–30, 1976.

60. *Aviation Noise Abatement Policy*, Department of Transportation, Federal Aviation Administration, November 18, 1976.

61. *Airport Noise Regulatory Process*, EPA NPRM 76-24, 41 FR 51522, November 22, 1976.

62. *Noise Abatement Landing Flap Amendment and Decision Not to Prescribe Two-Segment Approach Requirements Submitted by Environmental Protection Agency*, Federal Aviation Regulations Part 91 Amended, and FAA Notice of Decision Not to Prescribe, 41 FR 52388, November 29, 1976.

63. *Proposed Regulations Submitted by the Environmental Protection Agency to the FAA: Noise Abatement Minimum Altitudes for Turbojet Powered Airplanes in Terminal Areas*, FAA Notice of Decision Not to Prescribe, 41 FR 52393, November 29, 1976.

64. *Turbojet-Powered Airplanes Delayed Landing Flap Procedure*, FAA NPRM 76-26, 41 FR 52396, November 29, 1976.

65. *Proposed Regulation Submitted by the Environmental Protection Agency to the FAA: Fleet Noise Level Requirements*, FAA Decision Not to Prescribe, 41 FR 52398, November 29, 1976.

66. *Phased Compliance with Part 36 Noise Limits by Turbojets with Maximum Weights Greater than 75,000 Pounds*, Federal Aviation Regulations Part 91 Amended, and FAA Notice of Decision Not to Prescribe, 41 FR 56046, December 23, 1976.

67. *Noise Regulations for Propeller-Driven Small Airplanes Submitted to the FAA by the Environmental Protection Agency; Notice of Decision*, Federal Aviation Regulations Part 36 Amended (36-6), and FAA Notice of Decision Not to Prescribe, 41 FR 56056, December 23, 1976.

68. *Agriculture Operation and Fire Fighting Propeller-Driven Small Airplanes*, FAA NPRM 76-27, 41 FR 56065, December 23, 1976.

69. *Noise Limits and Acoustical Change Requirements for Subsonic Transport Category Large Airplanes and for Subsonic Turbojet Powered Airplanes*, Federal Aviation Regulations Part 36 Amended (36-7), 42 FR 12360, March 3, 1977.

70. *Civil Supersonic Airplanes Proposed Noise and Sonic Boom Requirements*, FAA Supplemental NPRM 77-23, 42 FR 55176, October 13, 1977.

10

Highway and rail traffic noise

Edwin G. Ratering
Director, Vehicular Noise Control
General Motors Corporation

HIGHWAY TRAFFIC NOISE

Scope

In any discussion of community noise control, the aspect of noise that is generated by the operation of motor vehicles on the streets and highways of the land can hardly be neglected. Indeed, the phenomenon of traffic noise has been the object of a great number of studies resulting in the tabulation of thousands of sound level readings by acousticians and engineers. Hundreds of compilations of attitudinal surveys and physical investigations to establish the bounds of noise interference with human activity have been accomplished by psychologists and biophysicists. Countless hours of philosophizing have occupied the time of sociologists and economists as they have considered methods to quantify the worth in dollars of improving the quality of life through noise control. Predictably, laws have been written; lawyers and administrators continue to theorize concerning enforcement schemes. And the process goes on.

The objectives that these investigators hope to attain are consistent with the new attitude toward the environment that emerged as we passed into the 1970s. Out of this work, certain broad generalizations have become apparent. One of these is the realization that we must think and act in terms of increments of immediate improvement and degrees of ultimate improvement when dealing with the ecosystem, rather than with assumptions of absolute-total-solutions-no-matter-what. Therefore, the engineer, the regulator, and the economist find themselves dealing in margins. In the case of traffic noise control, the engineers who design vehicles, highways, and noise-attenuation barriers, quite often find that the first increment of improvement is the largest and the least expensive when compared to subsequent increments. At each point, hard decisions must be made by the public, weighing the margin of cost, the change in functional character of the vehicle system, and the margin of environmental improvement. Each succeeding increment of improvement must then be accepted or rejected on the basis of whether it is marginally efficient.

In this chapter, we will attempt to set out the dimensions of traffic noise and to describe the means by which traffic noise reduction may be attained and the degree of environmental improvement that may be affected. In addition, the major aspects of locomotive noise and railcar noise are considered.

Population Exposed

In June 1974, the U.S. Environmental Protection Agency (EPA) published a report[1] that was intended to provide a first-order estimate of the numbers of people in the United States exposed to various levels of sound. The data used to generate noise exposure and population density estimates for urban areas, excluding freeway corridors and airports, were gathered from 100 sites located in 14 major cities. Sites were selected in situations where people live, ranging from nearby lightly traveled streets to points adjacent to major arteries (not freeways). The conclusions in the EPA report are in the form of noise-exposure projections for the national urban population, based upon statistical treatment of the data, which were mostly from the 100 sites. The results of this urban residential study are combined with the results of a previous study[2] in which the EPA estimated the number of people who live in freeway corridors and near airports, and the noise levels to which they are exposed. Then the noise-exposure condition of people living in rural areas was estimated and combined with the urban, airport, and freeway studies.

The resulting estimate of residential populations exposed to levels of outdoor day/night sound levels that exceed L_{dn} 60 is given in Table 10-1. (See Chapters 2 and 3 for introductory discussions of L_{dn}).

An estimated 62 million people live in circumstances in which the L_{dn} is over 60 dBA, presumably because of urban and freeway traffic noise. To put this into perspective, it is quite possible for the sounds that exist in one's dooryard

TABLE 10-1 People in the United States exposed to outdoor day/night sound levels. (From Eldred, K. M., Galloway, W. J., and Simpson, M. A. Population Distribution of the United States as a Function of Outdoor Noise Level. Prepared for U.S. EPA, ONAC under Contract No. 68-01-1886, June 1974, p. 30, Table 6-1.)

	PEOPLE IN MILLIONS			
L_{dn} EXCEEDS	URBAN TRAFFIC	FREEWAY TRAFFIC	AIRCRAFT OPERATIONS	TOTAL
60	59.0	3.1	16.0	78.1
65	24.3	2.5	7.5	34.3
70	6.9	1.9	3.4	12.2
75	1.3	0.9	1.5	3.7
80	0.1	0.3	0.2	0.6

in the middle of the night to reach the level of 55 dBA when no traffic exists. Such a level could be due to a gentle wind in the trees combined with the sounds of the crickets. The difference is, of course, that such night sounds are generally desirable whereas traffic sounds may be undesirable, notwithstanding that the actual level of the traffic noise may be in the same range as the natural night sounds.

Considering now the higher ranges of sound level due to urban traffic, it is apparent from Table 10-1 that, with any significant reduction in exterior residential noise exposure, the number of people exposed to L_{dn} exceeding 70 dBA will become negligible. Since physiological damage is not a factor at L_{eq} 70 dBA, it then becomes a proposition of determining the cost that society is willing to bear to reduce the number of people exposed to the band of L_{dn} between 60 dBA and 70 dBA. These costs will arise out of the development and manufacture of vehicle-component designs, which will reduce engine-related noise; additional costs will accrue due to development and production of tires, which will not only be quieter, but which must retain the other characteristics of safety, traction, and vehicle handling. Complementary noise reduction actions may include construction of roadside barriers, depressed roadways, and new routings for highways, rails, and mass transit systems. But these solutions, too, each carry some characteristic cost. In a real sense, we may say that the degree to which this band of community noise between L_{dn} 60 dBA and 70 dBA is resolved constitutes one quantitative expression of the national quality of life. It will be attainable in the degree that society is willing to pay for it.

Referring again to Table 10-1, the report shows an estimated 8.8 million people who live in circumstances in which L_{dn} exceeds 70 dBA due, in the main, to urban and freeway traffic. Even if these estimates were to be found in error by a factor of 50 percent high, it seems worthwhile to consider what may be done to improve the noise environment. We would add that in the process of making such consideration, it must be borne in mind that use of a relatively simple metric, such as L_{dn}, overstates the precision with which we can deal with a problem. For example, in many cases, annoyance is an incident event, and even large expenditures to attain equivalent energy reductions in terms of L_{dn} would still leave the probability of annoyance in localized incident event situations. Thus, in the localized situation it might have been better to solve the isolated problem of the annoying incidents rather than to contemplate the larger expenditure to reduce the whole plane of existing sound level, assuming, of course, that the level lies somewhere in the band of L_{dn} 60–70 dBA.

At any rate, if the estimates in Table 10-1 are taken at face value, they will show that if urban and freeway traffic noise were to be reduced by 5 dBA, the proportion of the population exposed to sound levels greater than L_{dn} 60 would be reduced by 57 percent while the proportion exposed to L_{dn} exceeding 70 dBA would be reduced by 85 percent. If, by some means, urban and freeway noise could be further reduced by another 5 dBA, the proportion of the population exposed to levels greater than L_{dn} 60 would be reduced from the present estimate by almost 86 percent. If the estimates are correct, then indeed a 10-dBA reduction of equivalent energy sound level, L_{eq} or L_{dn}, is a worthy goal. We will further explore these possibilities in this chapter.

Subjective Aspects of Annoyance

In 1971, research that had been conducted for the Motor Vehicle Manufacturers Association was reported, which described both the extensity and the intensity of annoyance due to motor-vehicle noise based on interviews with some 1200 urban respondents.[3,4] Of this number, 46 percent identified annoying vehicle-noise situations, while 54 percent were not annoyed. One interesting conclusion of the report was that when annoyance due to motor vehicle noise does occur, it does so in the context of a "situation" and therefore is not a human response related only to sound-pressure stimuli. Of the sample of 549 who were annoyed, one-third of the variance in overall annoyance was related to perceived loudness; however, the respondents felt that half of the time, drivers under 25 years of age were responsible for annoying operation of vehicles. Thirty percent of the annoyed respondents thought that drivers were operating their automobiles and motorcycles illegally; this attitude raised the intensity of annoyance one-half of one scale point relative to annoyance by vehicles thought to be operated legally. Half of the respondents thought that noise produced by vehicles was easily within the control of the driver, while another 15 percent thought that the driver could control noise output of the vehicle if he or she tried.

The reader should not be led to believe by these experimental data that measuring human reaction to noise is an exact science. The organized reaction to noise is complex indeed and is certainly far from being completely understood.

On the other hand, it seems worthwhile to begin an elementary tabulation of responses to noise in order that vehicle designers, urban planners, and regulators have a first order of facts as a premise for some determination of noise-control countermeasures that can be taken.

Table 10-2 delineates quantitatively the extensity and intensity of traffic noise resulting from the work conducted for the Motor Vehicle Manufacturers Association. The extensity number is associated with how often the annoyed sample of respondents subjectively identified sources. The intensity of annoyance shown in the table is the mean value of subjective expressions on a scale of 1 through 7. The gradings of annoyance are defined as: 3, "quite"; 4, "definitely"; and 5, "strongly."

Apparently, people have definite feelings about various vehicle sounds, and the intensity of their annoyance is affected by these feelings, in addition to the loudness. For example, considering the intensity of annoyance, which was expressed as mid-range for diesel-powered trucks, the extensity of annoyance is only 12 percent. From this, one is led to believe that people rate trucks as being noisy, but that their annoyance is partially ameliorated because they view truck operation as a useful element of commerce, and because they do not think that

TABLE 10-2 Annoyance due to motor-vehicle traffic. (From Bolt Beranek and Newman, Inc., Report 2082. Motor Vehicle Noise Identification and Analysis of Situations Contributing to Annoyance. Prepared for Motor Vehicle Manufacturers Association, June 1971, p. 24, Table II.)

	ANNOYANCE					
	Extensity			**Intensity**		
	RANK	NUMBER	PERCENT	RANK	MEAN	STANDARD DEVIATION
Busses	12	20	3	1	5.1	1.8
Diesel trucks: Auditor in transit	6	30	4	2	4.9	1.2
Automobiles: Tire screech	8	27	4	3	4.7	1.7
Automobiles: Exhaust noise	3	97	14	4	4.5	1.7
Motorcycles	1	160	23	5	4.5	1.5
Diesel trucks: Auditor at home or at work (or location is not relevant)	4	88	12	6	4.2	1.6
Automobiles: Noise origin irrelevant, brake squeal	8	27	4	7	4.1	1.5
Automobiles: Siren noise	11	21	3	8	4.0	1.8
Automobiles: Engine noise	5	44	6	9	3.8	1.4
Gasoline trucks	10	26	4	10	3.8	1.5
Traffic flow	2	115	17	11	3.7	1.5
Automobiles: Horns and other noises	8	27	4	12	3.3	1.7
Unclassified	-	11	2	-	5.6	0.6
TOTAL		693	100		4.2	1.6

drivers deliberately create noise. Referring again to Table 10-2, annoyance due to traffic flow is somewhat more extensive, but the intensity mean is shown as a lower value. Although individual vehicle noises in a stream of traffic are less identifiable, the researchers believe that traffic-flow-noise annoyance intensity is influenced by other aspects of the situation that directly affect individuals such as their locations and the degree to which the noise interferes with their activities. Thus, the dimensions of traffic noise that affect people tend to be more psychological than physical. These responses suggest that the efficiency of a traffic-noise-control program will be maximized by actions that take specific account of the psychological responses of people to "noise situations."

Predicting Traffic Noise for Environmental Impact Statements

Rapidly changing population patterns on the national scene—and a more highly developed public expectancy in terms of environmental effects—have generated the requirement for highway planners to furnish environmental impact statements whenever new highways are planned. A factor to be considered in writing environmental impact statements is the noise that may result from the traffic that will move along the new road or along the improved or widened section of an existing arterial road. In order to evaluate what the new condition will be, methodologies of traffic noise prediction have been developed by various agencies. The basis for one such prediction method is contained in the National Cooperative Highway Research Program Report 117 "A Design Guide for Highway Engineers."[5]

The authors of the report have systematically delineated the parameters that affect traffic noise and have described a procedure for collecting data. They present an analytical model with a calculation procedure, which results in estimated noise levels for the highway under consideration. A detailed narration of the procedures is not warranted here because the design guide is self-explanatory. However, a discussion of the parametric data required will be of interest to the reader who may need to become familiar with the terms and dimensions of the information, which are critical in any procedure for predicting the sound level along an arterial road that may be in the planning stage.

The problem of traffic-noise prediction is complicated by the fact that streets and highways are not flat, straight, or free from natural terrain variations that affect propagation. They are not of constant cross section, considering the number of lanes and the width of the median. Finally, the density and mix of traffic are not constant, nor is the vehicle speed. Each of these parameters must be taken into account. In order to do this, it is necessary to analyze the highway by examining elements of the section in question that have constant characteristics, through the use of a mathematical process known as finite element analysis. Of course, all the elements of constant character taken together in the model consitute the complete section of the roadway for which the planner wished to predict the noise level. The design guide contains charts and tables that must be used in conjunction with the traffic-design volume specifications of the highway, the surveyed terrain grade data along the right-of-way, the directional-plot plan for the road including distances and curvature, and the surface roughness rating of the roadway.

The charts and tables provided in the design guide may then be used to determine the statistical bounds of the sound levels resulting, in terms of L_{10} and L_{50} and other derived parameters (see also Chapter 2). The interrelation and organization of these data are illustrated in Figure 10-1. The resulting levels of L_{10} and L_{50} may thus be estimated for various hours of the 24-hour day.

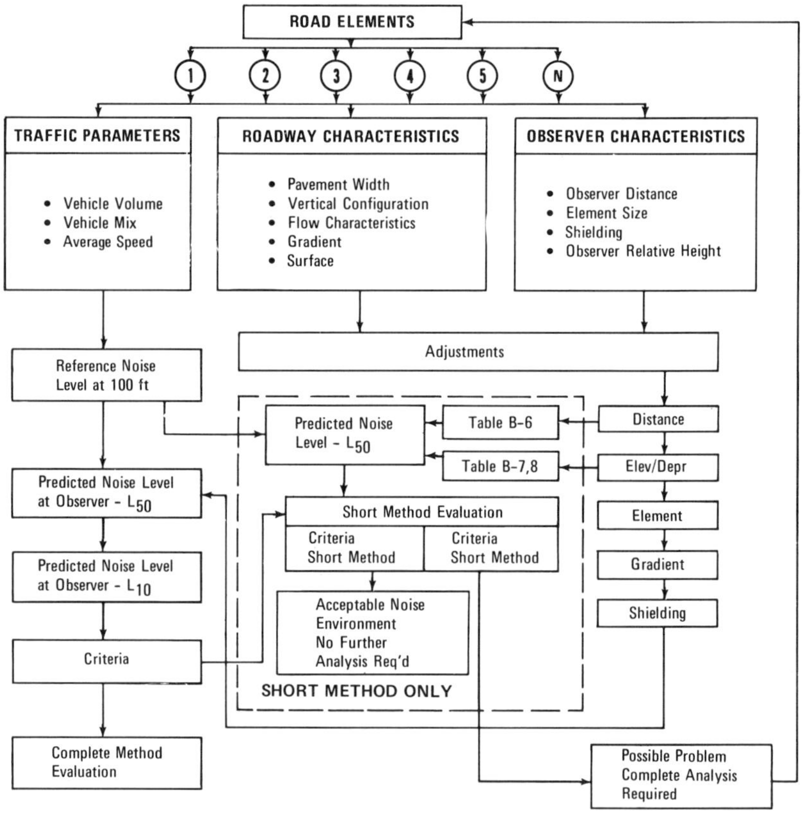

Fig. 10-1 Flow diagram of traffic noise methodology. (From Galloway, W. J., Gordon, C. G., Kugler, B. A., and Nelson, D. L. Bolt Beranek and Newman, Inc. *Highway Noise—A Design Guide for Highway Engineers.* **National Cooperative Highway Research Program Report 117[5], Sponsored by the American Association of State Highway Officials in Cooperation with the Federal Highway Administration, Highway Research Board, National Research Council, National Academy of Sciences, 1971, p. 40, Fig. 1.)**

TABLE 10-3 Summary of highway barrier costs and reported or expected effectiveness. (From Rackl, R., Sutherland, L., and Swing, J. Wyle Laboratories, Report WCR 75-2. Community Noise Countermeasures Cost-Effectiveness Analysis. Prepared for Motor Vehicle Manufacturers Association, July 1975, pp. 6–42, Table 6.7-4.)

TYPE	LOCATION	BARRIER HEIGHT (FEET)	COST PER LINEAL FOOT ($)	NOISE REDUCTION (dB)
Earth Berm	I-75, Madison Heights, Mich.	13		8 to 10
Earth Berm	I-94, US 131 Interchange, Kalamazoo, Mich.	10	12.50	7
Earth Berm	I-75, Lake Allatoona, Ga.	15	14.50	
Earth Berm	I-84, West Hartford, Conn.	10 to 15	83.50	10
Earth Berm	Oakdale Road, Georgia	15	27.00	10
Earth Berm	Route 157, Boulder, Colorado	7		6 (Cars) 0 (Trucks)
Earth Berm	Montgomery Co., Maryland	8 to 10		8 to 10
Earth Berm	I-182, Pasco, Washington	10	32.00	10
Earth Berm	S. Madison Beltline, Wis.	10	21.50	12
Earth Berm	North Freeway, Nebraska			9
Earth Berm	St. Paul, Minnesota			10
Earth Berm	Columbia, S. Carolina	11	17.50	7
Timber Wall	Lincoln Park, Mich.	6		0
Timber Wall	I-75, Allen Park, Mich.	14	66.00	10
Timber Wall	127, North Lansing, Mich.	8		0
Timber Wall	I-205, E. Portland, Oregon	10 to 15	46.00	7 to 10
Timber Wall	I-405, Bellevue, Washington	10 to 14	36.50 to 58.00	
Timber Wall	I-90, Tanner, Washington	4 to 12		5 to 10
Timber Wall	North Carolina	10	15.00	0 to 15
Timber Wall	US 59, Houston, Texas	8	7.00	10 at 50 feet
Concrete	I-205, E. Portland, Oregon	10 to 15	73.00	7 to 10
Masonry	Lordsburg, New Mexico	6	25.00	8
Concrete	Sepulveda Blvd., Los Angeles, Ca.	8	65.00	
Concrete	Sepulveda Blvd., Los Angeles, Ca.	9	23.00	
Concrete	Sepulveda Blvd., Los Angeles, Ca.	9	30.00	
Concrete	#92, Hayward, California	6	21.00	
Metal	Phoenix, Arizona	12	328.00	
Concrete	I-94, Minneapolis, Minn.	10 to 23		11 (first row of houses)
Earth Berm + Concrete	Nevada	8		6 to 7 (Trucks)
Earth Berm (+ Wall)	St. Paul, Minnesota			3.8 to 12.4
Earth Berm (+ Wall)	Roseville, Minnesota		68.00	5 to 11
Earth Berm (+ Wall)	Minnesota			2 to 8
Earth Berm + Polyester	Howard Co., Maryland	12	77.00	10
Earth Berm + Steel	Baltimore, Maryland	35		10
Earth Berm + Masonry	Tempe, Arizona	8	8.00	6
Earth Berm + Concrete + Wood	Sacramento, Ca.	5 to 19		10 to 15

In later research reported under National Cooperative Research Program Report 144, the design guide resulting from NCHRP Report 117 is further evaluated and modified by field studies of the noise-reducing characteristics of certain roadside constructions, such as barriers and elevated and depressed highway sections.[6] For a detailed discussion of the practical benefits and costs of roadside noise barriers, reference is made here to the excellent work by the Ministry of Transportation and Communications of Ontario[7] and the Minnesota Department of Highways.[8] The compendium of data found in Table 10-3 is from work done by Wyle Laboratories[9] for the Motor Vehicle Manufacturers Association of the United States.

Traffic Noise Reduction As a Function of Reducing Noise Levels of Vehicles

In the preceding sections, traffic noise has been related to characteristics of the roadway. Of course, other factors in community noise control are the capability and the feasibility of reducing vehicular noise by designing noise-control features into newly produced vehicles and by prohibition of deliberate vehicular modifications that produce increased noise. If these measures are taken, what quantitative improvement may we expect along the streets and highways in a community? General Motors engineers have developed a simplified mathematical modeling technique for predicting roadside noise from freely flowing traffic in order to assess the effectiveness of various vehicle-noise reduction strategies.[10] By entering sound-level data into the model, which represent various design possibilities relative to vehicular noise control hardware, changes in urban-street L_{dn} can be projected. Figure 10-2,[11] for example, represents the reduction in L_{dn} that can be expected if:

1. A statistically derived operational sound-pressure level of 70.5 dBA at 50 ft (15.2 m) for automobiles and light trucks is chosen as a future low-speed value. This assumes the elimination of higher vehicular sound levels resulting from exhaust-system modification by owners and from poorly maintained exhaust systems.
2. Medium- and heavy-truck low-speed sound levels do not exceed a standard of 83 dBA.
3. Tire noise from cars and trucks is idealized at the level consistent with best-known present technology.
4. The existing fleet of trucks is modified as required to meet a low-speed passby standard of 86 dBA.
5. The density of vehicles on the national street and highway system increases in proportion to the slope of the top line on the chart, as an extreme.

Figure 10-3 represents the reduction in L_{dn} that can be expected if the conditions of (1) through (4) are met, but the vehicular density of vehicles remains the same, as opposed to condition (5).

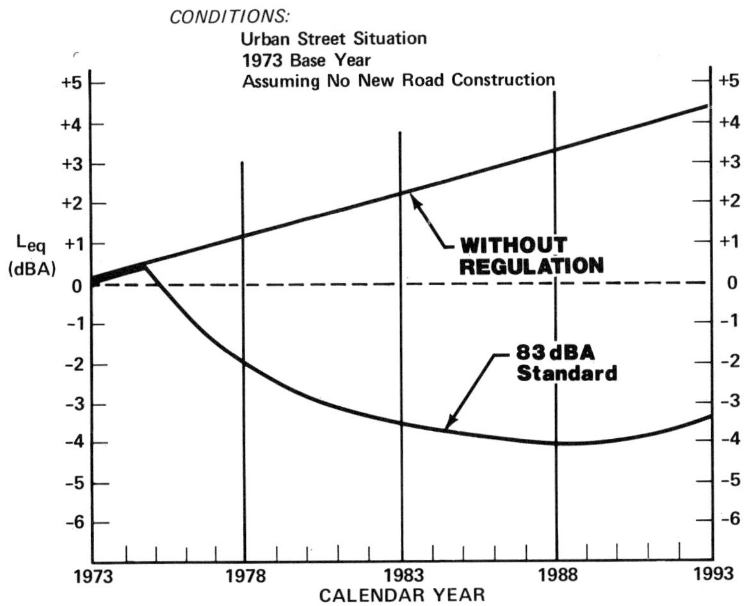

Fig. 10-2 Projected change in L_{eq} due to federal truck regulation conditions: Urban street situation, 1973 base year, assuming no new road construction. (From General Motors Corporation, Environmental Activities Staff, 1975.)

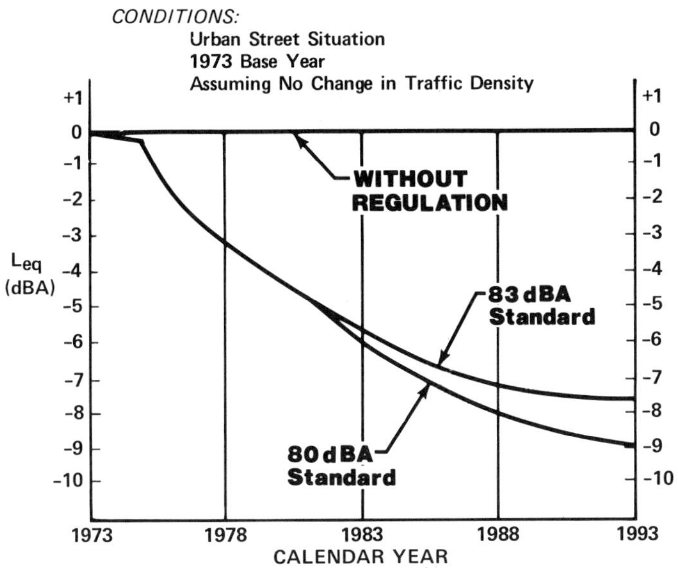

Fig. 10-3 Projected change in L_{eq} due to federal truck regulation conditions: Urban street situation, 1973 base year, assuming no change in traffic density. (From General Motors Corporation, Environmental Activities Staff, 1975.)

Of course, the reduction of L_{dn} can be further biased downward by choosing lower operational vehicular noise standards, but the estimate of incremental cost must then be developed for the attainment of each characteristic new vehicular sound level proposed. The results of successive L_{dn}-reductions and their associated costs may then be compared to the population-exposed goals discussed in relation to the values shown in Table 10-1. Studies in the field of cost for increasing these benefits have shown a sharp rise in the progression, beyond which the costs rise exponentially. It is not surprising that a point of diminishing returns exists in this relationship; social planners should not attempt to drive the benefit factor beyond that point.

Vehicular Component Hardware Development for Control of Engine-Related Truck Noise

It immediately becomes obvious to anyone who makes a traffic-noise survey that, aside from modified automobiles and motorcyles, the major source of noise on the streets is the truck. This is largely because truck engines are ordinarily used at near-full power output for a much higher proportion of operating time than are automobiles or light trucks with gross vehicle weight ratings (GVWR) of less than 10,000 lbs (4536 kg). This higher power usage from the engine arises out of the fact that medium and heavy trucks are relatively low power-to-weight-type vehicles as compared to cars and light trucks. Thus, to start the load of the truck and to accelerate it, the engine is usually operated near its governed speed and power rating, regardless of whether the truck is gasoline-powered or diesel-powered. Ordinarily, the sound one hears from trucks operating at speeds of 35 mph (56 km/hr) or less is engine-related, and not tire-related. The main contributors to engine-related noise are the engine cooling system fan, the engine exhaust, the internal mechanical reciprocating and rotating engine parts, and the ringing of the engine block as a product of combustion gas pressure pulses within the cylinders of the engine. The latter phenomenon is more apparent in the case of the diesel engine than in the gasoline engine.

If the truck engine noise is controlled to levels consistent with the existing Interstate Motor Carrier Noise Regulation, then at cruise speeds of 50-55 mph (80-88 km/hr), tire noise becomes a most significant contributor to total passby noise. Note that the truck manufacturers' responsibility is involved with low-speed, high-power-output, engine-related noise. To qualify the matter of who bears responsibility for truck tire noise, it must be understood that truck manufacturers furnish only original equipment tires, and do not control the aftermarket in tires. Of all truck tires being used in service at any point in time, it is estimated that 75 percent of them are the product of aftermarket purchases. Of the remaining 25 percent of tires on the road that are original equipment-truck tires, truck purchasers either specify or supply their own choice of tire for the truck in 25 to 35 percent of the cases. Finally, in the case of semitrailer rigs, the truck-tractor manufacturer has no control at all over the tires supplied for trailers. For all these reasons, the truck tire manufacturer is responsible for the noise performance of tires, not the truck manufacturer.

The truck engine-cooling-system engineer has a pair of options in terms of the methodology that may be used to control fan noise. The first approach is to attempt to adjust the various cooling-system design parameters so that the fan tip speed will be minimized, inasmuch as fan noise is a function of tip speed. The best combination is ordinarily the use of a large-surface-area liquid-coolant radiator with a large-diameter low-rotational speed fan. If there is insufficient space in the engine compartment for that combination, there may be room for a close-fitting fan shroud with sufficient axial length to improve flow conditions for the fan. If cooling system optimization is not possible in the degree required to sufficiently reduce fan noise, the second option is to drive the fan through a thermally actuated clutch. The truck accommodates this concept by the fact that ram airflow is of sufficient volume to satisfy medium to high vehicle-speed cooling requirements, even in quite warm air temperatures. The fan will be automatically cycled to the "on" condition when critical coolant temperatures are exceeded, and the fan will be automatically declutched when the coolant temperature falls below the critical level. In the likely event that the fan will engage during an engine idle period as the result of "hot soak" after a high temperature run, it is of little consequence from the standpoint of noise because of the relatively low rotational rate of the fan at engine idle. Thus, not counting engagement at engine idle, tests show that statistical expectancy of "fan on" time may be as low as 1 percent and as high as 10 percent.[12] This, of course, means that fan noise would be radiated less than 10 percent of the time.

Exhaust noise attenuation must take into account three aspects of noise generation resulting from the release of high-temperature gases under the pressure conditions existing within the engine cylinders due to combustion of the fuel:

1. gas-pressure pulse propagation through the exhaust-valve exit route from the interior of the cylinders
2. high-velocity viscous flow of expanding gas from the exhaust-pipe outlet
3. mechanical ringing of the exhaust piping and the muffler body, sometimes known as "shell noise."

Exhaust pulse noise is primarily attenuated through the use of chambered mufflers consisting of reactive and resistive elements that decrease the amplitude of the transmitted gas-pressure pulses. The ability of mufflers to accomplish this is quite largely a function of correctly proportioned chamber volumes, taking into consideration that excessive back pressure on the engine is intolerable. The exhaust-system designer does not have great freedom to use high-volume devices but rather has quite specific space restraints placed upon the system design. If the exhaust noise situation remains critical after the best muffler is applied, the usual alternative is to split the exhaust gasflow and to install a second muffler. Of course, dual exhaust systems add weight and cost.

Viscous flow noise is usually controlled by reducing gas velocity through the use of larger-diameter exhaust pipes. Shell noise radiation may be controlled through the use of double-wall exhaust-system pipes constructed so that the two layers of steel attenuate the transmitted sound by interfacial friction. This tends to dampen exhaust-pipe ringing. The use of large-diameter double-wall pipes adds weight and cost to the truck. In addition, the mufflers must be wrapped with some type of absorption material to dampen muffler-body ringing, and a protective layer of sheet steel must be used over the absorption material.

During the combustion process in diesel-cycle engines and otto-cycle engines, a high rate of gas-pressure rise is followed by a rapid pressure-decay rate. These pulses excite the many natural frequencies of the engine structure and cause it to ring. This is known as *combustion noise*, and, as it turns out, this source is more difficult to control than the other major sources. Much work has been done particularly to understand and to control diesel-engine noise, and research is continuing as more sophisticated analytical instrumentation becomes available.

Diesel-engine manufacturers have investigated the noise control possibilities of accoustic treatment of sheet-metal engine panels such as rocker-box covers. Commercial noise control packages attempting to exploit such treatment are available, but they are far from a total answer to engine noise. When it is required for engine noise to be further reduced, engineers have resorted to designing engine and transmission enclosures, which must be built into the truck cab and frame subassembly. These enclosures must be lined with acoustic absorption material and must be designed so that mating surfaces fit perfectly to provide a high-integrity absorbing barrier. Also, the enclosure must be designed for quick removal for engine and transmission accessibility when normal maintenance is required. Such enclosures may be envisioned as the counterpart of an aircraft engine nacelle and, in fact, quasi-aircraft design practice will be required if very low engine-noise levels are needed.

Development engineers have uncovered serious disabilities with the use of such enclosures, in addition to the expensive design inherent in their application. The enclosures impose a higher level of engine-cooling difficulty because truck engine compartments, as we know them, are quite open to cooling airflow axially as well as downward toward the roadway. In order to significantly increase cooling capacity, disproportionately large engine compartments are needed to accommodate large radiators, or else disproportionately small engines must be used with existing compartment envelopes. Full-scale model building and testing show that, to attain satisfactory engine cooling, either cargo space will be lost due to engine compartment-space claims, or prime-mover power capability will be lost due to use of smaller engines. This is contrary to best design practice for trucks in a period of petroleum shortages. Furthermore, because of the fire hazard that arises from accumulations of engine oil and fuel in the acoustic absorption material, particularly around the bottom of the enclosure, this engine enclosure approach to noise control must be considered marginal at best. Taken together with the maintenance difficulties inherent due to inaccessibility of the engine and transmission, the conclusion is that engine enclosures will change the functional character of the transport truck in a manner that is unacceptable. Development work has indicated that, expensive as it may be, a somewhat less severe application of engine enclosures may reduce the overall low-speed, full-power noise of a heavy diesel truck to the 80-dBA level, measured at 50 ft (15.2 m). The reader may recall that a significant reduction of noise is attainable in the national community with the use of trucks rated at 83 dBA (Figure 10-3). The impact of trucks controlled to that level is more significant to the spectator when one realizes that, in the recent past, uncontrolled trucks in commercial use could generate sound levels in excess of 93 dBA at 50 ft (15.2 m) under low-speed, full-power conditions.

Noise Test Standards for Trucks

The Society of Automotive Engineers has developed a method to rate the sound level of trucks; it is described in the SAE Handbook as Test Standard J366b.* The truck is accelerated at full power along a line of travel 50 ft (15.2 m) from the specified microphone and sound level measuring equipment. It must reach full-power-rated engine speed in a specified distance from the microphone. The truck noise is so measured from both sides of the vehicle, and the higher value becomes the sound-level performance number of that vehicle, expressed in decibels on an A-weighted scale (dBA). This test method is appropriate for trucks because truck engines are ordinarily utilized in that manner on urban streets and roads. In other words, the test level is a reasonable good measure of the sound level that a spectator might commonly experience in real-world conditions.

Noise Control of Existing Truck Fleet

As previously discussed, it is recognized that one of the most effective measures that can be taken to reduce urban highway and street noise is to control the existing fleet of past model trucks, assuming that new trucks will be manu-

*SAE Handbook, Part 2, 1976.

factured with appropriate noise control features. Accordingly, the EPA has published a regulation, effective October 1975, under the authority of the Noise Control Act of 1972, which sets a maximum passby sound level of 86 dBA at 50 ft (15.2 m) for vehicles over 10,000 lbs (4536 kg) operating in interstate commerce at speeds of less than 35 mph (56 km/hr). A level of 90 dBA measured at 50 ft (15.2 m) has been established as a maximum sound level for these vehicles when operated at speeds in excess of 35 mph (56 km/hr). The regulation is written to control the operational characteristics of trucks in service and not the manufacture of new trucks. Thus, it is the responsibility of truck owners and operators to make necessary changes to their vehicles in order to comply with the regulation. Recognizing that truck owners do not often have engineering facilities to determine the degree of retrofit necessary, the Department of Transportation has funded several studies to determine and demonstrate the type of new hardware that may be required to meet the requirement of the EPA rule. Out of this has grown interest on the part of truck manufacturers to establish centers in the field at which trained technicians and mechanics will perform truck-retrofit services to reduce engine-related noise. Some manufacturers have published standardized parts lists for retrofit of their various model categories.[13] Of course, truckers must institute periodic maintenance of noise-control systems to assure effective community noise reduction, and the federal rule must be actively enforced to realize appreciable lessening of traffic noise. In addition, the states may enforce their own noise laws in order to attain local improvement, as long as such laws are not preempted by federal law.

The Noise Control Act has separate provision for the regulation of new-product noise. These provisions are also preemptive over local law in order to avoid the counterproductivity of nonuniform rules operative upon truck manufacturers who distribute their products nationally.

Relating Light-Vehicle Sound Level Ratings to Actual Operation Levels

Engine-related noise of passenger cars and light trucks must be understood in an entirely different context than truck-engine noise. For this discussion, light trucks are defined as transport vehicles with GVWR of less than 10,000 lbs (4536 kg). These light trucks and cars are in general much lighter in proportion to their engine-power ratings than are medium and heavy trucks. The light vehicles are capable of being accelerated in traffic with the use of proportionately less power, and thus they generate much less noise.

General Motors engineers have performed an extensive study of the modes in which light vehicles are operated by conducting a survey in 12 cities in the United States.[14] An extensive analysis of the measured data is given in their report, but it is worth noting here some of the conclusions about the percentage

TABLE 10-4 Light-vehicle operating mode summary. (From Gray, Richard F. General Motors Proving Ground, Engineering Publication 6313. A Survey of Light Vehicle Operation. July 1975, p. 8, Table IV.)

MODE	TIME (min)	% OF TOTAL	MILES	% OF TOTAL
Non-Highway Driving:				
Idling	1376.98	13.54	0.0	0.0
Cruising*	5604.64	55.12	2803.76	74.32
Accelerating	1564.30	15.38	506.31	13.42
Decelerating	1621.88	15.96	462.50	12.26
All Driving:				
Idling	1411.14	10.54	0.0	0.0
Cruising*	8466.84	63.24	5248.39	81.96
Accelerating	1715.50	12.81	595.83	9.30
Decelerating	1794.52	13.40	559.31	8.74

**ABSOLUTE VALUE OF ACCELERATING IS LESS THAN 0.05 g.*

of time spent accelerating and the acceleration rates, so that we may relate these to noise output of light vehicles. Notice in Table 10-4 that during nonhighway driving (that is, along urban streets and arterials), light vehicles are accelerating about 15% of the time. In Table 10-5, we can see that the time spent accelerating light vehicles in excess of 0.25 g is extremely small, certainly less than 0.5% of the time.

In a separate investigation, GM engineers have established that the first row of light vehicles accelerating from stoplight situations in fast-moving traffic

TABLE 10-5 Vehicle driving time spent in acceleration bands. (From Gray, Richard, F. General Motors Proving Ground, Engineering Publication 6313. A Survey of Light Vehicle Operation. July 1975. p. 8, Table VII.)

	NON-HIGHWAY DRIVING		ALL DRIVING	
ACCELERATION (g)	TIME (min)	% OF TOTAL	TIME (min)	% OF TOTAL
0.05				
0.10	1027.13	65.66	1148.64	66.96
0.15	359.32	22.97	378.49	22.06
0.20	146.88	9.39	155.33	9.05
0.25	28.23	1.80	29.63	1.73
0.30	2.40	0.15	2.77	0.16
0.35	0.20	0.01	0.43	0.03
0.40	0.02	0.00	0.03	0.00
0.45	0.02	0.00	0.02	0.00
0.50	0.03	0.00	0.07	0.00
0.55	0.07	0.00	0.08	0.00
TOTAL	1564.30	100.00	1715.50	100.00

consistently move away so that they travel the first 100 ft (30.4 m) in 5 seconds. While the actual acceleration rate in that situation is not uniform, it is equivalent to a calculated average rate of 0.25 g. From these data, we may predict noise levels of accelerating vehicles in traffic by measuring the sound level of light vehicles during the performance of such controlled "urban accelerations" on the proving ground strip.

A sample of 1975 production light vehicles were tested to determine (1) the average wide-open-throttle sound levels within the sample, (2) the "urban accelerations" sound levels using the average acceleration rate criterion of 0.25 g, and (3) the 35-mph (56 km/hr) cruise-by sound levels of the sample. Figure 10-4 illustrates the results. It is apparent that these production-run vehicles typically operate in traffic in the sound level range of up to about 70 dBA. Somewhat contrary to these facts, the industry has historically adopted the Society of Automotive Engineers Test Standard J986a to rate the exterior sound level of light vehicles. This test method requires that the sound-rating number (dBA) be assigned on the basis of the sound level resulting from a controlled wide-open-throttle acceleration. While this is an acceptable engineering test from the standpoint of repeatability, it surely does not give an indication of the real character of the sound level produced by light vehicles in urban traffic.

Referring to Table 10-5 again, wide-open-throttle acceleration or any ac-

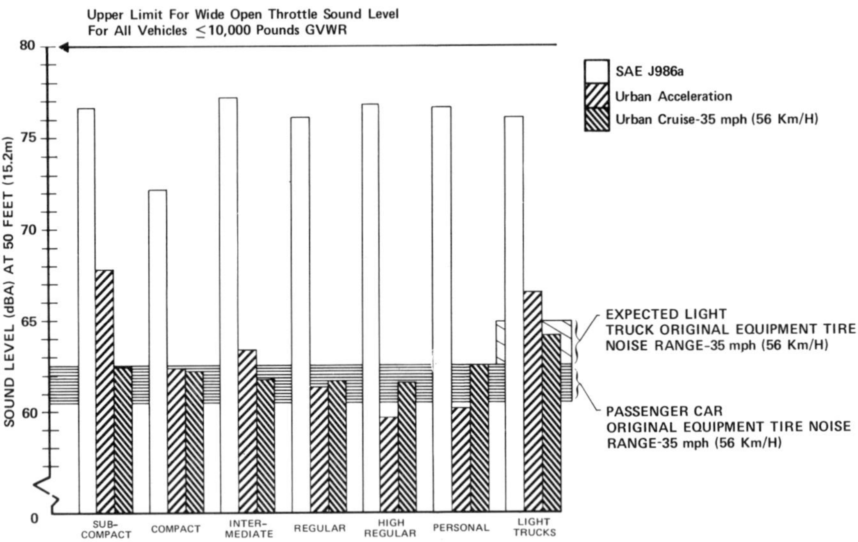

Fig. 10-4 Average exterior sound levels for 44 General Motors production vehicles (1975/76) under various operating modes. (From General Motors Corporation, Environmental Activities Staff, 1975.)

celeration greater than 0.25 g is used less than 0.5% of operating time. Wide-open–throttle acceleration rates can reach values ranging from 0.3 g to about 0.4 g, depending upon the engine power rating of individual models. One may note that this is quite the opposite of the situation with medium and heavy trucks, in which the SAE Test Standard J366b is quite well correlated with the sound level of the truck in traffic. Finally, the foregoing discussion of light-vehicle noise is confined to low-speed conditions. Sound level characteristics of light vehicles at speeds above 35 mph (56 km/hr) are essentially a function of tire noise (see p. 268).

Referring to Fig. 10-4, the sound levels given for wide-open–throttle acceleration (SAE J986a) are conditioned by the term *average*. The reader must not construe the values shown as absolute sound levels for a given car or class because considerable variability exists, even between automobiles manufactured to a set of identical specifications. This is true of light, medium, and heavy trucks as well. Figure 10-5 illustrates the result of noise-testing 277 nominally identical, new

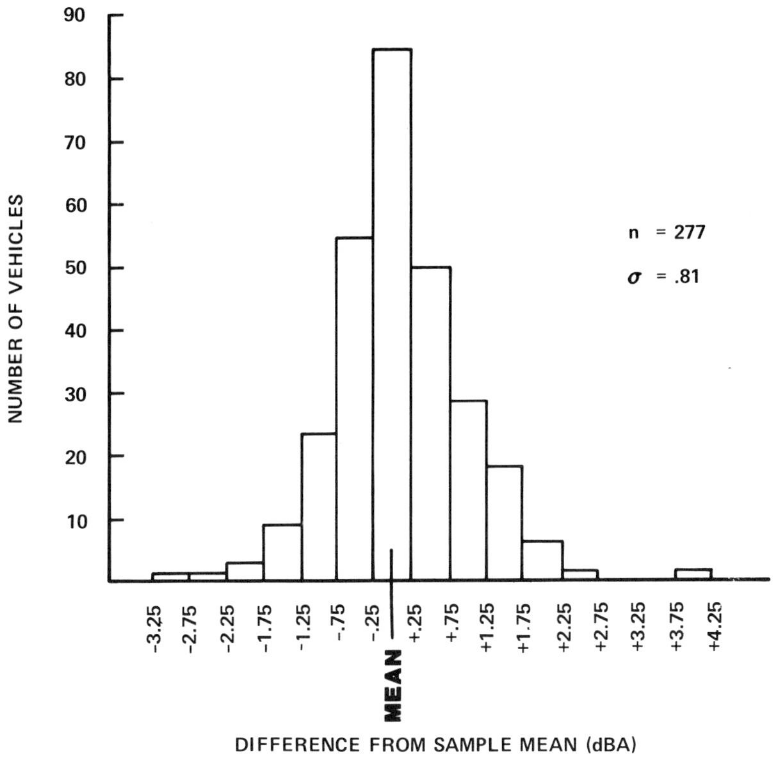

Fig. 10-5 Histogram of sound level for nominally identical vehicles–SAE J986 Test. (From General Motors Corporation, Environmental Activities Staff, 1973.)

production automobiles. If the single highest value is thrown out (defective muffler), the range of variability is 5.5 dB. The standard deviation is 0.81 dB. This means that the designer/manufacturer must target the sound level design point some 2 to 3 dB below the sound level specification if it is required that production vehicles must not exceed the specification. This explains why, in Fig. 10-4, the averages of the wide-open–throttle sound levels fall significantly below the upper limit specification of 80 dBA. Any lowering of such a specification implies that a redesign of the vehicle must take place to maintain the mean sound level some 2 to 3 dB below the new specification.

The work of conscientious vehicle noise control engineers is sometimes negated when owners modify the exhaust system of light vehicles in a manner that produces more noise. Noise control strategies that depend solely upon legislative action directed toward the designer-manufacturer of vehicles are inadequate unless they also provide for correcting the noise problem resulting from exhaust-system modification by owners or failure of owners to maintain vehicles. Various agencies have devised quantitative methods to determine the acceptability of individual vehicles on the street. A test method that may be used for this purpose is performed on the stationary vehicle and requires a noise measurement to be made with the sound level meter placed 0.5 m from the exhaust outlet, 45° from the axis of the exhaust-tailpipe outlet.[15] The procedure generally requires the engine to be operated at three-quarters rated speed with the transmission in neutral. A "pass-fail" noise level is adopted, which is related to subjective comparisons with vehicles known to have original equipment exhaust systems in good repair. A screening test such as this is simple, quick, and sufficiently quantitative that it may be used successfully for enforcement purposes.

Noise Control Hardware Development for Automobiles and Light Trucks

The components that are main contributors to the low-speed full-power noise level of light vehicles are the engine-cooling fan, engine exhaust, engine structure vibration, and the combustion air-intake system. High-speed noise is largely a function of tire noise. Cooling-system fan noise is controlled by refinement of the aerodynamic design of the cooling-system components in the manner previously described for trucks. If the vehicle is air-conditioned, recent practice is to equip the car or truck with a viscous drive fan clutch. The unit acts as a fluid coupling under high-temperature conditions, driving the fan to cool the engine. At reduced ambient temperatures, the fluid is internally diverted, and the coupling "declutches," no longer driving the fan. Furthermore, this type of viscous drive is usually designed as a torque limited coupling. This means that even when the fan is thermally engaged, its driven speed is limited to a value less than the input speed of the driven pulley. Since low-vehicle-speed cooling of the engine requires greater fan-generated airflow volume than high-speed cooling, then considering the availability of supplemental ram air, the speed-limiting feature of the viscous fan drive is attractive in preventing excessive fan speed, and thus noise. Other devices such as "flex" fans have been developed, which are molded from synthetic material that allows the blades to flatten out at high speeds under air load. This provides partial control of fan noise at lower cost but is not as effective from the standpoint of noise control as the viscous drive.

The same general technology used in truck exhaust systems is applicable to light vehicles, except that space is even less available. Ground clearance beneath the automobile is critical, and becomes a constraint upon the use of large exhaust plenums. Dual exhaust systems constitute a design option, but the cost-conscious engineer will think twice before resorting to such a solution. A more stringent noise specification on production cars has required increases in exhaust-pipe diameter to reduce high-velocity viscous-flow noise of expanding gases exiting from the tailpipe. Larger-diameter pipes, on the other hand, tend to reduce the attenuation capability of mufflers since the effectiveness of attenuating chambers increases with the area ratio of chamber to pipe. Depending upon the ability of the engine-compartment sheet metal to act as a noise barrier for engine-block and air-intake noise, the engineer will install supplementary noise-damping material on the hood and walls.

Under the full-throttle-acceleration SAE J986a Test Standard, for 1975 and newer model automobiles and light trucks, an exterior noise specification of 80 dBA is not uncommon. As shown in Fig. 10-4, this sound-level specification results in urban acceleration and cruise sound levels in the range of up to about 70 dBA. To reduce the engine-related sound level of automobiles and light trucks below those levels would appear to require the following:

1. exhaust systems with larger mufflers on virtually all vehicles
2. some provision for shielding exhaust systems to reduce radiated noise for most vehicles
3. fan noise reduction by means of a clutch-controlled fan
4. additional air injection reaction pump silencing on vehicles with such air pumps.

Also, some vehicles will require further air-intake noise silencing. A number of other vehicles may require considerably more control for engine noise or other mechanical noise sources. The real point of interest relative to further reduction of engine-related noise is delineated in Fig. 10-4; virtually no benefit exists for such reduction in the urban speed range because tire noise becomes the dominant component source. Of course, this is even more apparent at speeds greater than 35 mph (56 km/hr), since tire noise becomes increasingly dominant compared to engine-related noise.

Automotive Tire Noise

Much of the traffic noise radiated from moving vehicles is generated by the interaction of the tires with the roadway. The aspect of tires that is most important as it pertains to noise generation is the tread pattern. The most recognized test standard for tire noise has been developed by the Society of Automotive Engineers to rate highway-truck-tire noise. It is published as "SAE Recommended Practice; Sound Level of Highway Truck Tires, J57."

In a report of research to determine the mechanisms of tire noise, Richards postulates as contributors the following: tread vibration, tread squirm, carcass vibration, air pumping, and aerodynamic noise.[16] Richards believes that tread vibration is the primary contributor to tire noise. If a body vibrates in air, sound will be generated depending upon vibrational amplitude and time pattern. In the case of tires, the vibration can be a function of either tread pattern or road-surface texture; in actual practice, it is a composite of both.

Tire noise increases as the vehicle is operated at successively higher speeds. There appears to exist a relationship such that tire noise varies with the fourth power of vehicle velocity. However, the frequency spectrum (Fig. 10-6) of tire noise does not vary much with changes in vehicle velocity.[16] Figure 10-7 illus-

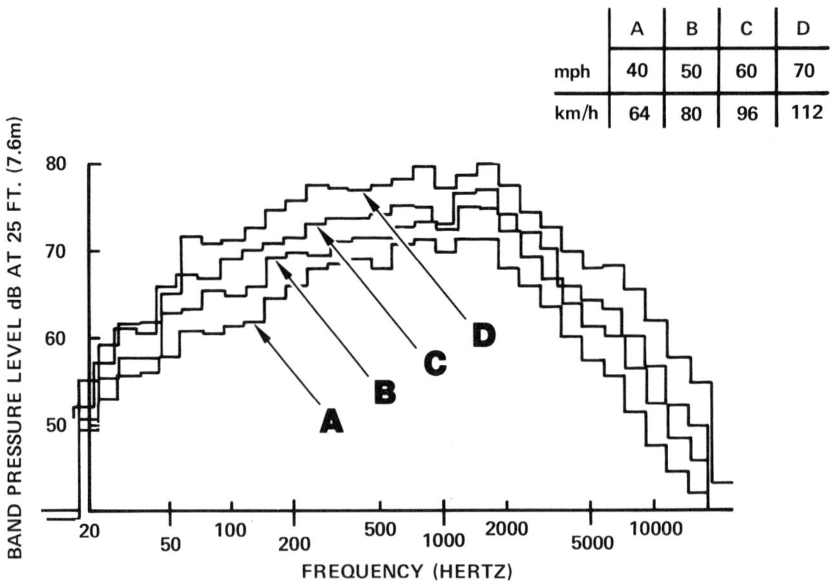

Fig. 10-6 Passby spectra of steel-belted radial tires. (From Richards, M. G. General Motors Proving Ground, Engineering Publication A-2973. Automotive Tire Noise—A Comprehensive Experimental Study. September 1973, p. 3, Fig. 4.)

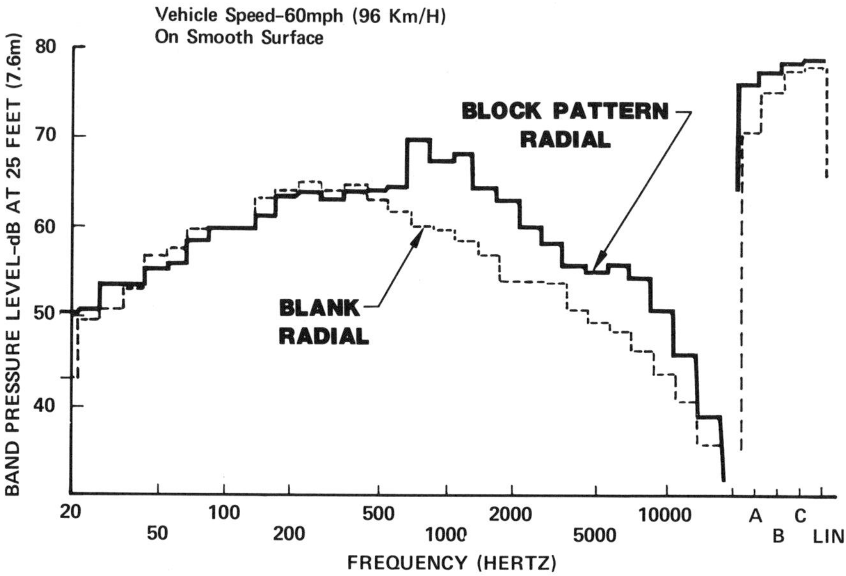

Fig. 10-7 Effect of tread pattern on passby spectra. (From Richards, M. G. General Motors Proving Ground, Engineering Publication A-2973. Automotive Tire Noise—A Comprehensive Experimental Study. September 1973, p. 5, Fig. 7.)

trates a comparison of a nominally aggressive block design pattern with a patternless blank tire. In general, a more aggressive tread design, such as a cross-lug type, is noisier than a rib-type tread design tire. Figure 10-8 shows the noise levels of various combinations and conditions of tires during operation of a loaded five-axle truck. Tire noise due to air-pumping results from the use of tires with tread designs that form unvented pockets. Ordinarily, commercially available new tires are not manufactured with such tread pockets. Thus, the condition can be avoided if all tiremakers and retreaders use vented tread designs.

In practical terms, automobile tires can be built with tread designs that balance the desirable aspects of safety, traction ability, durability, ride, and handling, with sound-level characteristics that are only 2 to 3 dB above the patternless or smooth tire. Figure 10-9 illustrates the comparison.[17] It is significant that tires that represent the noise floor of technology are not prototypes in the laboratory stage, but are commercially available. At this writing, the laboratory seems to have nothing further to offer relative to tire noise reduction. The future may hold improvement, but the shape and structure of tires is to a significant extent dictated by their functional requirements and, contrarily, further reduction in tire noise is probably dependent upon basic changes in form and structure.

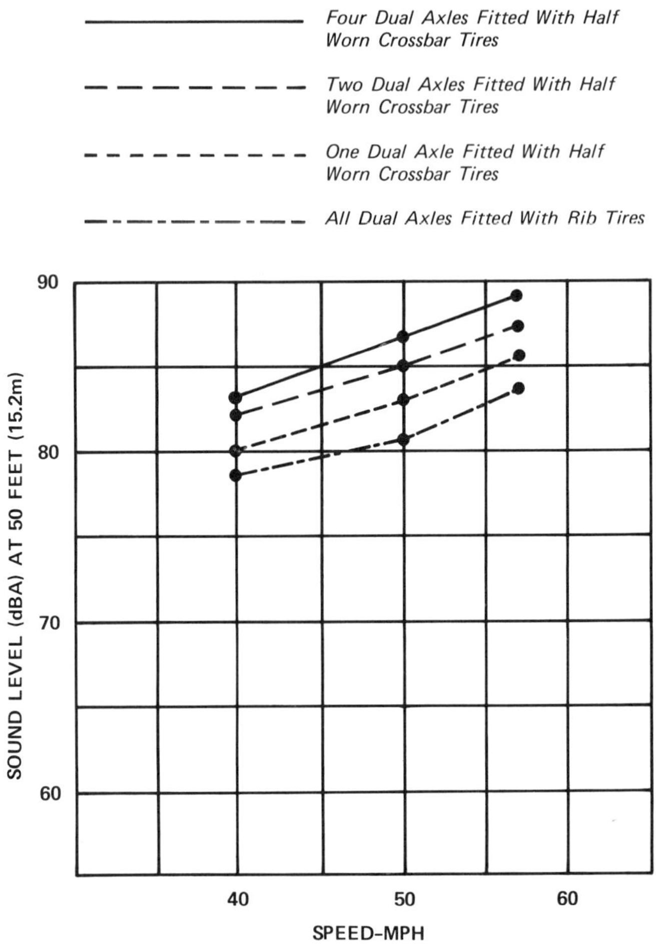

Fig. 10-8 Truck tire noise as a function of vehicle speed for a five-axle truck-trailer combination. (From Tetlow, D. Passenger car and truck tire noise. Proceedings: Conference on Motor Vehicle Noise. April 1973, p. 46, Fig. 24.)

RAILROAD NOISE

Community noise resulting from the operation of rail-transportation systems differs from highway and urban-street traffic noise, because passage of a train is a timed event with a discrete beginning and an endpoint whereas urban streets generally constitute a more or less continuing source of noise. Furthermore, urban and suburban railroad lines are far less in number than arterial streets and thus affect far fewer numbers of people. The obvious fact is that rail noise is an

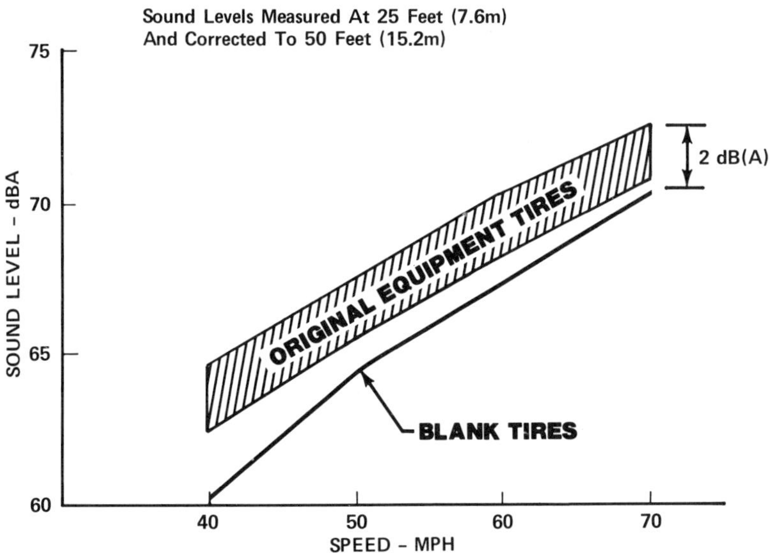

Fig. 10-9 Comparison of original equipment tires with a blank tire. (From Tetlow, D. Passenger car and truck tire noise. Proceedings: Conference on Motor Vehicle Noise. April 1973, p. 40, Fig. 4.)

annoyance to a given observer as an incident event; when the train has passed the point, the ambient sound level is restored. By itself, that fact does not ameliorate the need to consider control of rail-system noise, but it does impose a different psychological response, and thus quite possibly a different criterion of resolution.

In the following sections, we discuss two aspects of rail-system noise: the sound resulting from operation of the power plant of the diesel electric locomotive and the sound of car wheels rolling on steel rails.

Locomotive Noise

Prime mover power for line-haul railroad locomotives in the United States is, in most cases, generated by diesel engines. Such engines drive on-board electrical generator systems, which furnish electrical energy to drive-wheel motors. The sound resulting from operation of the power plant system of a diesel-electric locomotive is due largely to sound pressure waves emanating from the engine exhaust system, the cooling fans, engine structure vibration, and car body vibration.

Traditionally, exhaust-system noise has been attenuated to some degree by the incidental use of exhaust-driven turbochargers, and the effects of exhaust manifold spark arresters. Developmental work on the critical aspects of space engineering of locomotive components has resulted in the capability of the designer/manufacturer to incorporate exhaust muffling systems into the design of new-production locomotives. This is important because the full power noise of the individual locomotive representative of the existing fleet is controlled by the exhaust-noise level, when measured at 100 ft (30.4 m). Figure 10-10 illustrates the dominance of the exhaust noise by comparing the A-weighted octave-band sound levels measured at 3 ft (1m) from the exhaust outlet and at 100 ft (30.4 m) from the side of the locomotive during full-power operation. The cooling fans are not operating in order to eliminate their influence in this particular exhaust-system noise study. Good correlation is shown to exist between the

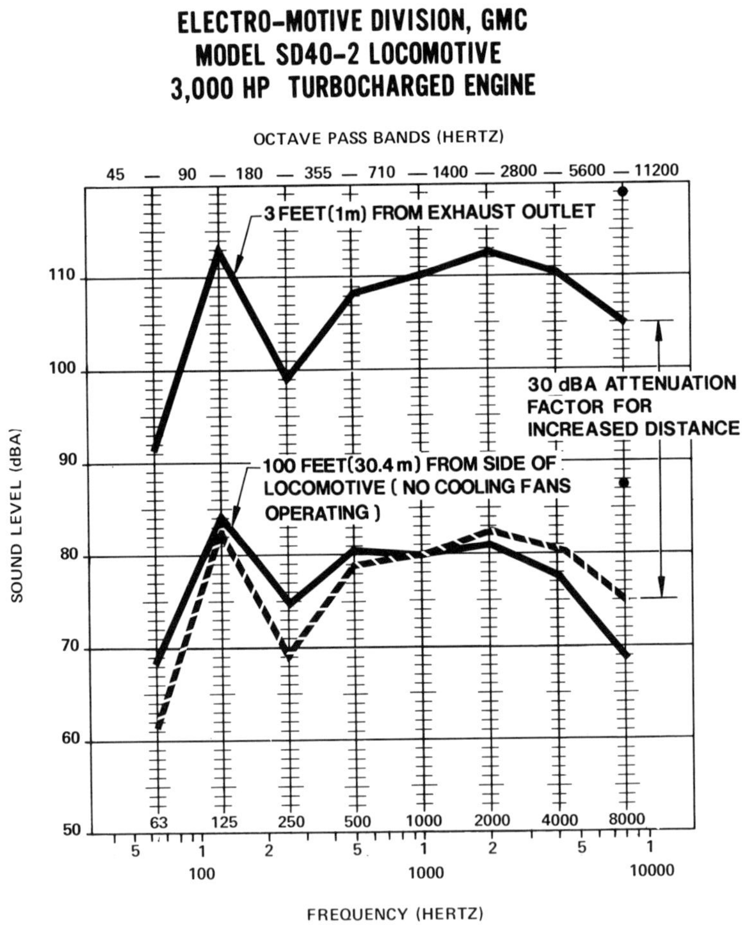

Fig. 10-10 Locomotive exhaust noise during eighth throttle, full load operation. (Courtesy of Electro-Motive Division, General Motors Corporation.)

full-power exhaust-noise spectrum at 3 ft (1 m) and the overall locomotive noise spectrum measured at 100 ft (30.4 m), when a 30-dB attenuation factor for hemispherical sound spreading is used to correct for the increased distance. At several points in the spectrum, the measured octave-band level at 100 ft (30.4 m) is somewhat less than that predicted using the 30-dB attenuation factor. This indicates some unaccounted for attenuation. But, on the other hand, wherever the measured points exceed the predicted level, it is believed that structurally radiated car-body sound is contributing to the overall locomotive noise. At idle, this latter condition seems even more pronounced (Fig. 10-11), demonstrating that under the idling condition, exhaust noise is not the major con-

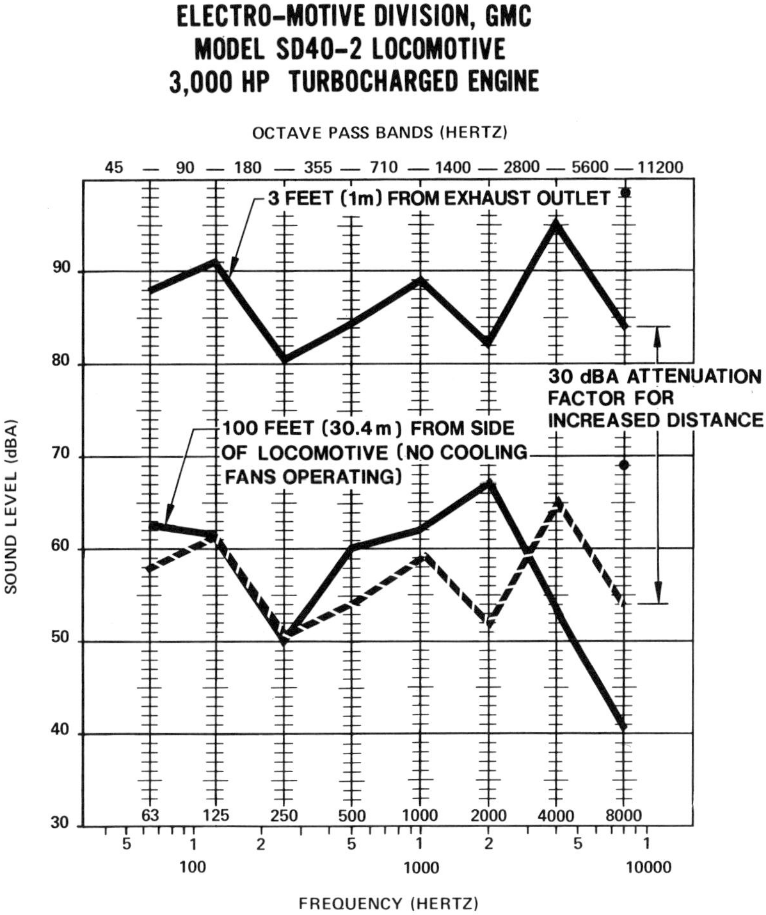

Fig. 10-11 Locomotive exhaust noise during idle operation. (Courtesy of Electro-Motive Division, General Motors Corporation.)

tributor. The significant point here is that at full power, exhaust noise muffling can significantly reduce overall locomotive noise; however, the effect of muffling on idle noise is less. Thus, noise from the idling locomotive does not yield much to the application of the exhaust muffler.

Obviously then, locomotive noise at the higher power settings could be reduced by the application of mufflers to existing locomotives that are in service. However, it must be understood that to redesign existing configurations so that muffling systems can be applied is indeed a complex matter. The power plant section of a locomotive is a space-intensive system of components. To undertake such redesign and rebuild of the existing fleet on a retrofit basis has implications of such magnitude that national railroad economics would be significantly affected. The situation is obviously much different when future new-production locomotive noise is considered, inasmuch as the space allocation for muffled exhaust systems can accommodate new designs in an orderly and logical manner.

Test procedures used to determine locomotive-engine-related noise ratings recognize that reproducible measurements can be made by channeling the electric-power output of the system to a trackside load-cell. This allows test personnel to conduct the tests with a stationary locomotive and provides engine operation and sound-measuring conditions that can be well controlled for all engine power settings of interest.

Developmental work for later designs of locomotives indicates that power-plant-related noise control can be applied that can limit the sound level to 87 dBA at any throttle setting, when measured at 100 ft (30.4 m) from the locomotive, using the stationary test procedure described.

Wheel/Rail Noise

In general, the standard roadbed configuration in the United States uses tie-and-ballast construction. The ballast is usually a crushed rock aggregate placed upon appropriately drained and graded earth. The ties are laid at calculated intervals in the ballast, and the rails are fastened to the ties. In most cases, the ties are cut from wood; ties formed as concrete castings have been used upon occasion. The primary function of the tie is to distribute load imposed upon the immediate section of rail that it supports.

Because tie-and-ballast construction appears to offer better sound attenuation than elevated structures or other nonballasted construction, the standard for comparing wheel/rail noise performance includes tie-and-ballast construction as a condition.[18] For measuring comparative levels of wheel/rail noise of differently engineered configurations, a distance of 50 ft (15.2 m) is commonly used. Obviously, this is not necessarily a standard distance to evaluate noise-exposure conditions for people situated along the right of way; each of those situations must be evaluated in terms of the real proximity of the trackage to the adjacent population.

Wheel/rail noise is generated as the product of the nonuniformities that exist at the contact point of the rolling wheels with the rail. These nonuniformities are (1) rail roughness, (2) flat spots on the periphery of the car wheels, and (3) gaps at the rail joints. In general, these conditions account for the noise of rail cars moving along straight sections of track. When the radius of a curved section of track is small, wheel/rail noise increases markedly, as much as 15 dBA.[18] This proves not to be a major problem on most surface trackage because curve radii are quite large and thus wheel squeal and curve howling are not so likely. Noise reductions, which may be attained when taking noise-control measures related to wheel/rail noise, will be as the result of the following action:

1. elimination of rail joint gaps by "continuous" welding of the rail joints

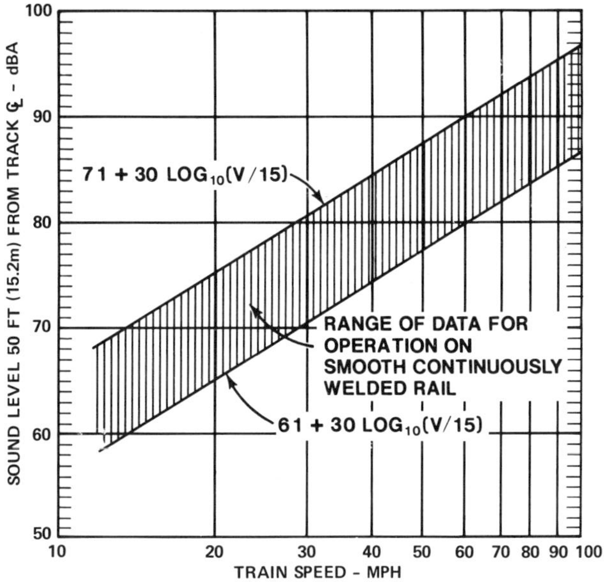

CORRECTION FACTORS*	
JOINTED RAIL	8 TO 10 dBA
WHEEL FLATS	8 TO 10 dBA
NEW OR ROUGH RAILS	3 TO 6 dBA
ROUGH WHEELS	3 TO 6 dBA
CORRUGATIONS	UP TO 15 dBA

**ADD ONLY ONE CORRECTION FACTOR - THE LARGEST APPLICABLE*

Fig. 10-12 Wayside noise for at-grade operations on tie and ballast track. (From Cann, R. G., Fredberg, J. J., and Manning, J. E. Cambridge Collaborative, Inc. Prediction and Control of Rail Transit Noise and Vibration—A State of the Art Assessment. Prepared under Contract to U.S. Dept. of Transportation, Contract No. DOT-TSC-643, April 1974, p. 39, Fig. 3.1.)

2. grinding wheels to eliminate flat spots and to restore relatively smooth contact surface
3. grinding rails to provide smoother, flatter surfaces (rails tend to wear in a nonuniform manner, which degrades quality of flatness)
4. car modifications, such as vibration damping devices in the interfaces of the wheel-to-truck-to-car suspension system.

Data gathered from rail systems in the United States, Europe, and Japan have been compared in order to establish some common band of wheel/rail noise performance.[18] These data are presented in Fig. 10-12 as a chart that shows a band of sound level plotted against train speed. For predictive purposes, correction factors are given for the various noise-generating conditions of wheels and rails. In general, the lower limit of the band represents a somewhat idealized noise and vibration controlled condition. If any of the conditions that are shown in the correction factor list exist, appropriate noise corrections must be added. Figure 10-13 illustrates data that describe the attenuation of wheel/rail noise as a function of distance of the observer from the railroad tracks.

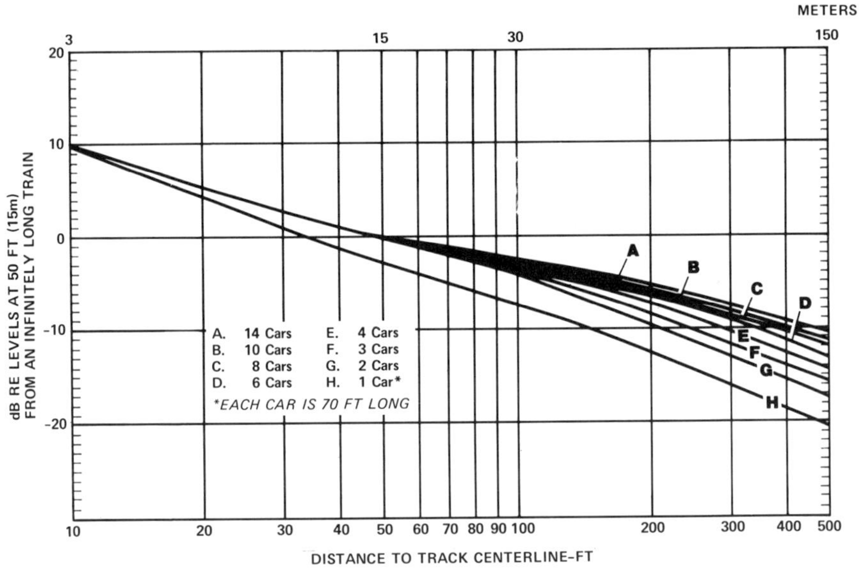

Fig. 10-13 Decrease in sound level with distance from the track. (From Cann, R. G., Fredberg, J. J., and Manning, J. E. Cambridge Collaborative, Inc. Prediction and Control of Rail Transit Noise and Vibration—A State of the Art Assessment, Prepared under Contract to U.S. Dept. of Transportation, Contract No. DOT-TSC-643, April 1974, p. 40, Fig. 3.2.)

A LOOK TO THE FUTURE

In a technically dynamic society, prophecies for the future are often little more than guesses. But relative to highway and rail traffic noise, we can say, at this writing, that significant improvements will be made in the future if noise control technology that already exists is properly applied in the mass production of land transportation vehicles. More community control will be exerted against the practice of automotive exhaust-system modification and the neglect of in-service maintenance of those systems. Finally, in a more technically sophisticated society, the public will become ever more sensitive to the fine balance between environmental improvement and the cost of the controls that will be required to attain the sought-after benefits. We will discover in the field of environmental improvement that it is not always necessary to plumb the technology to its full depth, with the attendant high costs to society, to attain real and satisfying gains. Finding the point of maximization of environmental protection for the least cost must become a science unto itself.

REFERENCES

1. Eldred, K. McK., Galloway, W. J., and Simpson, M. A., *Population Distribution of the United States as a Function of Outdoor Noise Level*, Prepared for U.S. Environmental Protection Agency, Report No. 550/9-74-009, June 1974.

2. *Report on Aircraft/Airport Noise*, Report of the Administrator of EPA to the Committee on Public Works, U.S. Senate, July 1973.

3. *Motor Vehicle Noise Identification and Analysis of Situations Contributing to Annoyance*, BBN Report 2082, Prepared for Motor Vehicle Manufacturers Association Inc., June 1971.

4. Bolt, Beranek and Newman, Inc., *A Study of Annoyance from Motor Vehicle Noise*, BBN Report 2112, Prepared for Motor Vehicle Manufacturers Association Inc., June 1971.

5. Galloway, W. J., Gordon, C. G., Kugler, B. A., and Nelson, D. L., *Highway Noise: A Design Guide for Highway Engineers*, National Cooperative Highway Research Program Report 117, 1971.

6. Kugler, A. B., and Piersol, A. G., *Highway Noise, A Field Evaluation of Traffic Noise Reduction Measures*, National Cooperative Highway Research Program Report 144, 1973.

7. Hajek, J. J., and Harmelink, M. D., *Noise Barrier Evaluation and Alternatives for Highway Noise Control*, Report RR 180, Ministry of Transportation and Communications, Ontario, Canada, September 1972.

8. Bouchard, T. J., Jr., and Lambert, R. F., *Experimental Evaluation of a Freeway Noise Barrier*, Minnesota Department of Highways, Report MN HW 5-180 74-6, January 1974.

9. Rackl, R., Sutherland, L., and Swing, J., *Community Noise Countermeasures Cost-Effectiveness Analysis*, Wyle Research Report 75-2, July 1975.

10. Bass, R. W., Orlosky, S. E., Pataky, P. P., and Whitney, D. R., *A Simplified Mathematic Modeling Technique for Predicting Equivalent Sound Levels Due to Traffic Noise*, General Motors Corporation, Environmental Activities Staff, Publication No. 6172, 1975.

11. General Motors Corporation, Response to EPA Proposal for Noise Emission Standards Medium and Heavy Trucks, USG-350/74-22, January 15, 1975.

12. International Harvester Letter, dated November 15, 1974 to Office of Noise Abatement, "Data Summary of Fan Clutch Field Test: Department of Transportation Quiet Truck Program with International Harvester."

13. Bullard, O. J., and Shaffer, G. M., *The Determination of Practical Noise Control Retrofitting of Pre-1970 Truck and Coach Models*, Transportation Systems Center, U.S. Department of Transportation, Report No. DOT-TSC-OST-75, October 1975.

14. Gray, Richard F., *A Survey of Light Vehicle Operations*, General Motors Proving Ground Engineering Publication No. 6313, July, 1975.

15. *Exhaust System Noise Measurement*, Art. 9, Title 13, California Administrative Code, November 14, 1975.

16. Richards, M. G., *Automotive Tire Noise–A Comprehensive Experimental Study*, General Motors Proving Ground Engineering Publication No. A-2973, September 1973.

17. *Proceedings: Conference on Motor Vehicle Noise*, General Motors Corporation, Environmental Activities Staff, April 1973.

18. Cann, R. G., Fredberg, J. J., and Manning, J. E., *Prediction and Control of Rail Transit Noise and Vibration–A State-of-The-Art Assessment*, Department of Transportation, Report No. PB 233 363, April 1974.

11

Preparation of noise control legislation

Janet F. Pawlak*

Environment Officer
International Council for the Exploration of the Sea
Charlottenlund, Denmark

COMPONENTS OF COMMUNITY NOISE CONTROL LEGISLATION

The recommended type of community noise control legislation is an ordinance that contains performance standards, that is, it specifies maximum noise emission levels for the operation of certain types of equipment and/or the conduct of specified activities. A performance standard community noise control ordinance is intended to replace the older nuisance provisions of community law, which prohibited "loud, unnecessary, or unusual noise."† Such "unnecessary noise" laws were not only difficult to enforce, because they required the

*At the time this chapter was prepared, the author, an attorney, was Noise Project Leader for the New Sources Branch of the Mobile Source Enforcement Division in the Office of the Assistant Administrator for Enforcement, Environmental Protection Agency, Washington, D.C.

†For a compilation of these laws, see *Laws and Regulatory Schemes for Noise Abatement*, pp. 1-99 to 1-104 (U.S. Environmental Protection Agency NTID 300.4, 1971).

enforcement officer to make a subjective judgment as to whether the sound in question was indeed "loud, unnecessary, or unusual," but also difficult to prosecute to an abatement of the noise, because the defendant could often prove that the noise was either not "unusual" or "unnecessary."

Some elements of the older, subjective noise ordinances can, and perhaps should, remain in a performance standard ordinance because the health and welfare impact of noise on people has many subjective elements. These subjective provisions should be written in such a way as to provide clear criteria for their application. Examples of such criteria will be provided later in this chapter.

A performance standard noise control ordinance may contain any or all of the following types of provisions, depending on the needs and goals of the particular community.

Maximum Noise Emission Levels for Equipment

1. *Use and operation.* The ordinance may specify the maximum sound levels allowed during the use or operation of the regulated equipment within the community. This provision is enforced against the owner or operator of the equipment. Types of equipment on which these standards may be used include motor vehicles, construction equipment, and refuse collection vehicles.

2. *License requirements.* The ordinance may specify the maximum sound level allowable for a product to be registered for use or operation within a community. This provision is enforced against the owner or operator of the equipment. Types of equipment for which these standards may be used include automobiles, motorcycles, construction equipment, and refuse collection vehicles.

3. *Sale.* The ordinance mav specify the maximum sound level for a product to allow that product to be sold or offered for sale in the community. This provision is enforced against the manufacturer or dealer of the product. Types of products for which the standards may be used include automobiles, motorcycles, power tools, and household appliances.

Operational and Zone Restrictions

1. *Curfews.* The ordinance may specify nighttime curfews for the operation of certain types of equipment or the conduct of certain types of activities. Such provisions are enforced against the owner or operator of the equipment or the person conducting the activity. Curfews are often used jointly with zone restrictions. Types of equipment or activities for which curfews may be used include domestic power tools, loudspeakers or public address systems, and construction.

2. *Zone restrictions.* The community may specify limitations on the conduct of noise-producing activities or the use of noisy equipment in those areas of the community that are exceptionally sensitive to noise, such as residential areas and

hospital, school, or library zones. Such provisions are enforced against the person conducting the activities or the owner or operator of the equipment. Zone restrictions are often used jointly with curfews.

Property Line or Receiving Land Use Sound Levels

A community may wish to retain certain subjective provisions from its nuisance from one land use and received by another. This type of provision will specify different maximum receiving sound levels depending on the sensitivity of the receptor to noise. Land uses are usually categorized into industrial, commercial, and residential, with residential uses receiving the greatest amount of protection, including lower nighttime maximum levels.

Land Use Planning Requirements

To prevent future noise problems from occurring, a community may specify ambient sound level guidelines for the planning of land uses and transportation corridors within the community. Planning requirements may include a specification of the maximum allowable exterior sound level on sites where residential, institutional, or recreational uses are planned.

Nuisance Law Provisions

A community may wish to retain certain subjective provisions from its nuisance noise law so that sounds that may be difficult to measure accurately using a sound level meter may still be covered by the ordinance. Although still retaining a subjective character, the criteria for determining when a particular "nuisance noise" violates the community ordinance should be as objective as possible. Nuisance noise provisions may be applied to such sources as barking dogs, televisions and radios, loading and unloading operations, vehicle repair activities, and vibrations.

For three of the above-mentioned types of ordinance provisions (sound emission levels for equipment, receiving land use sound levels, and land use planning requirements), the use of sound level measuring equipment is necessary to determine whether or not there is compliance with the ordinance. Therefore, in order to have a complete program, a community must develop an enforcement manual containing the appropriate measurement methodology to enforce each performance standard provision of the ordinance. Although this enforcement manual is not a law or regulation in itself, because of its close association with the law or ordinance it should be developed and reviewed concurrently with the ordinance. It also should be made available to the public in a manner equal to the availability of the noise control ordinance. However, unlike the noise control ordinance, revisions or amendments to the enforcement manual need not be submitted to the city council for enactment.

LEGAL CONSIDERATIONS

Public Health and Welfare

The legal authority of a community to control noise (as well as numerous other community problems) derives from the community's police power, which is granted by the state in which the community is located. As used in this chapter, the term *community* can mean any legally recognized political subdivision of a state, such as a county, municipality, township, or village. The total scope of authority granted by a state to its various types of political subdivisions varies from state to state. Also, the means by which such authority is granted differs, depending on the particular state constitution. Before developing a noise control ordinance, a community should make itself aware of its precise scope of authority in this area.

In general, the police power of a community or state is the basic power by which the government regulates the preservation of public order and tranquility, the protection of public health, safety, and morals; the use of land; and the prevention and punishment of crime. Environmental noise laws at the local and state level are based primarily upon the power to protect the public health and welfare, and also on the powers to control land use, to abate nuisances, and to preserve tranquility.

In developing a community noise control ordinance, it is important to keep in mind this underlying legal basis for the ordinance. As an initial step, the community should identify the public health and welfare goals it wishes to strive toward in the development and implementation of an environmental noise-control program. It would be helpful for the community to state these health and welfare goals in a formal document. An example of this type of document at the state level is "Explanation of Noise Control Regulations for Sound Source Sites," prepared by the New York State Department of Environmental Conservation in 1973 as the basis for its proposed noise control regulations. At the federal level, the U.S. Environmental Protection Agency document *Information on the Levels of Environmental Noise Requisite to Protect Public Health and Welfare with an Adequate Margin of Safety** serves as the basis for EPA noise control regulations; it can serve as a valuable source of information for communities on the health and welfare effects of noise.

As important as the statement of health and welfare goals of noise abatement is to the community noise control program, these goals should not serve as the sole basis for the determination of noise levels for the various provisions of the

*EPA Document No. 550/9-74-004 (March 1974).

ordinance. The choice of levels for noise emissions of products, for noise "spill-over" into receiving land uses, and for land use planning should also take into account the technological feasibility of such levels and the economic cost to the various members of the community and to the community as a whole associated with actual attainment of these levels. In order to be fully able to take these factors into account, an assessment of existing levels of noise in the community is important. Additionally, the ability of existing technology to abate noise emissions from sources that will be regulated must be assessed in order that the levels chosen will be realistic and enforceable.

Constitutional Issues

Several constitutional issues must be considered in the development of an environmental noise control ordinance. One issue involves the possible application of certain provisions of a noise control ordinance to activities involving the constitutionally protected rights of free speech and assembly; the other, which can almost completely be avoided by the use of a performance standard ordinance, involves the general constitutional issues of vagueness and overbreadth in a poorly defined "unnecessary noise" or "nuisance noise" provision.

Under the First Amendment to the U.S. Constitution, as applied to the states and their political subdivisions under the Fourteenth Amendment, all persons in the United States are guaranteed freedom from government infringement of their rights to speak and peaceably assemble. Noise ordinance provisions that impact the use of sound trucks and public-address systems on public streets or in public parks, or that prohibit "nuisance noise" generally or in hospital and school zones, can, if improperly drafted or applied, suffer constitutional infirmities under the First Amendment. Conversely, when such provisions are properly drafted and properly applied by law-enforcement officials, they can be valuable components of a community noise control ordinance.

In drafting an ordinance provision regulating the use of sound trucks, loudspeakers, and public-address systems, the community may wish to distinguish between the use of such systems for commercial advertising and their use for speaking on private opinions, such as political or religious views. The Supreme Court, in interpreting the First Amendment, has made a distinction between the use of speech for commercial advertising purposes and for private reasons, and has given the latter significantly greater protection. Thus, a community may place very stringent controls on the time, place, and manner in which sound trucks or public-address systems are used for commercial advertising; or, the community may simply place a complete ban on the use of these types of loudspeakers for commercial advertising in outdoor areas. These requirements will generally not constitute an infringement of a constitutionally protected freedom.

The community must be more careful, however, in the regulation of the use of sound trucks and public-address systems for private purposes. The Supreme

Court has said that a community may place reasonable regulations on the hours and places where public discussion with the use of loudspeakers may be conducted.[*] Thus, the community may, for example, place a nighttime curfew on the use of loudspeakers on any public right-of-way (e.g., a public street or sidewalk) or public place (e.g., a public park or plaza) if such use creates an annoyance or disturbance in a residential area or a noise-sensitive area such as a hospital, nursing home, or other institution where people reside. In the same ruling, the court also said that the community may regulate the sound level produced by loudspeakers and sound trucks. This type of regulation, although possible, would be more difficult to draft and to enforce than curfews and zone restriction provisions.

In drafting a loudspeaker provision relating to noncommercial speech, it is better for a community to avoid a system of issuing permits to allow the use of loudspeakers. Unless the community has very carefully and narrowly drawn the criteria for the grant or refusal of a permit, the ordinance could be struck down by a court as unconstitutional in imposing a "prior restraint" on the exercise of free speech, with no standards provided for the exercise of the administrative discretion involved in the permit procedure.[†] What the Supreme Court wishes to prevent, by finding unconstitutional permit systems with broad administrative discretion, is the possibility that the *content* of speech may be regulated. The court will not sanction a system in which a municipal official may, at his or her own discretion, grant a permit to use, for example, loudspeakers for a political speech, but refuse a permit for loudspeaker use for religious or racial equality speeches. A community should keep in mind these criteria not only when drafting loudspeaker provisions of their ordinance, but also when enforcing these provisions.

Finally, with regard to whether a community may completely ban the use of loudspeakers and sound trucks from any public right-of-way or public space, even though such use may be reasonably regulated in time, place, and volume, it appears that such ban would be found unconstitutional.[‡]

Another constitutional issue that must be considered in drafting a noise-control ordinance concerns the avoidance of "vague" and "overbroad" provisions. The issue of vagueness is important in criminal[§] and quasi-criminal[‖] statutes (into

*See *Saia* v. *New York*, 334 U.S. 558 (1948), p. 562

*Ibid., p. 560.

*See *Kovacs* v. *Cooper*, 336 U.S. 77 (1949), pp. 81–82.

†Criminal statutes are those that contain penalties that include the possibility of a jail sentence; they may also term a violation a misdemeanor or a felony, but such a specific statement is not necessary to the characterization of a statute as criminal. Some community noise ordinances contain the possible penalty of a jail sentence. This is not a recommended procedure, however, and should be avoided except, possibly, for repeat violators with several prior offenses of a similar nature.

‡Quasi-criminal statutes are those that specify as a sanction fines of a penal nature—that is, fines that serve to punish the offender rather than to provide restitution to a private person who has been wronged by the defendant.

which categories most community noise-control ordinances fall) because persons must be adequately apprised of what conduct would constitute a violation of the ordinance so that they may know that they should avoid such conduct.* Additionally, the statement of prohibited conduct must be clear enough so that law-enforcement officers may know what types of actions are prohibited in order to avoid arbitrary or discriminatory enforcement. For this reason, it is wise to avoid provisions that generally prohibit "loud, unnecessary, or unusual noise" or noise that "disturbs or annoys," unless such general terms are qualified in a way that informs citizens of what types of conduct or what levels of noise are prohibited. Communities that have had some type of nuisance noise statute† in effect for several decades will have a body of case law from local courts that interprets and more narrowly defines the meaning of the statute. This case law should be consulted when the community later seeks to replace the older statute with a new, performance-type community noise ordinance in order that the criteria of these cases may be included in the updated ordinance.

In addition to avoiding writing vague provisions for its noise control ordinance, a community must also avoid writing provisions that, although they constitute a constitutional use of the community's lawmaking authority, are so broad as to invade an area of constitutionally protected freedoms.‡ In the context of noise control ordinances, the constitutionally protected freedom of speech is in possible jeopardy. A possible example of such an overly broad provision could be a receiving land use property standard that prohibits *all* sounds in excess of 55 dBA from crossing a property line into a residential land use. Although this provision may generally be considered a constitutional exercise of lawmaking authority when applied to most sources§ of noise, it could be regarded as unconstitutionally overbroad if applied to the unamplified human voice. For this reason, it is recommended that a specific exemption for the unamplified human voice be included in any provision that could be unreasonably applied to human speech, such as receiving land use standards and general prohibitions against noise disturbances (as will be defined later).

In summary, to avoid the major constitutional problems that can impact upon a community noise control ordinance, a community should draft its ordinance to provide clear, unambiguous statements of the prohibited acts. The provisions should avoid unduly impacting the sensitive area of freedom of speech, but they may provide reasonable regulation of the volumes, times, and places of speech

*For a more detailed discussion, see, for example, *Lanzetta* v. *New Jersey*, 306 U.S. 451 (1939).

†For example, a nuisance-noise statute derived from the model published by the National Institute of Municipal Law Officers (in *Laws and Regulatory Schemes for Noise Abatement*, pp. 1-101 to 1-104 (U.S. Environmental Protection Agency NTID 300.4, 1971).

‡For a further discussion, see, for example, Note, "The First Amendment Overbreadth Doctrine," 83 *Harvard Law Review* 844 (1970).

§Other legal issues, which are beyond the scope of this discussion, may also impact this type of provision.

activities conducted on public streets and sidewalks and in public parks. Where a provision may be found to be overbroad due to a possible application of the provision to unamplified human speech, the use of the unamplified human voice should be specifically exempted from the operation of that provision.

Preemption

Although a community may traditionally have authority to control all of the various sources of community noise (except aircraft noise), recent legislation at the federal level and in some states has taken away, or preempted, the authority of communities to regulate certain types of noise sources.

At the federal level, the Noise Control Act of 1972 (49 U.S.C. Sections 4901 *et seq.*) contains provisions under which certain areas of state and local authority will become preempted after the effective date of regulations promulgated by the EPA under Sections 6, 17, and 18 of the act. In Section 2(c) of the act, Congress found that "while primary responsibility for control of noise rests with State and local governments, Federal action is essential to deal with major noise sources in commerce control of which require national uniformity of treatment." The purpose of federal preemption, thus, is to provide national uniformity of noise emission standards for products and equipment regulated by the EPA, which are shipped in interstate commerce or which are moving in interstate commerce.

Section 6 of the Noise Control Act gives the EPA the authority to develop and promulgate noise emission standards applicable at the time of manufacture and sale of new products that have been identified, under Section 5(b) (1) of the act, as major sources of noise and which fall into one of the following categories: (1) construction equipment, (2) transportation vehicles (including recreational vehicles), (3) motors or engines, or (4) electrical or electronic equipment. These regulations will be enforced against the manufacturers of the regulated products before or at the time of ultimate sale of the products (that is, a sale to a purchaser who does not intend to resell them immediately).

Subsection 6(e) (1) provides that after the effective date of an EPA regulation prescribing maximum noise emission levels for a particular new product or component, no state or political subdivision thereof may adopt or enforce, with respect to that particular new product or component, any law or regulation that sets a maximum noise emission limit on such product (or component) enforceable against the manufacturer of the product, unless such law or regulation is identical to the federal regulation. This is intended to provide uniform national standards for manufacturers to meet, so that their products may be shipped and sold throughout the United States without having to meet different local standards. Thus, by virtue of this preemption provision, state and local governments may not adopt or enforce any nonidentical new product noise emission limit regulations enforceable against manufacturers or dealers of new products regulated by the EPA under Section 6. In this context, the term *new product* means a product that has not yet been sold to the ultimate purchaser–in most cases, the first retail purchaser.

Under subsection 6(e) (2), state and local governments retain authority to control product noise by all other available means. This subsection states that nothing in the section precludes or denies the right of state and local governments to establish and enforce controls on environmental noise and sources thereof through the licensing, regulation, or restriction of the use, operation, or movement of any product or combination of products.

Thus, although a local government may not enforce a nonidentical local law specifying a maximum noise level for an EPA-regulated new product that affects the manufacture and sale of the product, the local government may regulate the noise impact of products in the community through regulations enforceable against the owners or operators of products. Types of regulations that are acceptable are:

1. maximum noise emission levels for the use or operation of a product
2. curfews on the operation of a product
3. restrictions on the zones, such as hospital zones, or areas, such as residential areas, in which certain products may be operated
4. requirements for periodic inspection and licensing of certain products.

Where a community wishes, under (1), to set maximum noise emission levels, enforceable against the owner or operator, for the use and operation of a product regulated by the EPA at the time of sale, the community should consider the EPA standards before determining what noise emission level to set. Although the community is not bound by the preemption provision to set product noise levels for use and operation identical to those of the EPA for manufacture and sale, the community must not, under the Interstate Commerce clause of the Constitution, create an undue burden on interstate commerce. A court could possibly hold that this had occurred if a community set operational noise level limits applicable to a federally regulated product that were so low that a product meeting EPA sale requirements could not be used in the community due to violation of the operational noise limits.

For communities that wish to set maximum noise emission level standards for the sale of new products that have not yet been regulated by an EPA Section 6 regulation, there is no preemption of this authority. However, for those products for which an EPA regulation is in effect, the Section 6 preemption provision requires that local laws be "identical" to the applicable EPA regulation. The EPA has interpreted the statutory term *identical* to mean that the local noise emission standard and the measurement methodology that defines the standard must be identical to those in the federal standard.* However, other elements of the local

*For a further discussion of the issue of identicality, see 41 *Federal Register* 2162 (1976).

law, such as the persons to whom the regulations are applicable, the sanctions for violations, the enforcement procedures, and any equivalent or correlatable "short test" measurement methods, need not be identical. Although this represents the present EPA interpretation of the scope of Section 6 preemption, the final determination rests with the courts of law that will hear and decide cases dealing with this issue.

The first two products for which preemption of state and local time-of-sale standards will occur are new medium- and heavy-duty trucks* and new portable air compressors.† Later regulations will apply to new buses, new motorcycles, new wheel and track loaders and dozers, new truck-mounted solid waste compactor units, and new truck transport refrigeration units. Scheduled for regulation at a still later time are lawnmowers.

Additional preemption of state and local authority with respect to noise control is found in Sections 17 and 18 of the Noise Control Act. Under Section 18, the EPA has promulgated standard‡ specifying the maximum noise emission levels allowed resulting from the operations of trucks and buses owned by motor carriers engaged in interstate commerce. The U.S. Department of Transportation, charged with enforcing the standards, has published compliance regulations§ containing a methodology for measuring noise emissions from such vehicles. EPA regulations‖ under Section 17 set maximum noise emission levels for the operation of locomotives and rail cars owned by surface rail carriers engaged in interstate commerce.

The preemption provisions contained in each of these sections are intended to eliminate the problems caused when trains, buses, and trucks, traveling in interstate commerce, are required to meet different noise emission level standards in various different states and communities. By virtue of preemption under Sections 17 and 18, vehicles and trains will need meet only one operational noise emission standard to be able to travel throughout the country.

The preemption provision pertaining to interstate motor carriers, Section 18(c) (1), provides that, after the effective date of EPA regulations applicable to noise emissions resulting from the operations of motor carriers engaged in interstate commerce, no state or local government may adopt or enforce noise emission standards applicable to the same operations of interstate motor carriers unless the standards are identical to those of the EPA. Section 17 (c) (1) provides a similar preemption with respect to noise emission standards applicable to the operations of equipment and facilities of surface rail carriers engaged in interstate commerce. However, Sections 17 (c)(2) and 18 (c)(2) provide that nothing in

*40 C.F.R. Part 205 (1977); effective October 1, 1977.

†41 *Federal Register* 2162 (1976), 40 C.F.R. Part 204 (1977); effective January 1, 1978.

‡39 *Federal Register* 38208 (1974), 40 C.F.R. Part 202 (1977); effective October 15, 1975.

§40 *Federal Register* 42432 (1975), 49 C.F.R. Part 325, as amended; effective October 15, 1975.

‖40 C.F.R. Part 203 (1977).

these sections shall diminish or enhance the right of state and local governments to establish and enforce standards or controls on levels of environmental noise or to control, license, regulate, or restrict the use, operation, or movement of any regulated product if the administrator of the EPA, after consultation with the secretary of the DOT, makes two determinations:

1. the state or local law is necessitated by special local conditions, *and*
2. it is not in conflict with regulations promulgated under the applicable section (Section 17 or 18).

The EPA has published interim guidelines to assist state and local governments to determine which types of laws and regulations would be impacted by these preemption provisions and to indicate the proper procedures to use in applying for a "special local condition exception."

The preemptive effect of these provisions can be divided generally into three categories, depending on the type of local law in question:

1. laws setting maximum noise emission levels on the operation of vehicles or trains
2. use or operational restrictions on the times or places of operation of vehicles or trains
3. general environmental regulations that could be applicable to certain operations of vehicles or trains.

The first case, a state or local noise emission standard applicable to the same operations of federally regulated rail and motor carriers as in Section 17 or 18 standards, will present very serious preemption problems if the local standard is not identical to the federal standard. Because the primary purpose of preemption is to enable vehicles and trains that meet the federal noise standards to travel from state to state without having to meet different noise emission requirements, any local noise emission standard that would require vehicles or trains to incorporate noise abatement measures in addition to those needed to meet the federal standard may be found completely in conflict with federal regulations. In this case, no exception would be granted, regardless of any compelling special local condition, due to the direct conflict with federal law.

In the second case (state or local use or operational restrictions on the times or places of operation of vehicles or trains), the decision regarding whether a special local condition exception should be granted will depend on several factors. One factor is the purpose of such regulation. If the primary purpose of the regulation is not to control noise, or if noise control is only one of many purposes for the regulation, then the regulation will not be subject to preemption and the community need not apply for a special local condition exception to be able to adopt or enforce it. An example of such a regulation would be truck routes that are designated based on several factors such as the safety of children, maximum load on streets, and noise. However, truck routes based solely on noise factors will be subject to preemption and must be reviewed by the EPA under the special local condition exception procedure. Other types of operational restrictions that must, if based solely on noise factors, be reviewed for a special local condition exception are curfews on hours of operation of vehicles or trains, restrictions on movement through residential areas or in noise-sensitive zones, and restrictions on the maximum daily cumulative noise level (for example, L_{eq} or L_{dn}) created by vehicles or trains. Restrictions of this type will be reviewed by the EPA to determine whether such restrictions are necessitated by special local conditions (for example, residential neighborhoods or schools and hospitals in close proximity to vehicle or train rights-of-way) and whether they are in conflict with the federal regulation.

The third case, general noise regulations that could be applicable to trains or vehicles, generally will not be affected by preemption. An example of such a general noise regulation would be receiving land use standards that are applicable to all sources operated on private property (see p. 312). Although vehicles operating on public rights-of-way are excluded from the application of this provision, federally regulated trucks and buses operating on private property (such as warehouse loading docks or passenger bus terminals) would be covered. A community wishing to enforce the provision against a warehouse or bus terminal that is causing a violation of the regulation would be advised to handle the situation carefully, however, to avoid a claim of preemption as a defense to the enforcement action. The enforcement action will generally not be considered to be preempted if there are reasonable and effective means of compliance other than controlling the noise emission level of the motor vehicle itself. Thus, if an effective means of compliance could be reasonably achieved by installing a noise barrier at the perimeter of the property owned by the warehouse or terminal, by moving noisy operations to other parts of the property, or by quieting machinery or equipment, other than interstate vehicles, that operate on the property, then preemption should not be a defense to this enforcement action because the federally regulated vehicles will not be directly impacted. If, on the other hand, the most effective and reasonable means of compliance would involve the installation of noise-attenuation equipment on federally regulated vehicles, this enforcement action would be preempted under Section 18 (c) (1).

The interpretation of federal preemption under the Noise Control Act can be expected to be refined and even changed as a result of judicial interpretations of the act due to cases arising under it. A community, in preparing a noise control ordinance, should ascertain the most recent interpretations of preemption at that time to ensure that the interpretations given above are still valid.

Preemption of community authority to enact controls on environmental noise can also occur at the state level. Several states, to date, have specified requirements for noise control ordinances in their political subdivisions. For example, in Minnesota, the community noise standards may not be more stringent than those at the state level.* Community noise regulations in Maryland may not be less stringent than state regulations.† In several states,‡ the state noise control law requires express authorization or approval of community noise ordinances before they may be effective. Before progressing too far in the development of a noise ordinance, a community should ascertain any relevant state requirements and also determine what types of assistance may be available from the state.

MAJOR COMPONENTS OF AN ORDINANCE

A community noise ordinance is a legal document and, as such, must contain certain provisions to meet appropriate criteria for legal documents. Additionally, a community noise ordinance will generally contain technical and scientific terms and standards, which must be understandable to the community as a whole, and to the enforcement officers, lawyers, and judges implementing and interpreting the ordinance. Thus, the ordinance should be prepared and reviewed by persons with expertise in various areas, such as lawyers, engineers, physicians, and environmental health specialists, to ensure that the ordinance is properly drafted and meets the requirements of the community.

A basic community noise ordinance usually contains provisions covering the following general categories, which are organized into articles of the ordinance:

1. definition of terms
2. grants of authority to administer and enforce the ordinance
3. prohibited acts (including performance standards and use and zone restrictions)
4. exceptions, variances, or permits
5. enforcement provisions, including penalties.

Beyond this bare outline of an ordinance, a community will choose what types of prohibitions and requirements to include in order to meet community goals. Additionally, depending on the community's requirements in noise control, other articles may be added. Such articles could include provisions requiring all municipal departments to participate in the goal of noise abatement in their programs and activities, land use planning standards utilizing noise exposure criteria, and building-code provisions specifying sound-transmission loss through exterior walls of buildings and interior walls of multifamily residential structures.

*Minnesota Pollution Control Agency Law sec. 116.07 (1974).
†Annotated Code of Maryland Art. 43 sec. 832 (1974).
‡Connecticut Public Act No. 74-328 sec. 7 (c) (July 1, 1974); Florida Statutes sec. 403.414 (8) (Laws of 1974, Ch. 74-110) (effective October 1, 1974); Revised Code of Washington Title 70, Chap. 183, sec. 6 (4) (1974).

In this chapter, the general outline of articles and certain major sections of a community noise control ordinance will be presented. For a detailed model community noise ordinance, with specific suggested wording for each provision, see the *Model Community Noise Control Ordinance** published jointly by the EPA and the National Institute of Municipal Law Officers (NIMLO).

Few communities will have the need or the desire to incorporate into their ordinances each type of provision that will be discussed. The choice of the actual provisions to utilize should be based on community health and welfare goals, the types of major noise sources and problems in the community, and the technical feasibility and economic impact of such provisions.

Definitions

The drafting of a noise control ordinance requires carefully bringing together scientific, engineering, and legal considerations to produce a document that is not only scientifically correct, but legally sound. Definitions of terms used in the ordinance play an especially important role in this process. The definitions of scientific terms, such as *sound*, *noise*, *decibel*, *sound pressure level*, and so on, must be carefully worded to impart a legal sense to scientific terminology. Additionally, such terms must be defined clearly enough so that lawyers and judges involved in cases arising under the ordinance will be adequately able to understand the concepts involved. For these reasons, it is not always possible to utilize directly the definitions of acoustic terms developed by technical groups such as the American National Standards Institute (ANSI).† Although ANSI definitions may provide a good basis for the definitions to be used in an ordinance, they must often be either simplified or expanded upon to provide a meaning appropriate for use in the ordinance. For examples of technical definitions that have been adapted for use in a noise control ordinance, see the EPA/NIMLO *Model Community Noise Control Ordinance*.

In addition to technical terms, the ordinance must define many other operative terms used in its substantive provisions. The most important of these terms are discussed in the following sections.

Noise versus *Sound*

The first issue to be discussed under this heading goes beyond the narrow topic of definitions: it involves choosing the operative word for the ordinance. The terms *noise* and *sound* are often used interchangeably with regard to *noise control* and *noise ordinances*. However, from the legal standpoint, with regard to the development and enforcement of an ordinance, the two words have very different and distinct meanings.

*EPA Document No. 550/9-76-003 (September 1975).

†ANSI Publication No. S1.1-1960 (Revised 1971).

The term *sound* is a purely physical phenomenon, and may be defined as

> An oscillation in pressure, particle displacement, particle velocity or other physical parameter, in a medium with internal forces that causes compression and rarefaction of that medium. The description of sound may include any characteristic of such sound, including duration, intensity, and frequency.

However, as used in the context of an ordinance, the term *noise* connotes a sound that has an adverse effect on a human being. Thus, *noise* may be defined as

> Any sound that annoys or disturbs humans or that causes or tends to cause an adverse psychological or physiological effect on humans.

In choosing between *noise* and *sound* for what might be considered the operative term of the ordinance, the community should consider the possible proofs and defenses associated with each word that may arise in an enforcement action brought under the ordinance. If the operative word chosen is *noise*, it is possible that a defendant cited for violating a performance standard provision may try to defend the action on the basis that his violation did not, in fact, disturb or annoy any human. Or, possibly, the court may require the prosecutor to prove that human annoyance actually resulted from the defendant's actions. To avoid these potential problems, a community may wish to use *sound* as the operative term of the ordinance. This usage will be further described in the later sections discussing suggested "Prohibited Acts" of the ordinance. By using this approach in developing sound emission standards for vehicles, equipment, and property lines, the community is in effect stating that sound levels from the regulated equipment that exceed the community limits are *per se* annoying or disturbing to humans or cause psychological or physiological effects in humans.

Note, however, that where violation of a prohibited act does not depend on exceeding a specified sound emission level, but rather depends on creating a "noise disturbance," as defined hereafter, the prosecution will have to prove the elements of a noise disturbance in order to make a case.

Person

In order to specify the persons to whom the ordinance is applicable, ordinances contain a definition of the term *person*. Such definitions are generally fairly standard. For example, *person* means

> Any individual, association, partnership, or corporation, and includes any officer, employee, department, agency, or instrumentality of a state or any political subdivision of a state.

In general, a community ordinance cannot be made applicable to any instrumentality or agency of the federal government, due to the doctrine of sovereign immunity. However, with regard to community environmental noise control laws (as well as air pollution laws), Congress stated that each department, agency, or instrumentality of the executive, legislative, and judicial branches of the federal government must comply with federal, state, interstate, and local requirements regarding control and abatement of environmental noise to the same extent that any *person* is subject to such requirements.

Although no cases have arisen under this provision in the Noise Control Act, federal courts of appeals, deciding cases arising under identical language in the Clean Air Act, have disagreed about the actual meaning of this provision. The Fifth Circuit Court of Appeals* has held that this provision applies to all state air pollution control requirements, whether substantive (e.g., emission standards) or procedural (e.g., construction permit requirements). On the other hand, the Sixth Circuit Court of Appeals, in *Kentucky* v. *Ruckelshaus*,† has interpreted the provision as requiring federal facilities and other federal activities to comply with substantive requirements only, and not procedural requirements. This latter case has been appealed to the Supreme Court, which will make the final decision on the issue. Although these decisions are interpreting the Clean Air Act, because the language in question is essentially identical to that in the Noise Control Act, the decisions will apply equally well to it.

Another problem arising with regard to federal facility compliance with state or local noise laws is that there is no authority under which state or local governments may sue the federal government for violations of their laws. Thus, although Congress has required federal compliance, state or local governments may not be able to completely assure such compliance because they cannot obtain a court order against a federal facility mandating compliance, nor can they prosecute the federal government to obtain fines for violations.

Due to these reasons, the definition of *person* suggested omits including the federal government as a "person" to whom the ordinance is applicable. Nonetheless, the community should keep in mind that, by virtue of Section 4 of the Noise Control Act, the federal government must comply with at least the substantive requirements of the community noise ordinance.

Noise Disturbance

In the substantive portions of the ordinance, the prohibited acts should be based wherever possible on sound level limits, applied to the operation of equipment or the conduct of activities. For those sources, however, for which measurement of the sound level is difficult or impractical (such as barking dogs, public-address systems, church bells, or loading and unloading operations), a more subjective criterion based on human annoyance must be used. In older noise ordinances, the criterion often used was whether the noise in question was "loud, unnecessary,

**Alabama* v. *Seeger*, 502 F. 2d 1238 (5th Cir. 1974).
†497 F. 2d 1172 (6th Cir. 1974).

or unusual." This criterion has been replaced, in the EPA/NIMLO model ordinance, with "noise disturbance" criteria, which are intended to provide greater objectivity to an essentially subjective determination. A *noise disturbance* is defined as

> Any sound that (1) endangers or injures the safety or health of humans or animals, or (2) annoys or disturbs a reasonable person of normal sensitivities, or (3) endangers or injures personal or real property.

Subpart (1) of this definition will draw upon the community and state common law with regard to protection of the public health and welfare from environmental and health hazards. Subpart (2) is based on the tort law standard of the "reasonable man"–in this case, how a normal person in a given situation would react to noise of the type in question in a particular enforcement situation. Finally, subpart (3) draws upon community and state common law with regard to protection of property.

The definition of *noise disturbance* should be included in any community noise ordinance that contains subjective or "nuisance noise" provisions.

Technical Terms

Depending on the precise terms utilized in the substantive provisions of the noise ordinance, the following scientific terms should be defined:

1. *A-weighted sound level:* the sound pressure level in decibels as measured on a sound level meter using the A-weighting network. The level so read is designated *dB(A)* or *dBA*.

2. *Day-night average sound level* (L_{dn}): the 24-hour energy average of the A-weighted sound pressure level, with the levels during the period 10:00 P.M. to 7:00 A.M. the following day increased by 10 dBA before averaging.

3. *Decibel* (*dB*): a unit for measuring the volume of a sound, equal to 20 times the logarithm to the base 10 of the ratio of the pressure of the sound measured to the reference pressure, which is 20 micropascals (20 micronewtons per square meter).

4. *Equivalent A-weighted sound level* (L_{eq}): the constant sound level that, in a given situation and time period, conveys the same sound energy as the actual time-varying A-weighted sound.

5. *Impulsive sound:* sound of short duration, usually less than 1 second, with an abrupt onset and rapid decay. Examples of sources of impulsive sound include explosions, drop forge impacts, and the discharge of firearms.

6. *Noise:* any sound that annoys or disturbs humans or that causes or tends to cause an adverse psychological or physiological effect on humans.

7. *Pure tone:* any sound which can be distinctly heard as a single pitch or a set of single pitches. For the purposes of this ordinance, a pure tone shall exist if the one-third octave band sound pressure level in the band with the tone exceeds

the arithmetic average of the sound pressure levels of the two contiguous one-third octave bands by 5 dB for center frequencies of 500 Hz and above and by 8 dB for center frequencies between 160 and 400 Hz and by 15 dB for center frequencies less than or equal to 125 Hz.

8. *RMS sound pressure:* the square root of the time-averaged square of the sound pressure, denoted P_{rms}.

9. *Sound:* an oscillation in pressure, particle displacement, particle velocity, or other physical parameter, in a medium with internal forces that causes compression and rarefaction of that medium. The description of sound may include any characteristic of such sound, including duration, intensity, and frequency.

10. *Sound level:* the weighted sound pressure level obtained by the use of a sound level meter and frequency weighting network, such as A, B, or C as specified in American National Standards Institute specifications for sound level meters (ANSI S1.4-1971, or the latest approved revision thereof). If the frequency weighting employed is not indicated, the A-weighting shall apply.

11. *Sound level meter:* an instrument that includes a microphone, amplifier, RMS detector, integrator or time averager, output meter, and weighting networks used to measure sound pressure levels.

12. *Sound pressure:* the instantaneous difference between the actual pressure and the average or barometric pressure at a given point in space, as produced by sound energy.

13. *Sound pressure level:* 20 times the logarithm to the base 10 of the ratio of the RMS sound pressure to the reference pressure of 20 micropascals (20×10^{-6} N/m^2). The sound pressure level is denoted L_p or *SPL* and is expressed in decibels.

14. *Vibration:* an oscillatory motion of solid bodies of deterministic or random nature described by displacement, velocity, or acceleration with respect to a given reference point.

Land Use Definitions

If a community wishes to have standards specifying maximum sound levels that may "spill over" into receiving land uses or environmental noise standards for land use planning, the particular definitions utilized by the community for the various land uses or zoning classifications (generally, residential, commercial, and industrial) must be included in the noise ordinance definitions. Even if a community does not intend to enact receiving land use standards or land use planning provisions, it should define the term *residential area*, because that is a key term in many other substantive provisions of the noise ordinance.

Noise-Sensitive Zone

Another key term is *noise-sensitive zone*. Designed to replace older terms such as *quiet zone* and *hospital zone*, the term *noise-sensitive zone* means all areas that contain sensitive activities and that are designated as noise-sensitive zones by the appropriate community official. Noise-sensitive activities include, but are not limited to, operations of hospitals, schools, libraries, nursing homes, and other institutions. Because the actual definition of *noise-sensitive zone* depends on which areas are actually so designated by the appropriate community official and may change from time to time, it is recommended that the definition merely refer to "any area designated pursuant to Section X of this ordinance for the purpose of ensuring exceptional quiet," rather than to actually define the term.

Public Space and Right-of-Way

Other terms that are used often in a community noise ordinance are *public right-of-way*, *public space*, and *real property boundary*. Definitions of these terms can usually be found in other community statutes; these other statutes should be consulted to ensure that the definitions of these terms are consistent with those in the noise ordinance. Suggested definitions follow:

1. *Public right-of-way*: any street, avenue, boulevard, highway, sidewalk, or alley or similar place, which is owned or controlled by a governmental entity.
2. *Public space*: any real property or structures thereon, which are owned or controlled by a governmental entity.
3. *Real property boundary*: an imaginary line along the ground surface, and its vertical extension, which separates the real property owned by one person from that owned by another person, but not including intrabuilding real property divisions. (This definition is intended to exclude portions of real property boundaries as they may pass through buildings, such as rowhouses or townhouses; it is also intended to apply regardless of height above the ground, in order to cover noise problems arising several stories above the ground.)

Motor Vehicles

A noise ordinance applicable to motor vehicles will need to define all terms applicable to the mobile sources included in the ordinance. The definition of *motor vehicle* may be taken from the state or community motor vehicle code, provided that it corresponds to the needs of the community noise ordinance; similarly, if appropriate, the motor vehicle code definition of *motorcycle* may be used in the noise ordinance. If the community wishes to apply the motor vehicle provisions to vehicles operated by interstate motor carriers, which are regulated by EPA noise emission standards,* certain definitions should be taken from these EPA regulations. The following are suggested versions:

1. *Motor carrier engaged in interstate commerce*: any vehicle for which regulations apply pursuant to Section 18 of the Federal Noise Control Act of

*39 *Federal Register* 38208 (1974), 40 C.F.R. Part 202 (1977).

1972 (49 U.S.C. §4917), as amended, pertaining to motor carriers engaged in interstate commerce.

2. *Gross vehicle weight rating (GVWR)*: The value specified by the manufacturer as the recommended maximum loaded weight of a single motor vehicle. In cases where trailers and tractors are separable, the gross combination weight rating (GCWR), which is the value specified by the manufacturer as the recommended maximum loaded weight of the combination vehicle, shall be used.

3. *Muffler or sound dissipative device*: a device for abating the sound of escaping gases of an internal combustion engine.

Construction

If the community intends to regulate construction noise itself or, in the context of a land use planning provision, the construction of residential structures in high-noise areas, the term *construction* must be defined. A suggested definition is "any site preparation, assembly, erection, substantial repair, alteration, or similar action, but excluding demolition, for or of public or private rights-of-way, structures, utilities, or similar property."

Because this definition may be used for land use planning provisions, the term *demolition* has been excluded from the definition of *construction*, so as not to include demolition activities within the application of land use planning provisions. *Demolition* may be defined as "any dismantling, intentional destruction or removal of structures, utilities, public or private right-of-way surfaces, or similar property." This definition may be combined with that of *construction* if the community does not intend to include land use planning provisions in its noise control ordinance.

Powers and Duties of the Administrator

In the next article to be discussed, referred to here as Article IV, the community must indicate which community agency or official will administer the ordinance and set forth the powers and duties required for its administration.

Smaller communities may have a small environmental protection office, which handles the administration of all environmental ordinances in the community, and which will also administer the environmental noise ordinance. Larger communities may have the resources and the desire to designate an office or official within an environmental protection office as the noise control office or officer. This choice should be indicated in the ordinance as the "lead agency" or "official" who will be responsible for carrying out the administration and enforcement of the ordinance.

This article should also enumerate the powers conferred upon the administrator in order to carry out the intent of the ordinance and the mandatory duties he or she must perform to administer it.

Among the powers that may be granted to an administrator are:

1. The power to conduct studies. Such studies can include, among others, studying the public annoyance due to noise and monitoring the sound levels in the community.

2. The power to conduct educational programs. This can include conducting public education programs regarding the health and welfare effects of noise, how to abate noise in various situations, and how to comply with the provisions of the ordinance.

3. The power to coordinate community efforts in noise control. This provision gives the administrator the power to coordinate the noise (and vibration, if desired) control activities conducted by all municipal departments. It also gives the administrator authority to cooperate or combine efforts with appropriate county and municipal agencies and to cooperate with the appropriate state and federal agencies.

4. The power to review actions of other departments. Under this provision, the administrator is empowered to request any other community department or agency responsible for any standard, regulation, or other action to consult with the administrator on the advisability of revising the action, if he or she has reason to believe that the action is not consistent with the ordinance.

5. The power to review public and private projects. By virtue of this provision, the administrator has the authority to review any public or private project (such as highway or power plant construction or a building permit for a private project) that is subject to mandatory review or approval by other community departments. Such review by the administrator would be for the purpose of determining whether the projects in question would be in compliance with the ordinance and would generally only be conducted where there was the possibility that the project would cause sound or vibration in violation of the ordinance.

The actual scope of this provision will depend on community policy. Some communities may wish to authorize the administrator of the noise ordinance to recommend to other departments appropriate modifications to projects to ensure that they will not violate the provisions of the noise ordinance. Some communities may even wish to give the noise control administrator veto power over projects that would significantly affect the noise environment.

Additionally, the community may wish to set criteria for determining which projects should be reviewed. Such criteria may either be stated in the noise ordinance provision or in administrative guidelines developed after the passage of the ordinance.

6. The power to conduct inspections. This provision delineates the authority given to the administrator and duly designated subordinates with regard to conducting inspections for the implementation and enforcement of the ordinance. This provision must be carefully drawn to avoid any constitutional problems under the Fourth Amendment (which prohibits unreasonable searches and seizures). To be constitutionally permissible, administrative inspections or

searches on private property must be made using a warrant procedure.* If private property owners refuse to allow community officials to inspect their premises or records kept on their premises, the officials must obtain a search warrant for the premises before they may inspect them. Additionally, because the Supreme Court has held that there is no distinction between the rights of a residential property holder and those of a commercial property holder with regard to searches or inspections,† both types of property may be treated the same in a community noise ordinance.

If the community desires, this provision may also grant authority to the administrator and the inspectors to stop moving vehicles to inspect them and, if necessary, to issue a notice of violation to the driver. Due to the greater element of risk to the inspector inherent in this type of situation, many communities prefer to use the police for this aspect of noise ordinance enforcement.

Suggested wording for the inspection provision is as follows:

> Upon presentation of proper credentials, enter and inspect any private property or place, and inspect any report or records at any reasonable time when granted permission by the owner, or by some other person with apparent authority to act for the owner. Whenever permission is refused or cannot be obtained, a search warrant may be obtained from a court of competent jurisdiction upon a showing of probable cause to believe that a violation of the ordinance may exist. Such inspection may include administration of any necessary tests.

7. The power to require that records be kept. Under this provision, the administrator may require the owner or operator of any commercial or industrial activity to establish and maintain records and make such reports as may reasonably be needed to ascertain compliance with the ordinance.

8. The power to require the owner or operator to make measurements. Similar to provision (7), this section gives the administrator the authority to require the owner or operator of a commercial or industrial activity to measure the sound level of (or the vibration level from) any source, according to methods and procedures specified by the administrator, and to furnish the administrator with reports of such measurements.

The advantage of this type of provision is that it allows a relatively small staff of noise enforcement officials to monitor the compliance of major industrial and commercial sources of noise by simply checking the self-monitoring results of the sources. Spot checks by the enforcement officers can ascertain the validity of the data.

9. The power to make recommendations regarding product performance standards. As part of a continuing program to add new provisions to the ordi-

*_Camara_ v. _Municipal Court_, 387 U.S. 523 (1967); _See_ v. _Seattle_, 387 U.S. 511 (1967).
†_See_ v. _Seattle_, 387 U.S. 511 (1967).

nance when a need is shown, this provision empowers the administrator to develop new provisions regulating the use and operation of any product, including a specification of the maximum sound emission level of such product.

One important use of this provision would be for the administrator to recommend appropriate noise emission levels for the use and operation of products that have been manufactured to meet specific noise emission levels according to EPA new product regulations promulgated pursuant to Section 6 of the Noise Control Act.

Another possible portion of this provision could empower the administrator to develop regulations prohibiting the sale of products that do not meet a specified maximum noise emission level. For products regulated by the EPA, the administrator must, of course, take into account federal preemption (see p. 286).

10. The power to make recommendations regarding noise-sensitive zones. Under this provision, the administrator has the authority to make specific recommendations to the appropriate community body regarding the designation of noise-sensitive zones within the community. Such a provision may read as follows:

> [The administrator shall have the power to] prepare recommendations, to be approved by (the appropriate authority), for the designation of noise-sensitive zones that contain noise-sensitive activities. Existing quiet zones shall be considered noise-sensitive zones until otherwise designated. Noise-sensitive activities include, but are not limited to, operations of schools, libraries open to the public, churches, hospitals and nursing homes.

After the enumeration of the powers of the administrator, the duties of the position should then be described. Among suggested duties are the following:

1. The duty to develop and recommend testing procedures. If test procedures and measurement methodologies to enforce each performance standard provision have not been developed at the same time as the ordinance, the first important duty of the administrator is to develop and publish guidelines or documents setting forth the appropriate measurement methodology for each provision. This type of document will be discussed later (see p. 323).

2. The duty to investigate and pursue violations. This provision places upon the administrator the duty to investigate, in accordance with the inspection provision above, and pursue according to the enforcement article (discussed on p. 000) the possible violations of the ordinance.

3. The duty to delegate authority. By virtue of this provision, the administrator is required to delegate functions, wherever appropriate, to personnel within his or her office or within other departments or agencies of the community.

4. The duty to review truck routes and transportation planning. Because of the importance of transportation noise sources in a community, the administra-

tor may be placed under a duty to assist in and review transportation plans and systems. A possible wording for this section is as follows:

[The administrator shall:]

a. Study the existing transportation systems, such as truck routes, within the community; determine areas with sensitivity to sound and vibration caused by transportation; recommend changes or modifications to transportation systems to minimize sound and vibration impact on residential areas and noise-sensitive zones.

b. Assist in or review the total transportation planning of the community, including planning for new roads and highways, bus routes, airports, and other systems for public transportation, to ensure that the impact of sound and vibration receives adequate consideration.

5. The duty to establish capital improvement guidelines. In order to assure that community building projects and other capital improvements meet acceptable standards of noise emission, the administrator is required to establish noise assessment guidelines for the evaluation of proposed improvements. The guidelines should assist in the determination of the relative priority of each improvement in terms of noise impact.

6. The duty to ensure consistency with state and federal laws and regulations. In this important provision, the administrator is charged with the duty to monitor developments in state and federal laws and regulations impacting noise control and to recommend changes or modifications to the community ordinance, as may be required. Additionally, the administrator is required to prepare and publish a list of products that have been manufactured to meet federal (or state or local) noise emission regulations, such as portable air compressors,* for which "tampering"† enforcement will be conducted. This will help ensure that products that have been manufactured to meet low noise emission standards will continue to meet such standards throughout the period of their use in the community.

7. The duty to administer grants, funds, and gifts. Under this section, the administrator is given the duty to administer noise program grants and other funds from private and public sources, including the state and federal governments.

Optional provisions that a community may wish to enact include:

8. The duty to develop plans to achieve long-term noise goals. If the community has sufficient funds, it may wish to direct the administrator of the noise program to develop a generalized sound level map of the community and utilize this map to prepare a long-term plan for achieving quiet in the community. He would then be required to take the appropriate steps for integrating this plan into the general planning process of the community.

*41 *Federal Register* 2162 (1976), 40 C.F.R. Part 202 (1977).
†For a discussion of "tampering," see p. 200.

9. The duty to prepare a periodic report. If a community desires to monitor its noise abatement program, it may wish to require the administrator of the program to evaluate its effectiveness and prepare a report periodically (for example, annually or biennially) in which recommendations are made regarding any legislative or budgetary changes necessary to improve the program.

The preceding list of powers and duties of the noise program administrator is not exhaustive, and a community may wish to delegate additional powers and duties to the administrator. Conversely, a community may feel that this list is too comprehensive and therefore may wish to excise those sections it believes are not needed to accomplish its goals. In this article, as in the entire ordinance, the community must tailor its choice of provisions to its needs and goals in noise abatement and to its resources to carry out the program.

Duties and Responsibilities of Other Departments

The next major portion, or article, of a community noise control ordinance may address the duties and responsibilities of other departments and agencies in the community to control noise in their own programs and activities. This article is based on the fact that many community departments can, in carrying out their programs, have a significant impact upon the noise environment of the community. Among the most important departments in this regard are the planning and zoning departments, the building permits department, and the transportation department. The provisions in this article, along with several from Article IV, require each community department to consider the noise impact of their programs. They also give the administrator of the noise abatement program the authority to coordinate and review the community efforts in this area. Due to the sharing of this responsibility, there is a potential for interdepartmental conflicts to arise. The community may wish to specify in the ordinance, or in administrative guidelines developed under the ordinance, a method for resolving such conflicts and arriving at a final decision. This may involve authorizing the city council, county board of supervisors, mayor, or other body to act as arbiter and ultimate decisionmaker.

Among suggested duties and responsibilities of other departments are the following:

1. Duties in carrying out programs. The first section in this article should place a general responsibility on all community departments and agencies to carry out their programs, to the fullest extent consistent with other law, in such a manner as to further the policy of the noise control ordinance.

2. The duty to cooperate. The second section should require all community departments and agencies to cooperate with the administrator of the noise control program to the fullest extent in the enforcement of the ordinance. This section refers particularly to those departments that will have a significant role

in assuring that the ordinance is properly implemented, such as the police, planning, and building permits departments.

3. The duty to comply with other laws. Under this section, community departments and agencies are required to comply with federal and state laws and regulations and the community ordinance regarding the control and abatement of environmental noise to the same extent that any person is subject to such laws. This section thus makes it clear that the community is not exempt from the application of noise control laws and regulations at any level of government.

4. The duty to consult prior to project approval. This provision requires all departments whose duty it is to review and approve new projects or changes in existing projects, that result or may result in the production of sound or vibration, to consult with the administrator of the noise program prior to project approval. This is intended to ensure that any adverse sound or vibration impact of community-approved projects will be minimized or avoided.

5. Duties with regard to contracts. In order to ensure that work done under contract to the community or equipment purchased by the community will be in compliance with the noise ordinance, this section requires that any written contract, agreement, purchase order, or other instrument, in which the community is committed to expenditures in excess of a specified amount of money in return for goods or services, contain provisions requiring compliance with this ordinance.

6. Duties regarding capital improvement programs. By virtue of this provision, all departments responsible for capital improvements budgets and programs must prepare an analysis of the noise impact of any proposed improvements in accordance with noise-assessment guidelines prepared by the noise program administrator (which guidelines are required in Article IV). Capital improvements covered by this section include land acquisition, building construction, highway improvements, and utilities and fixed equipment installation.

The next ordinance provisions to be discussed set forth the substantive provisions of the ordinance: the performance standards for equipment and land uses and the acts prohibited or required by the ordinance. For convenience, these provisions have been grouped into four major articles: general prohibited acts, sound levels by receiving land use, motor vehicle maximum sound levels, and land use provisions. As with all portions of the ordinance, a community must choose from the fairly comprehensive list set forth here those provisions that it wishes to enact based on community health and welfare goals, sources present in the community, and the level of resources that will be made available for the implementation and enforcement of the program. It is far better to enact a few most-needed provisions and enforce them well than to enact a comprehensive ordinance and subsequently enforce only a few provisions due to lack of resources. The latter course of action would not only be a disservice to respect for the law, but would also be a disappointment to citizens who had anticipated full relief from noise pollution.

Prohibited Acts

The first substantive article to be discussed, called here Article VI, sets forth a list of acts of various types that are prohibited by the ordinance. The article encompasses a variety of sources, from barking dogs and television sets to loading and unloading operations, which could not be placed into the more specific categories of the succeeding articles. For the most part, the provisions of this article are updated versions, containing quantitative standards wherever possible, of the older version of the NIMLO ordinance.* Because it is difficult to apply sound level performance standards to many of the sources covered in this article, the general criterion of a violation is the creation of a *noise disturbance* (see p. 294).

To begin with, the article may contain a general "catchall" provision to cover all sources and activities not covered in any of the more specific provisions of the ordinance. A section of this type is useful because it is impossible to foresee all sources of noise disturbances that may occur in a community for which legal relief should be possible. However, because so many potential sources are covered, this provision must be worded carefully to avoid the constitutional problems of vagueness, overbreadth, and infringement of First Amendment freedoms. The following wording is suggested:

> No person shall make, continue, or cause to be made or continued, any noise disturbance. Noncommercial public speaking and public assembly activities conducted on any public space or public right-of-way shall be exempt from the operation of this section.

In this, and in all the succeeding substantive provisions to be discussed, the community must make certain that the important operative words in the provision are identified and defined in the article containing definitions or in the provision itself. In the preceding provision, the terms *noise disturbance, public space*, and *public right-of-way* should be defined (in the article on definitions) in order to give the appropriate meaning to the section.

In order to avoid any First Amendment problems, noncommercial public speaking activities have been specifically exempted.

The remainder of this article should contain specific prohibitions, in which certain acts, and the causing thereof, are declared to be in violation of the ordinance:

1. Radios, television sets, musical instruments, and similar devices. To control noise disturbances caused by radios, televisions, phonographs, drums and other musical instruments, and amplifiers or similar devices that produce, reproduce, or amplify sound, the community may wish to

a. set a nighttime curfew on the operation of such devices where the opera-

*Reprinted in *Laws and Regulatory Schemes for Noise Abatement*, pp. 1-101 to 1-104 (U.S. Environmental Protection Agency NTID 300.4, 1971).

tion may create a noise disturbance across a real property boundary or in a noise-sensitive zone,

b. prohibit the operation of such devices in or on a motor vehicle on a public space or right-of-way, or in a boat on public waters, if such operation creates a noise disturbance at a distance of 50 feet (or more) from such device, and/or

c. prohibit the operation of such devices on common carriers (such as buses and trains) by any passenger if the operation creates a noise disturbance to any other person.

Although it is somewhat difficult to appropriately and narrowly define the terms, the types of devices covered in this provision should be distinguished from the sound truck and public-address system types of loudspeakers which will be discussed in (2). To attempt to avoid confusion on this issue and to avoid First Amendment problems, it may be wise to specifically exempt noncommercial speech, covered in the next section, from this provision.

2. Loudspeakers/Public-address systems. This provision covers sound trucks, public-address systems used outdoors, and other loudspeaker systems that can be used for amplifying speeches, commercial messages, and other such communications. To avoid First Amendment problems and obtain more control over the less-protected commercial speaking activities, the provision may be divided into two subsections. Suggested wording is as follows:

a. Using or operating for any noncommercial purpose any loudspeaker, public-address system, or similar device between the hours of 10:00 P.M. and 8:00 A.M. the following day, such that the sound therefrom creates a noise disturbance across a residential real property boundary or within a noise-sensitive zone.

b. Using or operating for any commercial purpose any loudspeaker, public-address system, or similar device (1) such that the sound therefrom creates a noise disturbance across a real property boundary or within a noise-sensitive zone, or (2) between the hours of ___ P.M. and ___ A.M. the following day on a public right-of-way or public space.

Note that the noncommercial use of loudspeakers is only prohibited during nighttime hours when it creates a noise disturbance in residential areas or noise-sensitive zones, whereas commercial uses are prohibited during the daytime if they create a noise disturbance on any category of land use and are banned at night in any public area.

3. Street sales. This provision bans street sellers, such as hot-dog vendors, from shouting about or "hawking" their goods within residential or commercial areas of the community. If the community feels that a complete ban is too stringent, the provision may be linked to a permit system and/or a curfew.

4. Animals and birds. This section prohibits owning, possessing, or harboring any animal or bird that frequently or for continued durations howls, barks, meows, squawks, or makes other sounds that create a noise disturbance across a residential real property boundary or within a noise-sensitive zone. The words *frequently or for continued duration* have been included to provide some criteria for determination of a violation.

5. Loading and unloading. In this provision, the loading, unloading, opening, closing, and other handling of boxes, crates, containers, garbage cans, or similar objects is prohibited during nighttime hours if it causes a noise disturbance across a residential real property line or within a noise-sensitive zone. This provision is intended to apply primarily to the sound associated with the objects being loaded and unloaded, not to the truck noise sources, such as idling engines or truck refrigeration units, which may also be present. If the provision is enforced against the engine or refrigeration-unit noise of vehicles covered by the Interstate Motor Carrier Noise Emission Standards* promulgated by the EPA under §18 of the Noise Control Act, the issue of preemption will arise (see pp. 288–289).

6. Construction. Among the several possible ways of controlling noise from construction and demolition sites, the present recommended means is a nighttime curfew on operations and, if desired, a ban on operations during the daytime on Sundays or weekends, where the sound from the operations creates a noise disturbance in a residential area or noise-sensitive zone. The section should provide an exception for emergency work by community and public service utility companies and should also allow the possibility of working under a special variance (see p. 000).

A nighttime curfew alone does nothing to control construction noise during the daytime period of operation. A specification of a maximum site sound level, however, is not a good solution because, if the level is high enough to allow operation of the noisiest equipment, the standard would be approving sound levels that clearly exceed health and welfare criteria. On the other hand, if a much lower maximum level were to be chosen, it could have the effect of prohibiting the use of certain types of equipment that must be used in construction. An alternative method is to utilize a performance standard based on a statistical level descriptor, such as L_{10}, or an energy average, such as equivalent sound level, L_{eq}. This method would allow all equipment to be operated for at least a short period of time each day, and, because the overall daily sound level is the basis for the standard, it provides an excellent incentive to use quiet equipment and/or add-on sound suppression devices. An ordinance provision of this type should set forth the sound level criterion (e.g., X dBA $L_{eq(8)}$), the point of measurement (e.g., at or across a real property boundary), and the period of

*39 *Federal Register* 38209 (1974), 40 C.F.R. Part 202 (1977).

measurement (e.g., the daily period of operation). It is recommended that the measurement location be based on an affected receiving land use rather than the perimeter of the construction site itself. This not only ensures that the provision is directed toward the health and welfare of the receiver, but also obviates the problem of determining the exact location of the construction site boundary. A full description of the measurement methodology and any other criteria to be used in the enforcement of this provision should be included in the community enforcement manual (see p. 323).

7. Vehicle or motorboat repairs and testing. Another type of noise a community may wish to control is noise caused by the repairing and testing of motorboats and vehicles in residential areas. A provision covering these sounds could prohibit repairing, rebuilding, modifying, or testing any motor vehicle, motorcycle, or motorboat in such a manner as to cause a noise disturbance across a residential real property boundary or within a noise-sensitive zone.

8. Places of public entertainment. Although the primary thrust of a community noise control ordinance is toward environmental or outdoor sounds, a more complete protection of public health and welfare may demand some regulation of interior sound levels in places of public entertainment—discotheques, nightclubs, and other places where music may be highly amplified. Even though the sound levels created by amplifiers in such places of entertainment can be high enough to cause a significant temporary threshold shift in patrons' hearing, and thus create a clear health and welfare issue, the problem can be difficult to regulate because patrons are voluntarily exposing themselves to these high sound levels.

One approach to the problem could involve a requirement for warning signs to be posted outside places of public entertainment when the sound levels inside would exceed a certain level at any point normally occupied by a customer. Supplemented by community public education programs, this type of regulation would ensure that potential customers understood the risk to their hearing involved in frequent attendance at such establishments. It may also, but not necessarily, bring about a decrease in the level of amplification that the clubs would choose to use. A suggested wording for this provision is as follows:

> [The following acts, and the causing thereof, are declared to be in violation of this ordinance:] operating, playing, or permitting the operation or playing of any radio, television, phonograph, drum, musical instrument, sound amplifier, or similar device that produces, reproduces, or amplifies sound in any place of public entertainment at a sound level greater than __dBA, as read by the slow response on a sound level meter, at any point that is normally occupied by a customer, unless a conspicuous and legible sign is located outside such place, near each public entrance, stating "Warning: Sound Levels Within May Cause Permanent Hearing Impairment."

Another approach, which is more effective but possibly more difficult to enforce, would be to set a maximum sound level standard, in L_{eq} over a period of time, which cannot be exceeded by places of public entertainment.

9. Powered model vehicles. A community in which engine-powered model vehicles, such as model airplanes, boats, cars, and rockets, are popular may wish to regulate noise levels and the use of such models to avoid noise disturbances. In a provision regulating powered model vehicles (which term should be defined), the community may wish to place a curfew on the use of those models that create a noise disturbance in a residential area, public space, or noise-sensitive zone. For the period of allowed operation, the community may set a maximum noise level caused by the use of the model, measured at a specified distance if used on a public space or across a property boundary if operated on private property.

10. Stationary nonemergency signaling devices. Communities wishing to control sound from stationary nonemergency signaling devices, such as church bells, factory whistles, and carillons, have a variety of control options, depending on how great the noise problem is from such devices and how much control the community wishes to exert over them. Among the options are:

a. complete prohibition of the sounding of all stationary nonemergency signaling devices
b. prohibition of the sounding of electronically amplified stationary nonemergency signaling devices
c. a night curfew on the sounding of all such devices
d. placing a maximum period of operation per hour on such devices.

Some communities may choose to place a blanket exemption on bells and chimes used in conjunction with churches and other places of religious worship.

11. Emergency signaling devices. Although they are necessary for emergency purposes and must be tested periodically, a community may wish to prohibit all nonemergency-related use of stationary emergency signaling devices, such as civil defense alarms, burglar alarms, and sirens used to alert volunteer firemen and rescue squads. In addition to prohibiting the intentional sounding of such devices for nonemergency purposes, carefully drawn limits may be placed on testing. Possible wording regarding testing is as follows:

a. Testing of a stationary emergency signaling device shall occur at the same time of day each time such test is performed, but not before ___ A.M. nor after ___ P.M. Any such testing shall use only the minimum cycle test time. In no case shall such test time exceed ___ seconds.
b. Testing of the complete emergency signaling system, including the functioning of the signaling device and the personnel response to the signaling device, shall not occur more than once in each calendar month. Such

> testing shall not occur before __A.M. nor after __P.M. The time limit specified in subsection (a) shall not apply to such complete system testing.

Additionally, for communities that experience frequent noise disturbances due to the sounding of burglar alarms on stores, houses, or motor vehicles, the noise ordinance may require that such alarms automatically terminate within a specified period of time after activation. The time limit should be chosen partially on the basis of police response time to burglar alarms in the community.

12. Motorboats. Communities that have lakes, rivers, or other waterways on which private boating takes place in their jurisdiction may consider regulating the maximum sound level of motorboats using these waters. The regulation may specify a measurement distance of 50 feet from the path of the boat or the nearest shoreline, whichever is less. This dual measurement distance can act to keep the noisier boats away from the shoreline, thus decreasing the noise disturbance to beach areas and waterside parks.

13. Domestic power tools. To avoid some of the noise problems caused by home use of power lawnmowers, other powered garden equipment, and power tools such as saws and drills used outdoors in residential areas, the community may wish to have a curfew on the operation of such equipment where a noise disturbance is caused at another residence or within a noise-sensitive zone. The hours of the curfew should be carefully determined so that ample time for use of the equipment is allowed while still protecting residents during sensitive hours, such as late evening on weekdays and early Sunday morning.

14. Tampering. In order to more fully implement and interface with the EPA new product noise emission standards under Section 6 of the Noise Control Act, a community may adopt an anti-tampering provision similar to that contained in Section 10 (a) (2) of the act. This would make it a violation of local law, in addition to federal law, to remove noise-control devices incorporated into new products to bring them into compliance with Section 6 noise regulations and to use such products after the noise-control devices have been removed. A suggested phrasing of such provision is:

> [The following acts or the causing thereof are prohibited:]
>
> a. The removal or rendering inoperative by any person, other than for purposes of maintenance, repair, or replacement, of any noise-control device or element of design or noise label of any product identified under Section — [state and federal laws and regulations (paragraph 6, p. 302)]. The administrator may, by regulation, list those acts which constitute violation of this provision.
>
> b. The use of a product, identified under Section __ [as in (a)], which has had a noise-control device or element of design or noise label removed or rendered inoperative, with knowledge that such action has occurred.

In addition, if the state has noise emission standards for products applicable at the time of sale, this provision may also be applied to tampering with these products.

As an additional subsection, the community may wish to declare it a violation of local law to tamper with any sound monitoring devices utilized by the community for environmental noise studies or enforcement of the ordinance.

15. Noise-sensitive zones. The provision relating to noise-sensitive zones should be carefully drafted due to possible First Amendment problems and because the activities protected by this provision demand lower sound levels than the other classes of activities protected by the ordinance. In the first instance, to avoid First Amendment problems, the prohibition should be directed toward sounds that actually cause or tend to cause a disruption of activities in the protected zone. Without such a relation to disruption of sensitive activities, a noise-sensitive zone provision could be found unconstitutional by a court, either on its face or as applied to activities, such as picketing or public speaking, protected by the First Amendment.*

Second, for both First Amendment reasons and to provide notice that noise levels entering and within noise-sensitive zones may need to be lower than in the general environment to protect patients recuperating in hospitals and students studying in schools, there should be a requirement that conspicuous signs be posted indicating the areas designated as noise-sensitive zones.

A suggested provision is as follows:

[The following acts, and the causing thereof, are declared to be in violation of this ordinance:]

a. creating or causing the creation of any sound within any noise-sensitive zone, designated pursuant to Section ___ [Article IV Section: Noise-sensitive zone recommendations], so as to disrupt the activities normally conducted within the zone, provided that conspicuous signs are displayed indicating the presence of the zone, or
b. creating or causing the creation of any sound within any noise-sensitive zone, designated pursuant to Section ___ [as in (a), above], containing a hospital, nursing home, or similar activity, so as to interfere with the functions of such activity or disturb or annoy the patients in the activity, provided that conspicuous signs are displayed indicating the presence of the zone.

16. Vibration. Communities who wish to control nuisance vibration in addition to noise may wish to add a provision prohibiting such vibration in this

*See, for example, *Grayned* v. *City of Rockford* (408 U.S. 104 (1972)), in which the Supreme Court upheld a municipal "quiet zone" provision which prohibited outdoor noises that caused a disruption of school activities. In the facts of the case, the defendants (persons picketing outside a school) were shown to have disrupted classroom activities.

article of the ordinance. Due to the difficulty of making quantitative measurements of vibration levels and the problems of actually correlating quantitative levels with human perception and attendant health and welfare effects, it is better to key violations to an affected person's "vibration perception threshold." For this purpose, *vibration perception threshold* may be defined as

> the minimum ground- or structure-borne vibrational motion necessary to cause a normal person to be aware of the vibration by such direct means as, but not limited to, sensation by touch or visual observation of moving objects.

Given this definition, the substantive provision may be as follows:

> [The following acts, and the causing thereof, are declared to be in violation of this ordinance:] Operating or permitting the operation of any device that creates vibration that is above the vibration perception threshold of an individual at or beyond the property boundary of the source if on private property or at ___ feet (meters) from the source if on a public space or public right-of-way.

This completes the list of substantive provisions regarding the use and operation of miscellaneous activities and sources. The next three articles to be discussed provide performance standards for sound emission or exposure levels in three major areas: receiving land uses, mobile sources, and land use planning.

Sound Levels by Receiving Land Use

A useful general means of regulating environmental noise, regardless of source, is to set performance standards specifying the maximum sound level (in dBA) that may pass from one parcel of land into another. There are numerous variations on this type of regulation and each community will have to decide which variation would best achieve the community's goals and remain within the resources allocated for the noise control program.

Performance standards for this type of provision are based on categories of land use. Generally, the three broad categories of land uses or zones are used: residential, commercial, and industrial. However, a community may further subdivide one or more of these categories, depending on the policies and needs of the community. The definitions for the categories chosen should be inserted in the definitional article of the ordinance.

After delineating and defining each land use, the health and welfare needs and the goals associated with them should be determined. In this connection, the community may wish to set down its health and welfare goals in a formal document or report.* In determining the needs resulting from the types of activities

*See, for example, the document published by the New York State Department of Environmental Conservation, *Prevention and Control of Environmental Noise Pollution* (1973).

associated with each class of land use, the following considerations are important: (1) for residential land use, the sound levels should adequately protect speech communication during the daytime and should allow rest, repose, and sleep during the night, (2) for commercial land uses, sound levels should adequately permit speech communication during all hours, and (3) for industrial land uses, the levels should take into account hearing conservation criteria.

Once goals related to health and welfare considerations have been decided, the community must study the impact of attaining these goals on the actual sources of noise in the community. First, the community should determine the technical feasibility of achieving such levels for major noise sources. Then, the economic impact on the sources in potential violation and on the community as a whole will need to be reviewed. Based on the results of these studies, the community will determine the practical feasibility of attaining the sound level goals identified to protect health and welfare. Very often, particularly in denser urban areas, these goals may not be immediately attainable due to technical and economic factors. In these cases, the levels actually chosen should reflect the levels best attainable, taking into account economic and technical considerations.

Another factor influencing the levels to be chosen is the duration, if any, over which the sound level must endure in order to cause a violation. The simplest, and most easy to enforce, way to handle the duration issue is to set a standard that specifies the maximum level that may be received into each category of land use at any time. This method is recommended by the EPA/NIMLO Model Ordinance* and is used by certain communities. Its advantage is that it is easy to enforce. Measurements need be taken for only a relatively short period of time in order to establish a violation and sophisticated statistical sampling methods or equipment need not be used. There are at least two disadvantages, however. Because the levels represent maxima with no duration factor, they should be set at a higher level than if duration is accounted for so as not to unduly penalize activities with relatively high-level sounds of short duration. This, on the other hand, will then legalize sound levels slightly below the standards regardless of the duration of the sound.

A sample provision of this type is as follows:

> No person shall operate or cause to be operated on private property any source of sound in such a manner as to create a sound level that exceeds the limits set forth in Table 11-1, when measured at or within the property boundary of the receiving land use.

Another approach is to set levels based on a statistical descriptor, such as L_{10} or L_{90}, or to specify a maximum level for continuous sound, with higher levels allowable for sounds of shorter duration. The State of Washington and several communities have standards based on this latter system. The advantage of this

*EPA Doc. No. 55/9-76-003 (September 1975).

TABLE 11-1 Maximum Sound Levels by Receiving Land Use.

Receiving Land Use Category	Time	Maximum Sound Level (dBA)
Residential	*X* A.M.–*Y* P.M.	A_1
	Y P.M.–*X* A.M.	A_2
Commercial	All times	A_3
Industrial	All times	A_4

type of standard is that the duration of sounds can better be accounted for, thus allowing relatively higher levels of sounds to be received for short durations and yet having a relatively lower allowable maximum for continuous sound than would be acceptable under the type of standard specifying a simple maximum. One disadvantage of this system, however, is that enforcement is more difficult. Measurements will have to be taken over a longer period of time and appropriate calculations made, tying up enforcement personnel for longer periods of time per inspection. In the alternative, relatively expensive equipment may be purchased, which may be used to automatically monitor sites for violation determinations.

Finally, there is another system in use by, for example, the State of Illinois, which utilizes the maximum sound level approach but determines the applicable standard not solely on the basis of the receiving land use category, but additionally taking into account the land use category of the source. This method is intended to balance the needs of the receiver with the ability of a source to abate property line noise. The standard in this case would consist of a matrix of the following form:

In this matrix, the lowest value in each vertical column is that at the top of the column (e.g., B) and the highest is that at the bottom (e.g., B″). The most protected category is Residential Receiver-Residential Emitter, whereas the least protected is Industrial Receiver-Industrial Emitter. The advantage of this system

TABLE 11-2 Maximum Sound Levels by Emitting and Receiving Land Use.

Emitting Land Use	Receiving Land Use: Residential	Commercial	Industrial
Residential	$A_{day}\,A_{night}$	B	C
Commercial	$A'_{day}\,A'_{night}$	B′	C′
Industrial	$A''_{day}\,A''_{night}$	B″	C″

is that it may reduce the burden on emitters without significantly compromising the interests of the receivers. On the other hand, the complexity could make it more difficult to enforce.

Regardless of which of the above types of standards a community chooses to enact, there should be a detailed description of the appropriate measurement methodology to be used in determining compliance set forth in administrative guidelines. Guidelines regarding other enforcement issues pertaining to the standards should also be specified. Among such issues are: tolerances on measurements, correction factors, and how to handle multiple sources and nonconforming land uses.

The measurement tolerance specified should be determined by the community in accordance with the type of equipment it will use in enforcement and the standard tolerances associated with such equipment. The community should also determine whether any correction factors should be used to account for particular circumstances (such as reflecting surfaces, certain site configurations, etc.). These should be set forth in detail in the enforcement guidelines. To assist enforcement officers in handling decisions arising from the presence of sources on several different properties which each cause (or the cumulative effect of which causes) a violation of receiving land use standards, the enforcement guidelines may set out measurement and policy criteria to help the officer determine how to handle the situation. Finally, the community will need to develop a policy with regard to nonconforming land uses and how this standard will apply to them. For example, if one home is surrounded by several industries, the community may determine that there should be no enforcement of the residential receiving land use standards in this predominantly industrial area. However, if the facts show that the home had been in a rural setting for many years before the industrial uses were developed around it, the community may decide that the equities lie in favor of protection of this residence from the sound emissions of the adjoining land uses. These policies should be set forth in a way that will provide a general rule for the majority of situations but will also allow flexibility for unusual circumstances.

Because this type of provision will apply to all sources of sound located on private property unless specifically exempted, it is important to study carefully the potential sources covered and determine whether any of these should be exempted and covered in another type of provision in the ordinance. One example is construction equipment. Application of even the least stringent levels to a construction site may severely restrict the activities at the site. Thus, it is a better course of action to exempt construction sites from the application of receiving land use standards and develop a provision specifically covering construction noise (see p. 307). Problems would also occur if these standards were applied to stationary emergency signaling devices, such as civil defense sirens and burglar alarms, and stationary nonemergency signaling devices, such as bells and chimes. These can be controlled more effectively with separate provisions developed specifically for them (see p. 309). For equipment such as lawnmowers and snowmobiles, which can be operated near the property boundary as well as in the center of the emitting property, the community should exempt these sources from receiving land use standards and handle them separately, particularly if the receiving land use standards are maximum levels rather than statistical levels. Lawnmowers can be handled with other domestic power tools (see p. 310), and snowmobiles should be covered by recreational motorized vehicle standards (see p. 319).

Two other sources should be exempted completely. Interstate railroad locomotives and cars should be exempted due to preemption under Section 17 of the Noise Control Act. The unamplified human voice should be exempted to avoid any conflict with the First Amendment guarantee of freedom of speech.

Finally, it is recommended that this provision cover only sources operating on private land uses; sources on public rights-of-way should be covered separately due to the nature of the different sources on public ways and to the problems that will arise with preemption under Section 18 of the Noise Control Act.

Motor Vehicle Sound Levels

All motor vehicles and motorized recreational vehicles will be covered in this next article of the ordinance. The most important section of this article is that which specifies maximum sound level limits for the operation of motor vehicles and motorcycles on public rights-of-way. Other provisions apply to mufflers, refuse collection vehicles, horns, and recreational vehicles.

1. Motor vehicles and motorcycles on public rights-of-way. This section specifies maximum sound levels for the operation of motor vehicles and motorcycles on public roads. The standard should indicate the appropriate breakdown of vehicle classes covered, the speed limit breakpoint if both low-speed and high-speed passby standards are to be incorporated, the standard measurement distance, and, finally, the maximum levels for each vehicle class and speed category. A sample provision is as follows:

> No person shall operate or cause to be operated a public or private motor vehicle or motorcycle on a public right-of-way at any time in such a manner that the sound level emitted by the motor vehicle or motorcycle exceeds the level set forth in Table 11-3.

In Table 11-3, a measurement distance of 50 ft (15 m) and a speed breakpoint of 35 mph were chosen because they are the criteria used in the Interstate Motor Carrier Noise Emission Regulations* promulgated by the EPA under Section 18

*40 C.F.R. Part 202 (1977).

TABLE 11-3 Motor Vehicle and Motorcycle Sound Limits, Measured at 50 Feet (15 m).

	Sound Level (dBA)		
Vehicle Class	Speed Limit ⩽35 mph	Speed Limit >35 mph	Stationary Run-up
Interstate motor carrier vehicle (GVWR/GCWR over 10,000 lb)	86	90	88
All other motor vehicles of GVWR/GCWR over 10,000 lb	A_1	A_2	–
Any motorcycle	B_1	B_2	–
Any other motor vehicle or combination of motor vehicles towed by any motor vehicle	C_1	C_2	–

of the Noise Control Act. The example incorporates the EPA standards in the first line of the table due to the preemption under Section 18(c) of all nonidentical standards applicable to vehicles over 10,000 lb GVWR or GCWR operated by interstate motor carriers. However, as indicated by the second vehicle class in the table, vehicles of the same weight rating that are not operated by interstate motor carriers are not covered by preemption; communities may, if they desire, set different standards for these vehicles. This may, however, complicate enforcement efforts because enforcement officers cannot always readily determine whether or not a truck passing them is operated by an interstate motor carrier. Thus, it is much simpler to apply the federal standard to all vehicles in this weight class.

In determining levels for the other two major classes of vehicles, motorcycles and all other vehicles (primarily automobiles), a community should first look to the state and to neighboring communities to determine whether they have noise control provisions of this type. Because automobiles and motorcycles travel from community to community and often from state to state, it is important to have uniformity of noise emission levels for this standard. Thus, unless there is a good reason not to do so, a community should adopt state sound emission levels for the operation of motorcycles and automobiles or, if there is no state standard, the levels of a neighboring community.

As with the receiving land use standards, the community should develop and publish administrative guidelines indicating the methodology by which the sound emission measurements are to be performed in enforcement. In addition to equipment and weather specifications, the methodology manual should specify any open site requirements, correction factors for measurements at different distances, operating mode of the vehicle, and any other criteria that are important to the enforcement of the provision. With regard to measurements on trucks, buses, and other vehicles covered by the Federal Interstate Motor Carrier Regulations, the DOT methodology* should be applied. However, once a violation of the community-adopted federal standard has been found, the community may utilize its own enforcement procedural requirements and sanctions on the interstate motor carriers, regardless of federal practice.

2. Adequate mufflers or sound-dissipative devices. This provision, which requires adequate mufflers or other types of sound-dissipative devices on all motor vehicles and motorcycles, is already in effect in one form or another in many states and communities. If the community finds that it already has such a provision, the existing provision should be reviewed to determine whether it is adequate for noise control purposes. The community should also decide whether the provision should remain in the code of law in which it was found (generally, the motor vehicle code) or whether it should be moved into the noise control code for enforcement by the government department implementing the noise ordinance.

3. Motor vehicle horns and signaling devices. Due to the large amount of annoyance which can be caused by motor vehicle horn honking, especially in dense urban areas, a community may wish to ban horn honking except for emergency purposes. However, before actually adopting a provision of this type, the community should determine how seriously it wishes to enforce it. Unless the community is to have a worthless law on the books, it will have to conduct a comprehensive public education program to alert all drivers to the existence of the law, and the police must be willing to apply consistent, nondiscriminatory enforcement. This approach has worked in many foreign cities, such as Paris and Vienna, but it must be applied consistently.

4. Refuse collection vehicles. There are several approaches to the control of noise caused in the collection of refuse. Because the noise produced by refuse collection and the operation of refuse compactor units on trucks is more disturbing during the nighttime, the simplest way to control the problem is to place a night curfew on refuse collection and/or the operation of truck-mounted refuse compactors in residential areas and noise-sensitive zones. Additionally, a community may wish to specify a maximum sound level for the operation of truck-mounted compactor units.

5. Standing motor vehicles. In order to control the noise caused by trucks and other vehicles idling or parked with auxiliary equipment (e.g., refrigeration units) operating, the community may wish to place time limits on the period for which such operation may be allowed in or near residential areas or noise-sensitive zones. Such time limits may be expressed as X minutes per hour during specified nighttime and early morning hours.

*49 C.F.R. Part 325 (1977).

6. Recreational motorized vehicles operating off public rights-of-way. This type of provision can set maximum sound level limits on the operation of all motorized recreational vehicles, whether subject to licensing or not, which are operating off a public right-of-way. Vehicles of this type include snowmobiles, trail bikes, racing vehicles, go-carts, dune buggies, and any other motorized recreational vehicle operating on land. Sound emission limits for the operation of each class of vehicle on private property can be set forth in a table similar to the table for vehicles operating on public rights-of-way (Table 11-3). The measurement distance can be specified as 50 feet or more from the path of the vehicle. This dual criterion for measurement distance is needed to give greater flexibility to the enforcement methods. Enforcement can be difficult due to the problem of determining actual measurement distance when there is no set path for operation of the vehicle.

This article on motor vehicles concludes the portions of the ordinance that contain use and operation prohibitions and standards for maximum sound emission levels of various specified sources.

Land Use Planning

In order to decrease and/or prevent the many environmental noise pollution problems that can result from poor land use planning, the community may wish to develop guidelines regarding noise criteria for planning. These guidelines may either be placed in the noise control ordinance or may be included in separate community land use planning guidelines. Regardless of where they appear, the planning guides to minimize environmental noise problems should provide criteria for the placement of residential structures and sensitive institutions (for example, schools, hospitals, nursing homes) and should eliminate incompatible use of land in all newly developing areas.

With regard to the construction of housing and structures to be used for noise-sensitive activities, the community should set guidelines for the maximum site exposure to sound. The sound levels should be measured or estimated for a reasonably long time period (e.g., 1 day) and expressed in L_{dn} (day-night energy average sound level). An example of such guidelines are those developed by the U.S. Department of Housing and Urban Development, which has site noise exposure criteria used to determine the acceptability of a site for use for public housing or other forms of HUD funding.*

In order to be enforceable in practice, this type of provision should not merely specify site environmental sound criteria for zoning or land use planning purposes, but should also include mechanisms for requiring applicants for building permits for these types of structures to set forth data indicating that the ambient noise on the site has been studied and that it falls within the criteria for residential or noise-sensitive land use purposes.

*U.S. Department of Housing and Urban Development Circular 1390.2.

Another type of sensitive land use is outdoor recreational areas and parks. Outdoor theaters and amphitheaters are the most noise-sensitive recreational uses, but others, such as children's playgrounds, picnic grounds, wildlife sanctuaries, zoos, and botanical gardens, are places where a quiet atmosphere is important to the full enjoyment of the recreational area. To ensure that all new recreational areas will meet minimum standards of ambient noise, the community may wish to include criteria for outdoor ambient sound levels using L_{dn} for all public and private outdoor recreational areas. As with residential structures, these criteria should be backed up with an application requirement which assures that the proponent of the project has conducted the appropriate sound level studies to determine compliance with the noise criteria.

Because commercial and industrial uses of land are not generally sensitive to exterior noise, the community does not need to develop ambient sound level site criteria for these uses. However, because such uses often contain activities that cause exterior sound, there should be some planning criteria controlling the amount of sound that a new or substantially modified commercial or industrial activity may be allowed to emit into the surrounding land uses. It is recommended that the standard for the maximum sound levels for receiving land uses (see pp. 313-316) be used as the planning criteria for new or modified commercial and industrial sources. Thus, a person applying for a building permit for a structure to be used for a commercial purpose—for example, a department store—must include information in the application to show that the activities conducted when the property will be in full operation will not cause sound levels on nearby land uses to exceed the applicable receiving land use standards for those uses. This requirement is intended to ensure that all necessary sound attenuation measures are incorporated at the earliest phases of site use planning and building design.

The land use planning guidelines should also incorporate criteria for new and expanded transportation systems, such as highways, rapid transit systems, and airports. These criteria should ensure that sound from the transportation system, when operating at its design capacity, will not exceed the values specified for residential areas and noise-sensitive zones.

The community may also wish to place requirements on its own activities in planning and zoning. Requirements can include an identification of existing and projected noise sources and associated sound levels for areas covered by a proposed zoning ordinance or comprehensive plan; these requirements should include a means to ensure that appropriate measures would be taken to avoid violation of any provision of the noise control ordinance. In addition, this provision may prohibit the approval of any zoning change application unless the required site feasibility study shows the impact of existing and projected noise sources on the site for a specified number of years and the projected noise impact of the completed intended use on surrounding areas. Such study must ensure the use of adequate measures to avoid violation of the noise control ordinance.

Another type of provision a community may wish to enact is a "truth in selling or renting" provision regarding the exterior sound level on sites containing structures or proposed structures for residential uses. This provision would require written disclosure to all potential buyers or renters that the site of the dwelling is exposed to sound levels in excess of the community criteria based on health and welfare effects. A suggested provision, from the U.S. EPA/NIMLO Model Community Noise Control Ordinance, is as follows:

> No person shall sell or rent, or cause to be sold or rented, any structure or property to be used for human habitation, where the structure or property is exposed to sound levels regularly in excess of (an L_{eq} in any hour of ___ dBA)/(an L_{dn} of ___ dBA), without making full written disclosure to all potential buyers or renters of the existence of such sound levels and of the nature of the sources. The administrator of the ordinance shall develop a standard format for written disclosures, which shall include information on the effects of noise on human health and welfare.

Because this provision may unfairly penalize private owners of single-family residences who bought homes in high-noise areas before the community passed this type of provision, the community may wish to develop a means of equitably handling this situation. The least acceptable means of dealing with this problem would be to simply exempt such persons from the operation of the provision; this would merely serve to continue the problem. A better way might be to pay the existing owner the difference between the fair market value of the house if it were in a quiet environment and the value in the noise-exposed environment obtainable with the "truth in selling" provision; this would, of course, assume that the owner had originally purchased the house for the former value. The best solution might be for the community to purchase houses in extremely high noise areas and sell the land for less sensitive uses, such as industrial activities, and provide community funds or a tax-incentive program for owners of homes in intermediately high noise areas to provide sound-attenuation measures for the exterior walls of their houses.

To ensure due process to persons applying to the community for building permits, zoning variances, or any approval in connection with the land use provisions discussed herein, the ordinance containing these provisions should specifically set forth the right of applicants to appeal adverse decisions in an appropriate court of law.

After all the substantive provisions of the noise control ordinance have been set forth, two general types of articles remain: (1) an article specifying exceptions from the ordinance, special variances, and/or permits, and (2) an article setting out the enforcement provisions.

Exceptions and Variances

Once the entire scope of the substantive provisions in the ordinance has been determined, it is then necessary to decide whether any of the activities or sources covered should be exempted from the application of the ordinance or allowed if certain requirements are met. Such special treatment may involve an exception, a variance, or a permit, and the provisions enumerating them should be set forth in a separate article.

An *exception* can be applied to any activity or source, or class thereof, to which the ordinance will not, under any circumstances, be applicable. Exceptions are generally only given to extremely important activities and/or sources, for which noise abatement measures would be extremely difficult or impossible to apply. It is recommended that only one exception be given in a noise control ordinance—an exception for sounds produced in response to an emergency or while doing emergency work. An emergency may be defined as "any occurrence or set of circumstances involving actual or imminent physical trauma or property damage that demands immediate action."

For other sources and activities that may not be able to comply with the substantive provisions of the ordinance immediately at the time the ordinance goes into effect, but which can be made to comply in time, the community should have a provision to grant *variances* for time to comply. A variance is simply an agreement by the community government not to prosecute a particular violation of the ordinance if the violator meets certain conditions imposed by the community. Thus, under a variance procedure, owners or operators of sources regulated by the ordinance, such as factories, construction equipment, and power plants, may apply to the community government for a variance for additional time to comply, showing that compliance within the required time limit would constitute a hardship to the applicant or other affected persons. If good cause is shown, the community may then grant the variance, specifying all necessary conditions and including a schedule for achieving compliance. The criteria that the administrative officer must consider in determining whether to grant or deny the variance should also be included in this provision.

Another type of procedure may be needed to grant special variances to allow certain other types of activities, such as construction activities at times other than the permitted hours of operation. The special variance procedure may be similar to the procedure to obtain a variance for time to comply, except that the special variance will apply only to discrete activities conducted over a relatively short period of time. In lieu of a special variance procedure, the community may choose to have a *permit* system for these types of activities, similar to community permit systems for parades or building additions. However, technically, the special variance procedure is the correct one to use. As with the variance for time to comply, a special variance can only be obtained for good cause shown, taking into account the applicable criteria concerning the grant or denial of a special variance.

In order to assure due process for applicants denied variances or permits, a section regarding appeals of adverse decisions should be included in this article.

Enforcement

The final article of the ordinance should set forth the enforcement requirements of the ordinance and miscellaneous legal provisions, such as the severability clause and the effective date.

The sanctions for violating the ordinance should preferably be civil money penalties, with a specification that each day of violation of any provision of the ordinance will constitute a separate violation. Although older community noise ordinances often contained criminal penalties, such as a short jail sentence, this is not recommended.

Because the primary purpose of a noise control ordinance is to abate noise sources in the community, an abatement order is an important alternative to issuing a notice of violation, which will result in assessment of a civil money penalty. Under the abatement order procedure, a person who has been cited for violating the noise ordinance due to a sound source that exceeds the ordinance performance standards (e.g., receiving land use standards or motor vehicle sound emission limits) may be ordered to bring the source into compliance within a certain time period. If compliance is properly achieved, no civil money penalty or other sanction will be applied to the violator.

With the completion of the article on enforcement, the community noise control ordinance is a complete document, ready for enactment into law by the appropriate legislative body and enforcement by the designated government department. However, as indicated several times previously, the legal document is not fully complete until an administrative document containing measurement methodologies for the various performance standards has been prepared. This document should include the appropriate measurement methodology for each of the following provisions, if they have been included in the community ordinance: (1) receiving land use sound level standards, (2) construction site L_{eq} standard, (3) motor vehicle sound emission levels, and (4) land use planning criteria.

Included in the measurement methodology document should be specifications for the minimum requirements for sound level measuring equipment, calibration requirements, minimum weather conditions, microphone height, and any other general conditions required for valid measurements. Additionally, for receiving land use sound levels, the manual should specify the minimum duration of measurement, the appropriate location of the microphone, and, if a statistical level standard is being used, the appropriate means of calculating the level. Similarly, for a construction site L_{eq} standard, the document should specify the minimum measurement time and the means by which the L_{eq} should be calculated. Motor vehicle sound level measurement methodology can be modeled after the DOT methodology for pass-by and stationary run-up tests for interstate motor carrier vehicles.* Additional criteria will have to be specified, however, for the measurement of sound emissions from off-road recreational vehicles. Finally, the land use planning enforcement criteria should include the measurement methodology for determining L_{dn} and L_{eq}, as well as guidelines for determining the impact of future noise sources, particularly transportation systems.

*See methodology set forth in 49 C.F.R. Part 325 (1977).

12

Control of noise in the home

Paul Jensen
Bolt Beranek and Newman, Inc.
Cambridge, Mass.

We live and work in a noisy world. Daily, we are bombarded by noise from cars and trucks, construction and industrial machinery, and office and household equipment. No corner of our continent is too remote to escape an airplane's drone, and once-quiet woodlands are readily invaded by snowmobiles. In 1974, a sizable group of Harlem residents were asked to list, by importance, the problems afflicting their section of New York; noise—the accumulated sounds of city living—was number 8 on their list.[1] A year later, and an ocean away from New York, a man shot his neighbor. The reason? He told the police he could no longer tolerate the noise of her incessant piano playing.

The message here is this: if we cannot find quiet elsewhere, we should make very sure that we can find it in our homes. Quiet buildings can be built, and even noisy houses and apartments can be quieted. They are, after all, buildings with acoustical qualities, qualities that are determined by the physical properties of materials within the space that is divided into house or apartment rooms. And these qualities can be thoughtfully planned to provide the desired acoustical environment.

In this chapter, we will discuss the noise sources (both outside and inside) that we are exposed to in our homes, how they affect us, how they are being dealt with through building codes and how we can contend with them.

NOISE SOURCES

Outside

Ask a noise control engineer to name the major source of noise that penetrates American homes from the outside. "Highway noise," he will answer and will go on to describe the many sources of highway noise: road surfaces, road conditions, the number of vehicles using the road, the kinds of vehicle using the road, what kind of tires those vehicles are riding on, length and width of the road itself, its distance from nearby dwellings, the landscaping along the road. All these, and more, are included in the term *highway noise*.

Second to highway noise in its impact on American homes is noise from airplanes and airports. A word of caution: the number of homeowners markedly disturbed by airport-related noise is small compared to the number of Americans whose homes are exposed to highway noise. Yet for several million people, airport noise is a special problem. Scientists at Bolt Beranek and Newman, a Cambridge, Massachusetts, consulting firm specializing in acoustics, estimate that the homes of 5.9 million Americans are *significantly* impacted by airport noise, and that 1.6 million of that number experience *severe* impact daily.[2] Why do they stand for this noise? Why don't they move to quieter neighborhoods?

There is no simple answer. Economics, social and cultural pressures, convenience, apathy, or inurement to noise are all present in neighborhoods where residents are subjected to noise that others could not or would not tolerate. These questions lead to another, which is the most philosophical in noise control engineering: When does sound become noise? A favorite anecdote of a Danish doctor concerns an elderly woman who had lived for many years in an apartment near an elevated railway structure. Trains thundered past her window so frequently that the doctor chose her apartment as ideal for conducting sound level tests. "May I," he asked politely, "bring my equipment into your home and measure the noise?" And the old woman, puzzled, asked "What noise?"

In terms of noise control in the home, however, houses in areas *significantly* or *severely* impacted by airport noise present special problems of insulation from noise, frequently more specialized than the more ordinary problems addressed in this chapter.

Inside

When it comes to noise inside our homes, we are often our own worst enemies—or our neighbors'. We have equipped our homes with vacuum cleaners, blenders, washers and dryers, dishwashers, disposals, hair dryers, power tools, and loudly flushing toilets. We entertain ourselves with radios, television sets, and a variety of musical instruments.

How do we sound to our neighbors? A study of neighbor noise in apartment houses brought forth some interesting facts.[3] More than 50 percent of the persons interviewed complained about some mechanical installation in the next or a neighboring apartment, but an equal number objected to noise from the stairway; more than 30 percent complained about banging of doors; more than 25 percent were being disturbed by neighbors playing musical instruments. Radios and televisions, which might have been expected to rate high in annoyance, bothered about 15 percent of the respondents, about the same number that objected to the sounds of neighbors arguing or fighting with each other.

In a study of consumer-product noise conducted by Bolt Beranek and Newman in 1972, 33 home appliances were judged for their mean sound level and their sound-level range.[4] Figure 12-1 lists the equipment and shows how sound levels varied. This figure shows clearly the considerable range in sound level between the quietest and the noisiest models, a range of from 15 to 30 dB. Remember, a 10-dB increase in sound level is considered by most listeners to *double* the noise of a piece of equipment, and a 20-dB increase is equal to a *quadrupling* in noise.

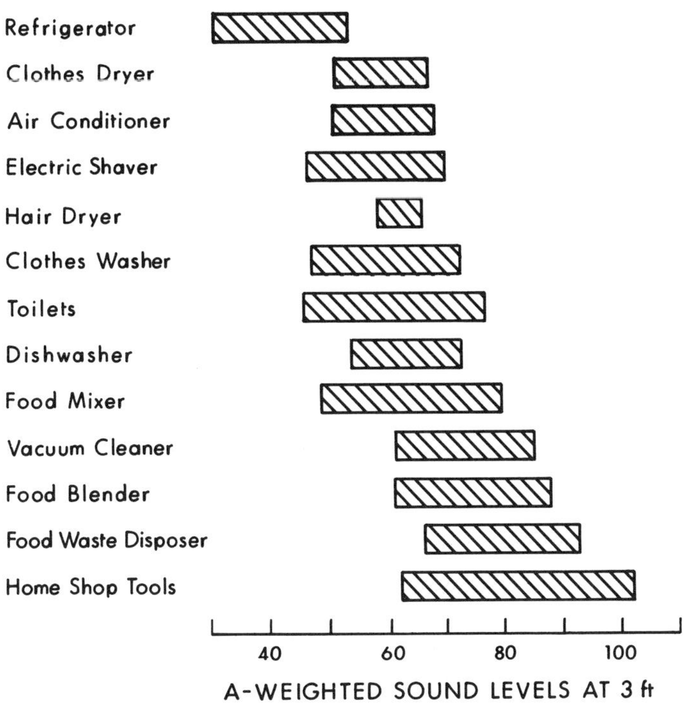

Fig. 12-1 Typical sound levels of appliances.

THE EFFECTS OF NOISE IN THE HOME

At home, humans tolerate noise in different degrees in different places. We accept higher sound levels in such areas as kitchens, family rooms, and laundries than we do in bedrooms. In our workshops, we are not annoyed by the noise of an electric saw or sander, although the much lower whir of a hair dryer may bother us if we are trying to listen to television.

Acousticians measure the annoyance of noise in the home by two methods: actual sound levels and sleep/speech interference. The simplest measurement is reported in A-weighted decibels; this single-number measurement can be made directly with a sound level meter, and the A-level correlates very well with most people's judgments of the loudness and/or annoyance of many common sounds (see Chapter 2).

In terms of A-weighted decibels, Americans rarely encounter really loud noises at home. The Occupational Safety and Health Act, which governs noise exposure in industry, allows for peak sound levels of 140 dB. The loudest sounds in a home acoustical environment are generated by power tools: electric saw (90–110 dBA), router (85–100 dBA), sander (85–105 dBA), gasoline-powered lawn mower (80–100 dBA).

Another means of measuring the annoyance noise in the home may be to judge it in terms of sleep interference, speech interference, or hearing damage.

SLEEP INTERFERENCE

Chief among annoying noises are those that interrupt sleep. Humans treasure sleep; if we miss a night's sleep, or if sleep is broken, we may be short-tempered and weary and may yearn for a midday nap. In fact, researchers have proved that sleep is essential for physical and emotional health.

Noise can interfere with sleep even when the sleeper is not consciously awakened. Canadian scientists, studying the impact of noise on sleepers, exposed test subjects to a recording of the noise from a passing truck, played at a selected peak level several times a night.[5] At the 40-dBA level, they reported, there was a 5 percent probability of awakening the subject, and at the 70-dBA level, a 30 percent probability. However, when significant changes in the electroencephalogram records of the sleeping subjects were taken into account, the probability of a shift in sleep level (up to and including awakening) rose to 10 percent at 40 dBA and to 70 percent at 70 dBA (see Figure 12-2).

SPEECH INTERFERENCE

A measurement has been worked out to determine the level at which noise interferes with speech. The (Preferred) Speech Interference Level (PSIL or SIL) was

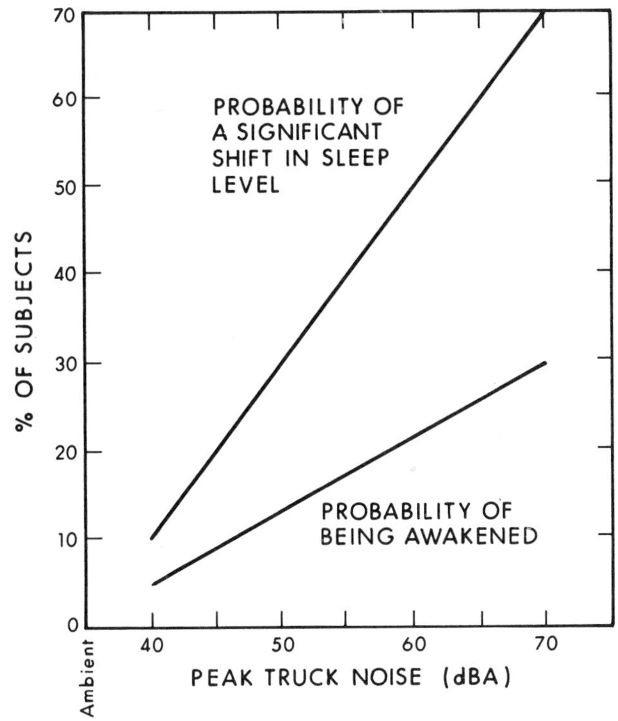

Fig. 12-2 Sleep interference.

developed to evaluate cabin noise in aircrafts. The PSIL is the arithmetic average of the sound pressure level in the 500-Hz, 1000-Hz, and 2000-Hz octave bands. Although it is still used in aircraft, it is also used to evaluate the acoustics of rooms in homes, classrooms, auditoriums, and offices. Although the PSIL is a useful technical tool, noise in homes can be evaluated equally well with A-weighted noise ratings.

HEARING DAMAGE

Any noise greater than 85 dBA can damage hearing if the human ear is exposed to it over an extended period. The Occupational Safety and Health Act defines as compensable hearing loss a 25-dB average hearing loss at 500 Hz, 1000 Hz, and 2000 Hz. The Occupational Safety and Health Administration (OSHA) regulation is designed to minimize the risk of permanent hearing impairment from exposure to hazardous levels of occupational noise. But while OSHA protects most workers, there is no protection for people at home, who may be incurring permanent damage by using home workshop tools or recreational vehicles, or pursuing hobbies such as rifle shooting. (The National Bureau of Standards, investigating the noise generated by the firing of popular guns, reports that they typically exceed safe levels of sound by considerable margins.[6]) And hearing loss far below that named as dangerous by OSHA will block out high-frequency sounds in music, bird songs, and certain speech sounds.

Workers on noisy jobs must be made aware of the cumulative impact of noise at work plus noise at home. At work, by OSHA regulation, no worker should be exposed to sound levels in excess of the limits illustrated in Fig. 12-3. For noise exposure consisting of several periods of exposure at different levels, the daily noise dose, D, should not exceed unity (see Chapter 7). Since the daily noise dose of a worker exceeds or approaches unity, the worker may be risking permanent hearing damage. Thus, if the worker comes home and uses power tools, or drives a snowmobile, or plays an instrument in a rock band, the additional noise exposure heightens the risk of hearing impairment. This factor underscores the importance of considering total exposure of individuals—not only occupational experience. Hearing conservation is a 24-hour concern.

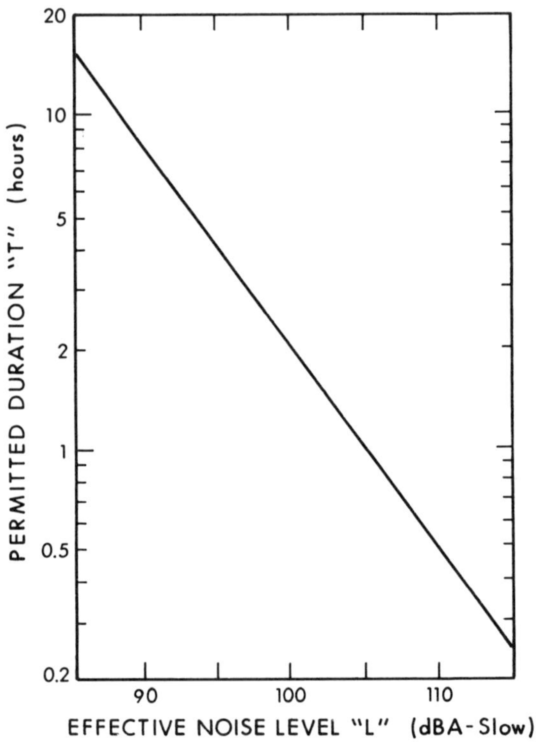

Fig. 12-3 OSHA noise exposure limits.

GOVERNMENT HELP IN ACHIEVING QUIET HOMES

In the 1960s and 1970s, both federal and community governments took the first steps toward government-regulated noise control for homes and equipment in and around homes. The Department of Housing and Urban Development (HUD), as described in Chapter 6, established guidelines for evaluating house sites and set conditions for building sites whose developers seek federal assistance.

By 1976, approximately 20 American communities had added some type of acoustical requirements to their building codes. The most comprehensive is the building code for New York City; however, although it is at the moment the best code, it also shows how far short of effectiveness even the best code falls.

The New York City building code was adopted in 1968 and took effect on December 6, 1969. It applies to all residential structures of more than two apartments. Clause 1208.0, "Noise Control in Multiple Dwellings," states that:

1. Walls, partitions, and floor-ceiling constructions separating dwelling units from each other or from public halls, corridors, or stairs shall have a minimum sound transmission class (STC) rating of 50, and dwelling unit entrance doors shall have a minimum STC of 35.
2. Floor-ceiling constructions separating dwelling units from each other or from public halls or corridors shall have a minimum impact noise rating (INR) of 0.*

In addition to these two requirements, provisions are also made to limit the noise impact from mechanical equipment located within the building. Mechanical rooms such as furnace spaces must be separated by constructions that provide a minimum STC rating of 50.

After the building is ready for occupancy, the builder is required to engage an independent firm to conduct measurements to determine whether or not the building code has been upheld.

Good acoustical separation or privacy between adjoining apartments requires that adequate attenuation has been obtained in all the possible paths by which sound from one room enters another. Even though the New York City code is the most comprehensive in the United States, it is not completely effective for two reasons:

1. It is not at this date (1976) enforced.
2. The specifying of the STC rating for walls and ceilings in no way guarantees privacy.

It is important to make absolutely clear that there is a difference between the transmission-loss values of a partition and the resulting or achieved noise reduction between rooms. It is naive to believe that requiring a party wall with a laboratory-determined STC rating will ensure adequate sound isolation in the

*See Chapter 6 for a discussion of STC ratings and of INR.

finished building. Even the call for a field-performance test of the party wall does not allow for the possibility of flanking paths not involving the party wall at all.

What is needed is a two-step certification procedure. Before an application for a building permit is granted, the building blueprints should be reviewed by the building inspector and a determination should be made as to whether the selected building components have the proper STC rating. Only then, if construction ensures privacy, should a permit be issued. And this permission should indicate only tentative approval; after the building is finished, field tests should be performed to determine if the expected privacy has been achieved.

A two-phase certification system would not be much different than the system now being used in Scandinavia and other European countries. Different rating systems or more complex measurement procedures may be used abroad, but tests of finished buildings show that after the building codes have been in use for about 4 years, about 50 percent of the investigated buildings comply with the noise-reduction requirements, only 2 percent of the buildings had noise-reduction values 10 dB below the requirement $I_a = 52$ dB.

Since it has been shown by many investigators in the past that the A-weighted sound level in dBA gives good correlation between absolute level and annoyance, Schultz has suggested that the privacy index I_p should be used in new multi-dwelling units.[7]

The privacy index is determined by the sum of two numbers: (1) the A-level difference L_A, between the source and the receiving rooms, and (2) the A-weighted level N_A of the background noise in the receiving room. Measurements in the completed building must demonstrate a value of I_p of at least 75 as a minimum requirement.

$$I_p = L_A + N_A > 75.$$

The privacy index is an efficient procedure; it also has the advantage of being simple enough to be measured by a building inspector using basic equipment. Schultz also proposed a stepwise approach to quieting housing, so that disruption would be minimized. First, he suggests, determine how much isolation is ultimately desired for housing, and express this isolation as a certain value of L_A. For the first year or so after the code is in effect, approve drawings only of construction that usually would yield somewhat better performance than the ultimate goal, for example, $L_A + 5$. Also, at first, when tests are made in the finished building, approve the building for occupancy even if it fails to meet the desired goal by, for instance, 5 dB ($L_A - 5$). In other words, allow, at first, a 10-dB margin for error during construction.

Then, gradually, as the builders learn how to improve their assembly technique to avoid leaks and flanking, narrow the permitted margin. The result, in 5 to 7 years should be significant improvements in achieved privacy in all kinds of dwellings. This system will require diligence, hard work, dedication, and cooperation, but it is the only way to achieve a better indoor environment.

THE NOISE CONTROL ACT OF 1972

In the Noise Control Act of 1972 (Public Law of 92-574), Congress declared that "it is the policy of the United States to promote an environment for all Americans free from noise that jeopardizes their health or welfare." Through the act, Congress required the U.S. Environmental Protection Agency (EPA) to order manufacturers of products that "emit noise capable of adversely affecting the public health or welfare," or which are "sold wholly or in part on the basis of . . . effectiveness in reducing noise," to give notice of the noise level of those products to the consumer.

Do these requirements make sense? Yes! Americans tend to think that noise equals power—that a noisy motorcycle, lawn mower, or vacuum cleaner is more efficient than a quiet one. Consumers, seeking quiet equipment, have been offered noisier and noisier products. The EPA effort in labeling equipment is the beginning of a reverse trend—cutting back noise to tolerable sound.

NOISE CONTROL PROBLEMS AND SOLUTIONS

Existing Noise Inside the Home

The most efficient way for homeowners to control noise in their houses is often to do it themselves. Following are a range of the most common noise-related problems and their solutions.

Problem: Buying a new appliance. How to ensure that it is quiet?

Solution: Check its tags or label for a noise rating, and check the rating against the levels listed in Figure 12-1. Have a salesperson run different models for you, so that you can compare them. Check with your friends; have they bought similar appliances? do they find them noisy or quiet? If you find a quiet model—or if you *don't* find a quiet model—write the manufacturer. Technology *is* available for noise control, but quiet appliances will be produced only in answer to consumer demand.

Problem: Kitchen-appliance noise, from disposal, dishwasher, electric can opener, electric mixer.

Solution: Disposal. First quiet the kitchen sink, if the underside is not coated with a damping material. Trowel on a damping treatment, or glue a bitumen-based sheet directly to the steel. To cut disposal noise, install a 0.5-inch-thick neoprene gasket between the device and the sink. Even more reduction in noise can be achieved by fastening 1-inch-thick, medium-density fiberglass board on the cabinet walls around the disposal.

Solution: Dishwasher. There are three sources of dishwasher noise: splashing water, motor noise, and structural vibrations. Effective treatment first demands a space larger than the machine, so that

1. The drum can be damped with a sound-deadening material.
2. Neoprene (soft waffle pattern or double-ribbed) can be installed under the legs in place of the usual nonskid pads.
3. Air space between the dishwasher and surrounding cabinets can be packed with fiberglass.
4. Flexible connectors and air chambers can be installed in the water-supply lines.

Solution: Small appliances (can openers, blenders, mixers). Never connect rigidly to walls or counter tops; mount with neoprene sleeves or neoprene pads for vibration isolation.

Problem: Room-air-conditioner noise.

Solution: The sources of room-air-conditioner noise are the high-speed blower and the compressor. High-frequency noise is radiated through the grills along with the air; low-frequency noise is structureborne. To some extent, high frequency noise can be reduced by installing sound-absorbing materials inside the air conditioner. (Where? Impossible to answer in a general text; as of 1976, more than 60 manufacturers are offering more than 1600 models for sale.) Make sure that air conditioners are vibration-isolated from the wall or window in which they are installed by placing a neoprene sleeve around the case.

Problem: Noisy fans in kitchen or bathroom.

Solution: Line the ducts with acoustical material. Or, if possible, substitute a centrifugal fan for a propeller fan and use 0.5-inch-thick acoustical lining in the duct.

Problem: Noisy water pipes.

Solution: This problem of noisy water pipes is widespread and particularly annoying in apartment buildings. It is generally caused, not by water flowing through the pipes themselves, but by water passing through the constrained spaces of valves, faucets, and other narrowed spots. When the water reaches such a spot, the velocity increases and cavitation results, causing noise and vibration, which is transmitted through the pipes to the building's structures where the pipes are supported. The building structure, in turn, acts as a sounding board, radiating the noise generated in a valve or faucet.

There are two solutions to this problem: (1) separate the pipes from the radiating structure or (2) minimize the source of noise.

Solution 1: Where pipes are supported from walls, use an oversized pipe clamp, and insert a neoprene layer between the pipe and the clamp. Where pipes penetrate walls, the opening should be large enough to allow a padding of fiberglass around the pipe. Finish the job by sealing off the padding, at the face of the wall, with a thermoplastic caulking compound.

Solution 2: Reduce the water pressure by installing a pressure regulator in the water pipe and maintain a water pressure of less than 45 psi. Another approach would to be to insert a 3-foot-long piece of 0.25-inch copper tubing between the water supply and an especially noisy faucet.

Existing Noise Outside the Home

Problem: Noise from passing vehicles and airplanes.

Solution: Research has shown that the best way to control outside noise from penetrating a home is to improve the sound-isolation values of the exterior walls of the house. Recent studies proved that the most efficient method is to add a pane of glass to each *window*, fitting it inside the present windowpane. Both the existing window and the new pane should be set in a neoprene gasket. If air leaks are noticed between the window frame and the wall during installation of the new glass, the homeowner can cut noise and prevent energy leakage by filling the leaking area with a nonstiffening thermoplastic caulking compound.

All *doors* opening into the house from the outside should have neoprene gaskets. Note that the gasket must fit well, or noise-quieting performance will be poor. If the door already has weather stripping, the neoprene gasket will add to its efficiency. In homes where doors open directly into the living room, a more elaborate way to seal off noise—and also inclement weather—is to build a little hallway around the entrance. Such a little hall will improve the acoustic performance of the door, preserve heat in winter, and keep in air-conditioned cool air in the summer.

Roof/ceiling sound isolation can be improved by blanketing the attic with 6-inch fiberglass covered by 0.5-inch gypsum board.

Will heavy draperies seal out noise? No! Noise intrusion from the outside is mostly in the low to middle frequencies; drapes have few acoustical values in that frequency range. What drapes—and carpets, upholstered furniture, and so forth—will do is to help absorb high-frequency sounds, but not necessarily because they are placed in front of windows.

Problem: How to prevent noise by good planning.

Solution: Plan ahead. When it comes to noise, prevention is far cheaper than cure. If you are lucky enough to be able to plan a dwelling, clump the rooms in the house into three categories: noisy, seminoisy, and quiet. Noisy rooms include kitchen, bathroom, furnace room, laundry room, music room, and playroom. Seminoisy rooms: hallways, family room, and dining room. Quiet rooms: living rooms, bedrooms.

In planning the house, place noisy rooms adjacent to each other, if possible, and bridge the area between noisy and quiet rooms with seminoisy rooms. Plan the electrical system so that plugs are *not* opposite each other in the walls; separate them by 16 to 24 inches. Pay particular attention to the walls and floors between noisy or seminoisy and quiet rooms. Such walls should consist of wood-stud construction with two layers of gypsum board on both sides and a blanket of fiberglass between the studs. Remember: noise can travel along flanking paιns, either through air leaks directly between rooms or through common hallways.

Floors between noisy and seminoisy rooms can be of normal wood flooring improved with a layer of gypsum board fastened to the underside. Use the same treatment between seminoisy and quiet rooms. Between noisy and quiet rooms, however, add *two* layers of gypsum board, suspend it on resilient hangers underneath the floor beams, and place a thermal isolation blanket between the beams.

Acoustical ceilings with a rating of NRC = 0.75 should be installed in furnace or laundry rooms, and doors to such noisy rooms should be solid-core wood doors with neoprene gaskets. This gasket should also close the air gap at the floor, and, if necessary, a threshold may have to be installed.

If the fresh air supply to the furnace room has to be drawn in through another room, it can be taken in through the furnace room wall duct. The duct can be made of either sheet metal or sheet rock; the lining should be 1-inch-thick fiberglass.

Bathroom doors should also be furnished with a neoprene gasket, and, if necessary, a threshold must be built up under them.

Grouping noisy rooms, separating sleeping and living areas, using clothes closets to improve the performance of the walls, gasketing doors—what do all these techniques add up to? *Sound-conditioning*. See Figure 12-4 for the layout of a house with sound conditioning.

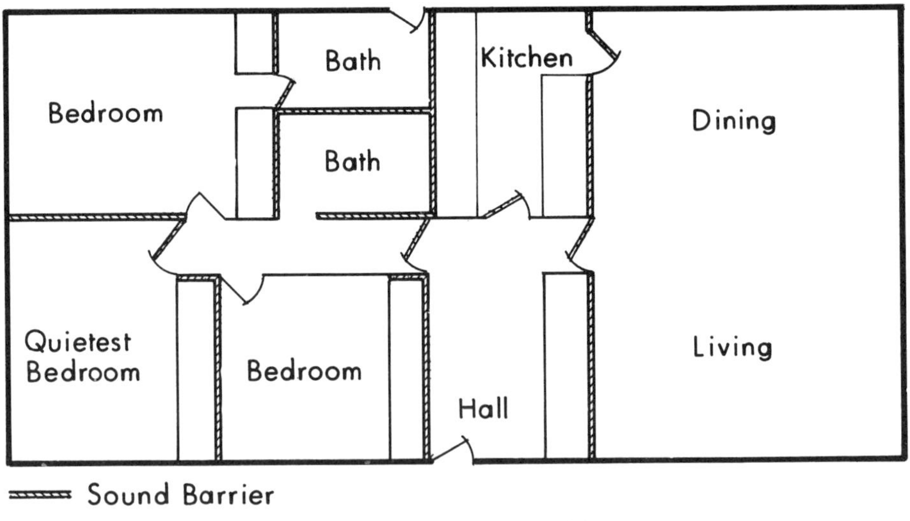

Fig. 12-4 Sound-conditioned house.

Problem: Noisy neighbors.

Solution: You may think your neighbors bother you in more ways than you can count. Acoustically speaking, however, the noise they make can be divided into two categories: (1) airborne sound sources and (2) structureborne sound sources. Airborne sources include speech, musical instruments, stereo systems—the kind of sources that transmit sound energy through the air. Insulation against these sources is called *airborne sound isolation*. Structureborne sounds include footsteps, plumbing, and banging doors. Since the character of such sources is impact against the building structure, the insulation against them is called *impact sound isolation*.

Airborne sound isolation: The New York City Building Code requires that the dividing walls between multifamily dwelling units should have an STC rating of 50. Assuming that the building is well constructed—no air leaks, no flanking transmission—then what does the rating of 50 mean? Figure 12-5 indicates the subjective impressions of various levels of insulation values and contains comments on the ease of improving poor situations. As Figure 12-5 shows, the New York City requirements are quite reasonable.

In the control of airborne sound, there are certain rules of thumb:

1. The heavier a material is, the more isolation it will provide.
2. If two materials have the same average weight, the one that is thicker and has low-average density will probably provide less acoustical isolation than the high-density "limper" material.
3. If the wall is well constructed, yet sound isolation is not good, suspect air leaks.

(Air leaks are difficult to spot visually. They can be detected, however, by listening or using sound measuring equipment. During the author's early years in acoustics, his primary job was field-testing dwellings. He and his assistant would enter separate rooms and try to talk to each other through the partition. Where they heard each other most clearly was the sound path; thus, they would stuff that area with fiberglass and seal with a thermoplastic compound.)

Flanking transmission through corridors or facade walls will seldom cause a problem between individual apartments unless the dividing wall exceeds the requirements (STC 50) and the flanking walls are very light.

If you determine that the sound isolation between two rooms is not as high as desired, yet there appears to be no holes in the wall, your best solution will be to cover the entire wall with an impervious 0.625-inch-thick gypsum wallboard skin. Details of construction depend on your degree of annoyance. If you are only marginally annoyed, the solution is to nail 1 × 2 inch furring strips on the wall about 16 inches apart, then nail a 0.625-inch gypsum board to them. More sound isolation can be obtained by placing a fiberglass blanket in the wall cavity. In cases where the noise intrusion is very severe, the gypsum board skin should

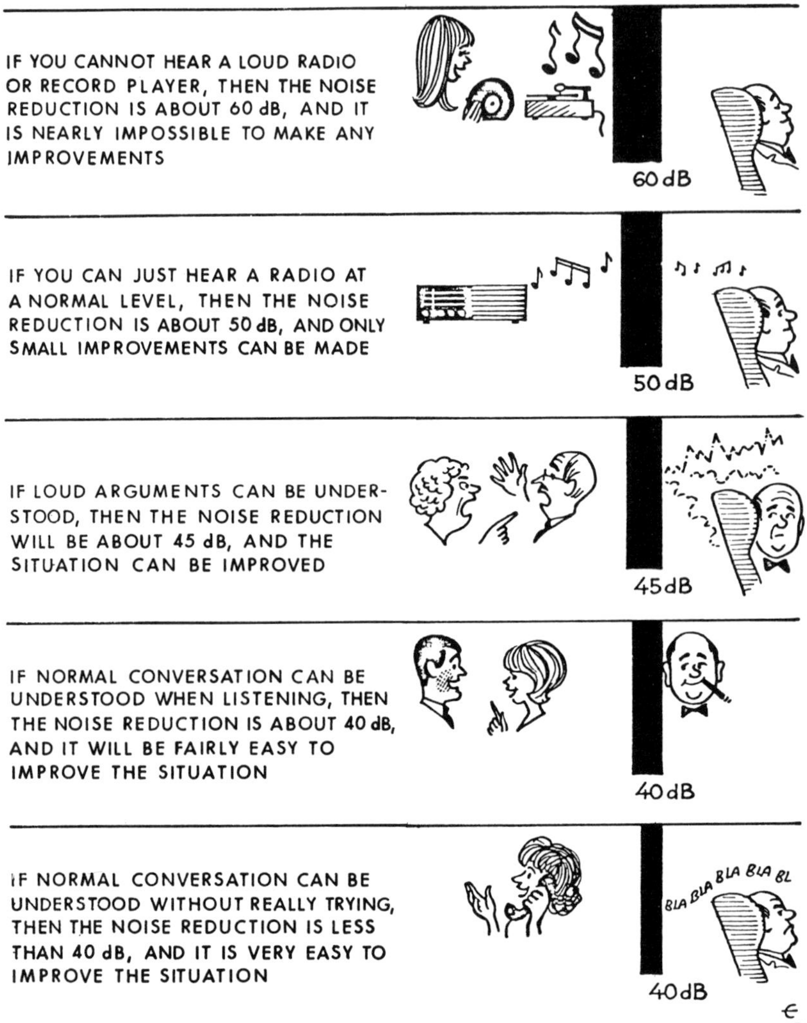

Fig. 12-5 Noise reduction between rooms.

be supported by a resilient channel, and a fiberglass blanket must be placed in the cavity.

Airborne noise coming through the ceiling can be reduced by hanging a resilient ceiling 4 to 12 inches from the existing ceiling. Consider, too, a fiberglass blanket in the cavity. The ceiling material can be a 0.625-inch gypsum board or a combination of a 0.625-inch gypsum board with acoustical tiles glued to the underside.

STRUCTUREBORNE SOUND ISOLATION

The requirement for structureborne sound isolation in the New York City Building Code is that INR equals zero. This rating has to be determined with a special measurement procedure that uses a standard tapping machine.

Several authors, and especially Schultz, have expressed some doubt as to the adequacy of the standard measurement procedure used to determine the INR. Gösele stated in 1949 that a measurement procedure is required for an acoustical evaluation of different floor construction. The test results obtained through this procedure should lead to the results that footstep noises under two different floor constructions should give test results that correspond to the loudness as perceived by the human ear. Unfortunately, the standard tapping machine that is used for the test does not give results that are in good agreement with subjective measures of intruding footstep noise. Schultz therefore suggests that the standard tapping machine be modified so the impacts on the test floor resemble normal footsteps more closely.[8]

Meanwhile, the most efficient way of reducing footstep noise is by installing carpets. Unfortunately, the carpets will normally have to be installed by your neighbor, who may not be cooperative. One solution that always works is to suspend a ceiling as described previously.

CONCLUSION

It is evident, then, that there are many sources of noise that can cause annoyance. You have the least control over out-of-doors noise sources; the only way to escape them effectively is to move to a quieter location, rather than wait for the outcome of federal or state planning in controlling environmental noise. With sensible planning, however, you can improve your indoor noise environment, and the noise from your neighbors can be controlled by the suggestions offered in this chapter. Finally, speaking about neighbors' noise, remember: *You* are *your* neighbor's neighbor.

REFERENCES

1. *New York Times*, November 21, 1974.
2. Bolt Beranek and Newman, Inc., Report No. 3186, *Massachusetts Airport Noise Study*, January 1976.
3. Jorgen, G. Ø., *Noise in Dwellings*, NBI Report No. 16, 1955 (in Norwegian).
4. Bolt Beranek and Newman, Inc., *Noise From Construction Equipment and*

Operations, Building Equipment, And Home Appliances, EPA Report No. NTID 300.1, December 1971.

5. Thiessen, G. J., and Olson, N., Community noise–surface transportation, *Sound and Vibration* 2(4): 10 (1968).
6. National Bureau of Standards, Now hear this, protect your ears . . .or don't shoot. *Dimensions*, October 1975.
7. Schultz, T. J., Acoustical privacy and noise control requirements in building codes, *Proceedings, Inter-Noise 75*, Sendai, Japan, 1975.
8. Schultz, T. J., A proposed new method for impact noise tests, *Proceedings, Inter-Noise 75*, Sendai, Japan, 1975.

The following references are suggested for further reading:

Beranek, Leo L., *Acoustics*, McGraw-Hill Book Company, New York (1954).

Beranek, Leo L., *Noise and Vibration Control*, McGraw-Hill Book Company, New York (1971).

Jensen, P., and Sweitzer, G., *How You Can Soundproof Your Home*, Lexington Publishing Company, Concord, Massachusetts (1974).

Sound Control Construction, United States Gypsum, Chicago, Illinois (1972).

Newman, Robert Bradford, Acoustics, in *Time-Saver Standards*, McGraw-Hill Book Company, New York (1974).

13

Control of noise from recreational activities

Robin Harrison, P.E.
Forest Service
U.S. Department of Agriculture
Equipment Development Center
San Dimas, California

INTRODUCTION

This chapter will consider methods and strategies to control noise generated by recreational activities. It will deal with only those sources that are peculiarly associated with such activities. It will not consider such things as barking dogs, lawn mowers, and so on, which are recreational only in the sense that they are not industrial or work-oriented. The first step, then, is to identify the noise sources with which we will be concerned in this chapter. Next, the effects of recreational noise will be considered. Finally, control strategies applicable to the particular problem will be reviewed.

The most basic prerequisite of any efficient noise control program is to have as complete an understanding of the noise problem as possible in all its aspects—technical, sociological, psychological, and so on. Many of the recreational activities that give rise to noise problems, particularly those that involve internal combustion engines as their prime movers, are focal points of subcultures with which the noise control official (unless the official is a participant) is not likely to be familiar. To help the reader to speak the language of the various groups represented, the following is a brief description of noise sources considered in this chapter along with some observations about the group involved.

Motorcycles

Motorcycles are probably the number one recreational noise problem in America (and perhaps the world) today. It's very difficult to draw the line between recreational use of motorcycles and transportation (highway) use of the same machines. Perhaps 50% of all new motorcycles sold in this country are intended exclusively for highway use. However, one-half of the total (estimated 10 million) units nationwide are in either off-road/on-road combination use, or in strictly off-road use.

Because of the magnitude and impact of the motorcycle noise problem, this chapter will deal most thoroughly with these popular vehicles. Many of the comments made here are equally applicable to other types of off-road vehicles (snowmobiles, boats, dune buggies, etc.).

Motorcycles can be divided into three classifications: highway motorcycles, combination motorcycles or trail bikes, and competition motorcycles. Typical motorcycles of these different types are shown in Figs. 13-1 through 13-3.

Highway Motorcycles. Highway bikes are the most numerous, but one of the easiest problems with which to deal. These bikes, which are designed for use strictly on the road, are equipped with adequate mufflers when sold new. All

Fig. 13-1 Highway motorcycle.

Fig. 13-2 Combination motorcycle.

Fig. 13-3 Competition motorcycle. (*Courtesy of Kawasaki Motors Corporation, USA.*)

enforcement philosophies discussed (except zoning) can be effectively applied to highway motorcycles.

Enforcement will be relatively easy because an effective enforcer already exists—the local police. Enforcement officers require training, motivation, and equipment, but in localities where adequate antinoise direction has been received from City Hall, effective programs have been carried out with a minimum of fuss and bother. Principles applying to highway motorcycle noise control are treated in Chapter 10, but because of the similarity between enforcement methods and philosophies used for enforcing highway and other motorcycle noise regulations, some policies will also be mentioned here.

The clientele, highway motorcycle riders, will probably believe—mistakenly—that louder motorcycles go faster and have more power, which will complicate enforcement. Although the old "Hell's Angels" image is disappearing from the highway motorcycle scene, the "one percenters" or "outlaws" are still well represented in the biking population; these people are likely to remove or modify mufflers under the mistaken impression that this will increase the performance of their machines. There is ample evidence that the mere removal of the muffler from a motorcycle, particularly from a two-stroke motorcycle, not only fails to enhance performance but in some cases actually degrades it. Also, there are numerous high-performance exhaust systems produced by reputable aftermarket manufacturers that do indeed increase motorcycle performance without increasing noise level. This will be further explored under Enforcement Schemes.

Combination (On/Off-Road) Motorcycles. Combination bikes are generally sold subject to the same regulations as highway bikes; however, they are designed to be used off-road as well as legally on the highways. Because of their more sporty flavor, they may be somewhat louder than a highway version of the same basic bike. Trail-bike riders often think of themselves as budding racers and are even more subject to the temptation to remove their mufflers than highway-bike riders.

While some states specify noise levels for highway motorcycles, most do not specify noise levels for off-road motorcycles. Thus, in most states, a combination bike is subject to noise level regulation when used on the highway, but not when used off-road. California is a notable exception, but even there, the off-road allowable level is somewhat higher than for the highway (86 dBA versus 83 dBA when evaluated by SAE Standard J331a at this writing). Few riders are likely to be conscientious about replacing mufflers, which they removed for a trail ride, when they take a trip downtown.

The combination-bike or trail-bike rider is likely to be more easily reached than the highway-bike rider. A trail bike is relatively expensive, and so average trail-bike riders use their cars or pickups for transportation. They are probably in a higher socioeconomic sphere than highway bikers and, thus more amenable to educational approaches.

One problem with combination bikes is that they are often used by young riders in areas close to urban or suburban population centers. Since the bikes are used off-road, noise provisions of the state vehicle code do not apply in most cases. Also, the operation is localized and concentrated. Anyone who lives near an area of extensive unmuffled motorcycle operation is likely to become a bona fide motorcycle hater and constant complainer to the local noise control official.

The motorcycle industry acknowledges that noise from this type of operation is one of their most significant problems. They also recognize that this problem can be alleviated somewhat through education. Far from being indifferent to the problem, the industry can be your best ally in the battle against this annoying noise source. National organizations are listed at the end of this chapter. Many have programs for noise control education, and information about such programs is available upon request.

Competition Motorcycles. Most major manufacturers sell motorcycles designed purely for competition use. Currently, the most popular, competitive, pure competition motorcycles are motocross racers. A motocross is a closed-course competition over a track that may include water crossing, ruts and bumps, jumps, many turns, and rough and smooth terrain. It is a highly developed sport, one which draws large crowds, both in Europe and the United States. It takes tremendous dedication on the part of the competitor and highly developed engineering on the part of the motorcycle manufacturer. Light weight, high power, and good handling are essential to a good motocrosser. Because of this, most of these vehicles have in the past been sold without any vestige of silencing treatment. This is slowly changing, however, and exhaust mufflers as well as other silencing treatments are beginning to appear on motocross machines.

Motocross racers would probably not provoke too much of a noise problem if their use were confined to competition tracks, except for those tracks located in residential areas. But, the qualities that make a good motocross bike also make an excellent play bike; thus, a large number of these machines find their way into the hands of riders who do not use them on closed tracks, but for general off-road riding. These machines are considerably louder than their trail-bike counterparts and are often used in the same location. The youngster who has a specialized motocross bike is going to humiliate the youngster with a trail bike when they face off in an impromptu race. Excellent motocross bikes are available off-the-shelf at very reasonable cost. Research on this subject is nearly nonexistent, but the author's observations are that 80 percent or more of all motocross bikes sold in the last 5 years are used at least part-time in areas where noise can be a serious problem.

Many genuine motocross racers are dedicated and intelligent individuals, and many of these people are members of organized racing clubs and associations. Most of these people realize that noise is a real problem. At the club level, at least, they can probably be convinced that they need mufflers, at least when not racing. An exhaust muffler, if properly designed, is generally adequate to reduce the noise level of a motocross machine to nearly that of a stock trail bike. The young person who thinks he or she is a motocross racer will present more difficulty.

Snowmobiles

Snowmobiles present a noise problem similar to that of motorcycles; however, there is no highway snowmobile corresponding to a highway motorcycle. Snowmobiles are designed and used strictly in off-road situations. Most snow-belt states have regulations that limit the noise from snowmobiles, and all snowmobiles (with the exception of highly specialized factory racers) are sold with adequate silencing. Snowmobile riders are less likely to remove or modify their exhaust systems than are motorcyclists. One important reason is that almost all snowmobile engines are of a two-stroke design, in which the muffler must be carefully matched to the engine to obtain maximum performance. Removal of a snowmobile muffler will almost always result in a degradation of performance. Also, since the muffler of a snowmobile is hidden from view by the cowling, the aesthetic "improvement" obtained by removing the muffler from a motorcycle is not seen when snowmobile mufflers are removed.

The snowmobile rider is more likely to be a member of a club or organized group than the motorcyclist. This is particularly true with regard to the adult user. The snowmobile industry has successfully promoted an image of the responsible snowmobiler as one who uses his snowmobile on organized trail rides. Where snowmobile clubs and associations are well organized, peer pressure has been effective in several areas of snowmobile safety and regulation.

Younger snowmobilers who see themselves as racers, however, are likely to cause problems. They will operate their snowmobiles at what seems to be breakneck speed on frozen lakes, in backyards, in parks, and so on, and often with total disregard for the safety and acoustical rights of others.

Boats

Outboard motorboats are likely to present noise problems only if the engines are modified. All of these motors are quiet as sold. However, some boaters who still believe that "noise is power" often go to great trouble and expense to defeat the effects of this excellent technology, often to the result of degrading the performance of the motor. Inboard ski boats can present a different problem; usually built, at least partially, by the operators, they are often assembled without mufflers. The competitor image is operating here again. An unmuffled ski boat is horrendously loud (well over 120 dBA at the boater's ears, in some cases) and presents a totally unacceptable noise source when operated within a mile, or even more, of other lake users not particularly interested in high-speed ski boats.

The owner of such a ski boat is likely to be quite knowledgeable mechanically and correct in believing that the installation of an effective muffling system on the boat would at least increase weight, cost, and complication, if not detract from performance. In this problem area, appropriate regulations, strictly enforced, seem to be indicated. Because of the diversified nature of the boating industry, less help from manufacturers is to be expected, particularly with regard to inboard boats, than can be obtained from motorcycle or snowmobile industry sources.

Other Off-Road Vehicles

To a certain extent, jeeps, dune buggies, hovercraft, and custom-built off-road vehicles present noise problems similar to those presented by motorcycles and snowmobiles. These vehicles are best dealt with on an individual basis, however, because of their varied nature. A noise limit reasonable for four-wheel-drive pickups, which are operated on public roads as well as off-road, could be too restrictive to be reasonable for such high-performance vehicles as go-karts. It is essential that the noise-regulation writer thoroughly investigate the technical difficulties involved in silencing any particular class of vehicles before noise limits are specified. A "blanket" noise regulation for all off-road vehicles is bound to be unduly lenient for some classes and too restrictive for others. It will meet strong resistance from some quarters, and thus may be unenforceable and, therefore, a failure.

Organized Races

Motor racing is here to stay. Statistics are conflicting, but it has been estimated that it is the second (or third) most popular spectator sport in the United States, after baseball (and possibly basketball). Devotees of racing are among the most loyal of all sport fans. Most motor-racing enthusiasts become so emotionally involved in the sport that many participate to at least a limited degree themselves. A large portion of the individuals viewing a drag race have "occasionally burned off a fast quarter" on the local back roads, even if they have not participated in organized drags. Many members of the audience at any motorcycle race arrived on motorcycles, which are a major factor in their lives. Spectators as well as competitors enjoy the color, pageantry, speed, and thrills of the whole competition scene. A number of them, unaware of the effects of noise on their hearing and general well being, even claim to like the noise. Motorized competition provides a healthy outlet for energy and interest for a large percentage of the population. Motor racing in all forms has contributed significantly to technology, even to silencing and emission technology. Advances made on the racetrack often show up in future production vehicles.

On the other hand, those who live or work near racetracks are likely to be severely annoyed by the noise issuing therefrom. Several tracks have been closed because of action against the noise problem. Managers of some racing facilities now require that competition vehicles be silenced, at least to a certain extent.

People who disapprove of motor racing will often seize upon noise as the factor to legitimize their disapproval and give them something to complain about. Complaints of noise are very commonly received from residents in the areas of sporting arenas. Often these complaints persist on evenings when the arena is unused; some signal (such as the turning on of the parking-lot lights) stimulates the complainers to believe that normal activities are going on there.

Nonmobile Sound Sources

A wide variety of recreational activities carried out at fixed locations gives rise to significant noise pollution. Some of these, such as shooting ranges, are not amenable to treatment of the individual noise sources within the recreational activity since silencers on guns are illegal. Area acoustical treatments (noise barriers, etc.) must be applied here. A discussion of such approaches can be found in Chapter 6.

Other recreational noise can be abated in two ways: (1) silencing of individual noise sources within the recreational activity, and (2) zoning or acoustical treatment of the recreational area. Examples of noise sources in this category are model airplanes, rock concerts, and so on. With regard to the latter, only the effects on listeners and community are of interest here; the hearing loss suffered by performers is best considered under industrial noise control since this exposure relates in some ways to occupational conditions (see Chapter 7).

Other items that could be loosely considered recreational noise are high-level radio and television use, power tools used in and around the home, and so on. These are considered in Chapter 12.

EFFECTS OF RECREATIONAL NOISE

In general, there are two major areas in which recreational noise effects are felt—setting and health (Fig. 13-4). Settings include those locations where humans are considered present and those where they are not. Settings with humans, and in some cases animals, can be further broken down into

1. urban and suburban, where the presence of the noisemaker is objectionable, and the mere detection of recreational noise is a factor
2. wilderness, where the same human elements are at work
3. designated or *de facto* recreation areas, where the presence of a noisemaker is not so objectionable, unless it creates "excess" noise.

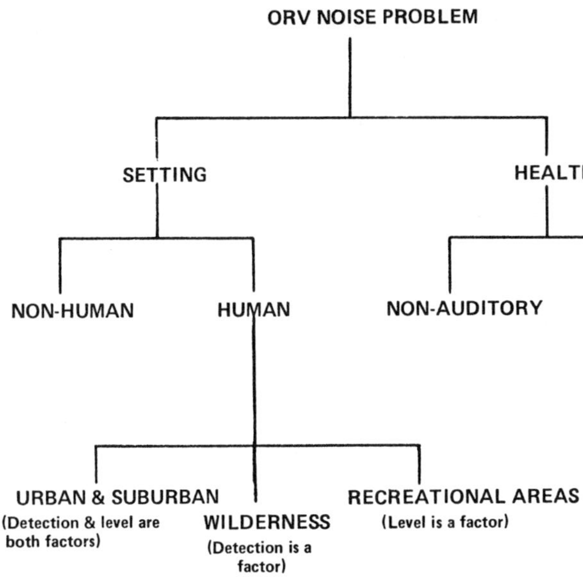

Fig. 13-4 Overview of difficulties caused by ORV noise.

On the health side, there are auditory and nonauditory effects; these effects are felt by participants and bystanders alike. Other noise issues involve the ability of participants in a recreational activity to communicate verbally and to perceive acoustical information.

No doubt many participants would prefer that their activities exposed them to lower noise levels, which would allow them to communicate more easily. There are many off-road vehicle (ORV) operators who harbor the mistaken impression that they can better monitor the operation of their engine, transmission, and so forth, if these components are excessively loud. This is a specious argument because, in most cases (such as a completely unmuffled motorcycle or dune buggy) when exhaust noise is very loud, it will mask any other mechanical noise information that may be of great value to the operator.

Various Settings

Nonhuman Settings. There is a great deal of speculation that recreational noise, particularly snowmobile noise, is dangerous if not downright fatal to certain species of wildlife or to some domestic animals. Although some well-controlled studies have been carried out on this specific subject, thus far, all have been inconclusive. Assertions of the destruction of fragile landforms due to sonic booms have been made, but the research supporting them is far from conclusive. So far, no one has seriously proposed that recreational noise causes any adverse effects on plants. So it seems that the "receivers" most affected by recreational noise are humans.

Humans. At this point, only the nonauditory effects of recreational noise are considered. The major effect, annoyance, can range from a mild feeling of being imposed upon by the recreationists to a feeling of unreasoning rage against them, depending upon the affected individual. Annoyance is generally considered to cause some impairment of health; this is dealt with in the discussion of nonauditory health aspects of noise.

Urban and Suburban. Where detection (annoyance) and level (hearing hazard) are both factors, ORVs (particularly motorcycles and snowmobiles) operated in areas near residences can cause significant noise problems. Many people still do not accept these machines as a desirable artifact of our way of life. Most motorcycle and snowmobile operation in urban and suburban settings is by younger cyclists who are likely to be less sensitive to the noise they cause than are the homeowners who live adjacent to the area being used as a motorized playground. Physicians tend to associate motorcycles and motorcyclists with injury, and some local police officers see motorcyclists in an unfavorable light.

Thus, the mere detection of the noise is sufficient to cause significant annoyance in a certain portion of the population. Perhaps 5% of the population would complain to local authorities if the noise level reaching them were loud enough for detection (about 45 dBA at their location), but not loud enough to cause speech interference. However, if speech interference is caused, a much larger percentage will complain. This is one of the strongest arguments for requiring silencing equipment on all ORVs, particularly if they are to be operated within detection range of homes.

Wilderness (or Wild Areas) Where Detection Is a Factor. The term *wilderness* is used here in its commonly accepted sense—that is, *any* area not developed by humans, and not only a *declared* wilderness area. Under wilderness (as opposed to urban or suburban) conditions, a much lower noise level is likely to give rise to considerable annoyance in any given person. As has so often been stated, most people who go to the woods go there to escape the hubbub of civilization. Even people who consider the noise of an ORV near their homes an inevitable adjunct of city life probably would not accept the noise after they have spent considerable time, effort, and money to get away from it. However, motorized recreationists also enjoy "getting away from it all," and some of the choicest ORV moments are experienced in the wilderness. Thus, conflicts are inevitable.

The *forest detectability distance* (the distance at which a machine will be barely audible above the typical forest background noise) for four types of vehicles in wild areas, is summarized in Table 13-1. The large variation in forest

TABLE 13-1 Summary of Forest Detectability Distances.[a]

	Vehicle Sound Level					
	Quiet		Typical		Loud	
	Detectability Distance (ft)	Sound Level at 50 ft (dBA)	Detectability Distance (ft)	Sound Level at 50 ft (dBA)	Detectability Distance (ft)	Sound Level at 50 ft (dBA)
Dune buggy	2000	77	4000	86	15,000	100
All-terrain vehicle	1500	74	4600	79	6200	83
Snowmobile	2800	73	6200	83	11,000	92
Motorcycle	1400	74	2600	83	3900	93

[a]From Harrison, R. T. Off-Road Vehicle Noise—Effects on Operators and Bystanders. Society of Automotive Engineers, Paper 740687, presented September 9, 1974.

detectability distances, especially for the louder ORVs, shows how important the spectra of both ORV and background are in determining the distance an ORV may be heard. It also shows the necessity of considering ORV-use impact on a case-by-case basis. The numbers shown in Table 13-1 are most useful as comparative, and not absolute, data. Furthermore, Table 13-1 demonstrates how futile it is to try to predict detectability from A-weighted levels alone. The forest detectability distances for snowmobiles are greater than those for other equally loud ORVs because of the lower background levels encountered in the winter forest. Background levels as low as 11 dBA have been measured near Lolo Pass, Montana, a popular snowmobiling area. Lower-than-average background noise, downwind propagation, the absence of intervening mountains and trees, and so on, can, if present, expand the distances shown in Table 13-1.

Recreational Areas Where Level Is a Factor. In campground-type areas where the nonmotorized recreationist or other user feels that the presence of ORVs is inevitable, level becomes a definite factor in annoyance. Tests conducted at the Oregon Dunes National Recreation Area at Siuslaw, Oregon, indicated that to experienced Forest Service recreation technicians, a dune-buggy level of 85 dBA at 50 ft was acceptable. These Forest Service employees are in constant contact with recreationists, both ORV and nonmotorized, and therefore are probably as accurate in their judgment of acceptability as anyone. Also, the repeatability and consistency of their judgment day to day, dune buggy to dune buggy, and person to person was remarkably high. It seems safe to conclude that in an area where the presence and use of ORVs themselves are not objectionable, but where their noise may be, 85 dBA at 50 ft is an acceptable level.

Health Effects

We will now consider both auditory (hearing loss) and nonauditory effects of recreational noise.

Nonauditory Effects. Nonauditory effects, briefly mentioned earlier, are difficult to define. There is no question that noise causes unpleasant physical reactions. Cardiovascular, digestive, and neurological disturbances have all been ascribed to noise of a lower overall level than that necessary to cause hearing damage. However, the literature is generally unclear as to whether these noise-caused reactions are due to annoyance and resentment at the noise sources, or are direct results of the noise exposure. The majority of scientific studies indicate that the annoyance and distress caused by noise, rather than the noise itself, cause the physical symptoms to appear. One notable exception to this is the startle response. Since noise created by a recreational activity is probably not startling, unpleasant, annoying, or distressful to the participant, the nonauditory effects of recreational noise are largely confined to spectators who did not voluntarily submit themselves to the noise. It is likely that psychological reaction, with its attendant physiological problems, far overshadows any singular physiological effects of the startle reflex alone.

Sleep interference is one of the most often-mentioned nonauditory recreational noise health problems. It has been amply shown that one need not be fully awakened by a noise to experience serious sleep disruption. There can be no question that this is a significant health problem in the United States today and that recreational noise does indeed contribute to this problem.

Auditory Effects

Spectators. In most cases, the noise reaching the spectator or bystander will not be great enough to cause any hearing damage; however, this is not true at organized motor races. Measurements made at motorcycle races indicate that spectators near the start-finish line of some tracks will be subjected to no less than 90 dBA during the races and may receive levels of up to 120 dBA during some of the events. In the final analysis, hearing hazard depends on the combination at exposure level and duration.

Operators and Passengers of ORVs. There is no question that noise from certain ORVs can cause permanent hearing damage to both operator and passengers. Table 13-2 summarizes the permissible exposure times for various classes of ORVs. This calculation is based upon OSHA regulations, which are liberal, at best. A more protective "floor" of 85 dBA, rather than the currently used 90 dBA, is employed. These permissible exposure times allow for a comparison of the probable hearing-damage potential of the various vehicles.

Some will not consider this a completely realistic criterion for recreational vehicle operation, since it is based on levels that, for a large population of typical

TABLE 13-2 Permissible Exposure Times, Various Types of ORVs (hr/day).[a]

	Vehicle Sound Level		
	Quiet	Average	Loud
All-terrain vehicles (ATVs)	3.2	1.9	0.5
Snowmobiles	1.2	1.1	0.9
Motorcycles	4.5	2.6	0.4
Dune buggies	6[b]	3[b]	0

[a]From Harrison, R. T. Off-Road Vehicle Noise—Effects on Operators and Bystanders. Society of Automotive Engineers, Paper 740687, presented September 9, 1974.
[b]Estimated.

working individuals, would not lead to significant hearing damage in 85% of the population for a long-term exposure of 5-day weeks, 50 weeks per year. Two or three 4- to 8-hour exposures per week are the maximum likely with most recreational vehicles. However, especially in the case of motorcycles and snowmobiles, many ORV recreationists ride 1 or 2 hours a day during the week. Industry sources indicate that the average snowmobile user operates 5 hours per week for 14 weeks of the year.

Comparison between noise measurements and measurements of the temporary threshold shift or temporary loss in hearing ability experienced by the operators of the test snowmobiles indicated that even this limit may be too liberal. Although experts do not agree on the correlation between noise-induced temporary threshold shift and noise-induced permanent hearing loss, there is universal accord that temporary threshold shift is perhaps the most adequate indicator that hazardous noises are being encountered.

Hearing Protectors for ORV Operators. To date, research has indicated that motorcycle helmets are not effective as hearing protectors for ORV operators. ORV operators often argue that if they wore earplugs under their helmets, they could not hear their machines or any warning signals such as sirens or bells. This is a specious argument, as the detection of a warning signal in noise is a function of the signal-to-noise ratio. Thus, the hearing protector attenuates the overall level of background noise the same amount that it attenuates the wanted signal. In some cases, signal perception is actually enhanced because frequency components of the background noise that interfere with hearing are reduced.

CONTROL STRATEGIES

There are four basic philosophies applicable to the control of noise from recreational activities. Three of these take the form of *source* regulations, while the fourth is a *zoning* concept. Source regulations are best applied to mobile noise polluters, while zoning works best for stationary sources. These four philosophies will be discussed in some detail with regard to recreational vehicles, and their applicability to other types of recreational noise will be mentioned briefly.

Source Regulations

With regard to recreational vehicles, source regulations can take three forms: sales or certification regulations, in-use regulations, and prohibitions against modification.

Sales or Certification Limitations. It is within the province of states to prohibit the sale or licensing of a recreational vehicle unless it meets certain requirements.* An example of such a regulation is provided by the California Vehicle Code, Section 38370, which states in essence that no person shall "sell or offer for sale a new off-highway motor vehicle . . . which produces a maximum noise exceeding the following noise limit at a distance of 50 ft . . . 86 dBA." The test procedure used by California to enforce this program is SAE Standard J331a for motorcycles, and SAE Standard J 192a for snowmobiles (which have a lower limit of 82 dBA). Presumably, other types of ORVs, such as dune buggies, are also covered. Note that this regulation provides for limits somewhat higher than those provided in California Vehicle Code, Section 27160, which covers noise standards for new highway vehicles. Roughly half of the states have similar provisions applicable to snowmobiles, and somewhat less than half of the states have provisions applicable to motocycles.

While the limit given in the California law is a reasonable one, some other jurisdictions are considering snowmobile noise limits of as low as 73 dBA. In the author's opinion, this is unwarranted. Most new snowmobiles currently marketed produce no more than 78 dBA when measured according to SAE J192a protocol. As pointed out earlier, most complaints about snowmobile noise are received from people in urban centers and directed at the "outlaw" snowmobilers who are operating where they probably shouldn't be in the first place. A snowmobile capable of producing no more than 78 dBA used on legitimate snowmobile trails has been shown to be an insignificant contributor to the suburban or rural noise environment. It therefore seems unfair to penalize the legitimate snowmobile user for the misdeeds of the irresponsible few, particularly when snowmobile engineers are generally agreed that lowering the noise level much below 78 dBA would seriously compromise the basic nature of the vehicle, increasing its cost and decreasing its safety and mobility. A similar argument

*At least, as of this writing it is. If the EPA sets a new product noise emission standard, such state action will be preempted. A standard for motorcycles is scheduled for proposal in 1977; off-highway motorcycles will be included.

applies to motorcycles. Because of the basic differences between motorcycles and snowmobiles, reducing the SAE J331a level of off-road motorcycles much below 86 dBA seriously burdens the basic design.

A word about the test procedures used is in order. The SAE Standard procedures are designed for new vehicle certification and are not applicable for enforcement. The procedures are rather complicated and difficult to run, but they do give excellent repeatability and do yield results compatible with the purpose of Section 38365 of California's vehicle code. That is, they indicate the maximum noise the vehicle is capable of producing under normal conditions. Using this test philosophy, it is not necessary for the state to actually run any tests. It can accept a certification, from vehicle manufacturer, importer, or distributor, that the limit is met. The SAE test methods are widely known and well standardized.

In-Use Noise Standards

Many states regulate in-use noise of *highway* vehicles. For instance, California Vehicle Code Sections 23130 and 23130.5, Vehicular Noise Limits, provide in part that

> no person shall *operate* . . . motor vehicles . . . at any time or under any condition of grade, load, acceleration, or deceleration in such a manner as to exceed the following noise limits . . . (on highways with a speed limit of 35 mph or less), 82 dBA; (on highways with a speed limit of more than 35 mph), 86 dBA (measured at a distance of 50 ft from the center of the lane of travel, and further) within a speed zone of 35 mph or less on level streets or streets with a grade not exceeding plus or minus 1 percent . . . 77 dBA. (These limits are applicable to highway motorcycles.)

Although effective for the enforcement of highway vehicular noise regulations, this control philosophy is difficult to apply to ORVs, because of the difficulty in locating a measurement site. Therefore, an operational noise limit test procedure is needed. Until recently, such a test has not been available. The Motorcycle Industry Council has developed an easily run stationary test, which admirably fulfills this purpose. The step-by-step test procedure appears in the appendix to this chapter. New-vehicle certification limits of 86 and 82 dBA would correspond to approximately 103 and 94 dBA when measured by the MIC procedure.

An in-use noise regulation might be worded as follows:

> Off-road vehicle noise limits. No person shall operate on any public lands within the jurisdiction of — any snowmobile that produces a noise level in excess of 94 dBA when measured according to noise test procedure appended to these regulations nor any other type of off-road vehicle that exceeds 103 dBA when

measured by the same procedure. Any person who operates an off-road vehicle upon any public lands in the jurisdiction of — shall be deemed to have given his or her consent to a noise test performed by such procedure, when requested to so submit by any peace officer. If any person refuses the officer's request to submit to such test, he or she shall be deemed to have failed such test. (See Chapter 11 for a detailed discussion of code preparation.)

Experience shows that the most effective penalty for failure of such a test is (for at least the first or second offense) to allow the owner to rectify the difficulty and bring his vehicle to the enforcement agency for a retest. After one or two violations, however, a fine commensurate with the seriousness of the infraction is indicated. This is called the *soft fuzz approach*.

One way to prevent the unauthorized use of loud ORVs near residences on private land is to require, as some localities in southern California currently do, that any operator of any ORV on private property within the locality be required to possess, from the landowner, written permission to be using the property. Many landowners, who don't mind the use of ORVs on their property without their knowledge, will not provide such permission for obvious legal reasons.

In-use regulations are also applicable to other internal-combustion-engine-powered forms of recreation. Muffler technology exists for engines as small as those for model aircraft, and indeed the Academy of Model Aeronautics requires exhaust mufflers on many of the classes of airplanes engaged in their competitions. Go-karts can be a particularly annoying problem, and effective mufflers are readily available for them.

Prohibitions Against Modifications. A provision similar to the California Vehicle Code, Section 27151, can be effectively used to enforce noise limits on ORVs. Section 27151 provides that ". . . no person shall modify the exhaust system of a vehicle in such a manner which will amplify or increase the noise emitted by the motor of such vehicle above that emitted by the muffler originally installed on the vehicle . . . no person shall operate . . . with an exhaust system so modified." A provision similar to this is essential in any ORV noise-control plan. It gives the arresting officer the option of ticketing a loud vehicle even if the officer does not have a sound level meter on hand, and therefore cannot run the test suggested previously in the operational noise limit section. It prevents the use of exhaust cutouts, or other noise-increasing appliances.

It should be pointed out that, while some aftermarket accessory exhaust systems are ridiculously loud, some maintain the original vehicle noise limits while providing a considerable increase in performance. It is inevitable that ORV enthusiasts, particularly motorcyclists, are going to modify their vehicles for aesthetic reasons, even if no technical reasons exist. Therefore, the use of well-designed, efficient, quiet aftermarket exhaust systems is to be encouraged. A section similar to the California Vehicle Code Section 27150.1 can address this end. This section provides that ". . . no person . . . shall offer for sale, sell or install a motor vehicle exhaust system, or any part thereof, included but not limited to a muffler, unless it (does not increase the noise level of the vehicle above the statutory limits shown above)." Whether or not such a system, once installed, does indeed increase the noise above the statutory limit is best determined by a test, as discussed in the section on operational noise limits.

Pure Competition Vehicles. Most states provide that competition vehicles are exempt from any noise limit. This is reasonable because vehicles singularly for competition are intended to be used out of earshot of everyone except enthusiasts. Also, most motor racing is held on private property, and there are obvious legal difficulties in regulating private activities on private land.

The sample language is drawn again from the California Vehicle Code, Section 38275.

> The provisions . . . (with regard to noise limits) . . . shall not be applicable to vehicles being operated off the highways in an organized racing or competitive event upon a closed course and which is conducted under the auspices of a recognized sanctioning body or by permit issued by the local governmental authority having jurisdiction.

This exemption for competition vehicles does not imply that they are not a noise problem. Indeed, some of the knottiest noise problems are presented by competitions held on closed courses near residential areas.

A "zoning" regulation, discussed in the next section, may be the best approach because it forces the *promoter of the race* to enforce the silencing of the individual noise sources. In a few cases, because of the nature of some vehicles, silencing puts a real technical burden on the competitors. Almost all forms of motorcycle racing, including motocross, scrambles, flat track, speedway, and drag racing, are amenable to exhaust treatment, at least. The same applies to boats and automobiles, with the possible exception of automobile drag racing. Here, perhaps the unlimited professional competitors should be exempt from silencing requirements for safety and technical reasons. If the track is remote from residences or other noise-sensitive areas, there is no need to insist on silencing, and noise-control efforts are better spent in other areas.

The silencing of 1-mile professional flat-track motorcycles and motocrosses has been successfully done at several racetracks throughout the country. The American Motorcycle Association is currently insisting that all competitors in their trans-AMA professional motocross series meet a 115-dBA limit when measured by the MIC test procedure. They also offer an award for the "quietest competitor," and, at one of the race meets, the racer who won the meet also won the quietest competitor award (this is rather conclusive evidence that fast vehicles need not be loud).

A word of caution, however. Because a particular brand of highway, enduro, combination, or trials bike might meet a certain noise level, does not mean that a motocrosser or short tracker of the same basic design can. To insist that a special-purpose off-road vehicle meet the same low noise limits that its highway counterpart can achieve would be unreasonable, because such limits would seriously compromise the basic nature of the vehicle, and counter-productive, because the owner-enthusiast would not stand for it. Rather than accept an unreasonable mandate, the owner-enthusiast would reject the entire concept of silencing, and pull the muffler and other silencing equipment off and throw it away. This is definitely not in the best interest of a peaceful environment. In any event, the noise-control officer will usually find both racetrack management and serious competitors anxious to cooperate with local authorities, and both of these groups should be involved in any contemplated regulation of competition noise.

Zoning

The concept of noise zoning, as discussed in Chapter 11, is quite applicable to recreational pursuits. However, certain caveats are in order. First, the time frame of the recreational activity must be considered. Very often, a noisy activity occurs only once or twice a year at any particular location. While infrequent use should not be a license to kill, it should be considered in setting noise-limit zones. Rather than an L_{eq} or L_{dn} approach, perhaps a maximum level (dBA) at the property line or 50 ft from the property line is more realistic. For example,

> No person shall operate any motor vehicle racetrack, discotheque, or other activity of a recreational nature that emits noise levels in excess of 85 dBA for more than 5 minutes in any hour, 1 hour in any week, or 10 hours in any year. Such activity shall not emit noise in excess of 65 dBA for more than 4 hours on any one day, or 2 days in any one week, and in no event, after 10 P.M. and before 7 A.M. This provision shall apply whether or not the recreational activity is held for profit, and whether or not an admission charge is made. Measurements shall be made outside the property line, using a sound level meter meeting the provisions of ANSI S1.4-1971, type 2.

This provision would allow for an occasional noise excursion into the normally unacceptable area, which, despite the best efforts of everyone involved, is inevitable. It would further permit a reasonable number of race meets, dances, and so on. It would, however, insure that noise from the recreational activity would not be so loud as to interfere with speech communication attempted by the neighbors, and would prevent interference with television and radio listening. Such a standard would not eliminate complaints. As mentioned previously, annoyance due to noise is related to the connotation the listener attaches to the noise. Very often, the idea of other people having fun is the most annoying of all connotations.

A word about acoustical treatment of noise sources. It has been conclusively shown that, as acoustical measures, planting trees around a racetrack, erecting noise barriers, and so forth, are futile. However, a tree screen that shields the activity from the view of unwilling bystanders often has the effect of reducing complaints, even if the acoustical efficiency is low, because of the elimination of the visual impact from the activity.

FURTHER RESOURCES

In summary, control of noise from recreational sources will challenge the ingenuity of noise control officials. They will be most successful if they take the time to acquaint themselves with the social, as well as the technical, aspects of each individual noise problem. Although source control will, in most situations, present the most promising possibilities for a successful noise control program, such a scheme must be thoroughly researched to ensure that it is both reasonable and enforceable. As with any source-control program, intelligent enforcement is the key to success. In addition, most recreational noise problems should be considered from the "noise zoning" point of view before a final control strategy is selected.

Investigations of the cause and effects of various ORV noise have been carried out by several branches of the federal government. The U.S. Environmental Protection Agency report, "Transportation Noise and Noise from Equipment Powered by Internal Combustion Engines," (NTID 300.13, December 31, 1971) is an excellent accumulation of basic engineering information on recreational vehicular noise sources. The USDA Forest Service has published reports on motorcycle, snowmobile, all-terrain vehicle, and dune-buggy noise, as well as ORV-noise measurement techniques. These publications are available directly from

USDA Forest Service
Equipment Development Center
444 East Bonita Avenue
San Dimas, California 91773.

An indispensable reference in the development of any vehicular noise ordinance is the State of California Vehicle Code, which is available for $1.00, including tax, from

California Department of Motor Vehicles
Sacramento, California 95801.

Finally, any noise-control ordinance aimed at a specific source, particularly an off-road recreational vehicle, should be discussed with representatives of vehicle manufacturers and users, such as the following:
For motorcycles

Motorcycle Industry Council
Suite 101
4100 Birch St.
Newport Beach, California 92660

American Motorcyclist Association
P.O. Box 141
Westerville, Ohio 43801

For snowmobiles

International Snowmobile Industry Association
Suite 850 South
1800 M Street, NW
Washington, D.C. 20036

For all-terrain vehicles (3- and 6-wheel type)

National Association of All-Terrain Vehicles
New Bethlehem, Pennsylvania 16242

Go-karts

International Kart Federation
444 West Foothill Boulevard
Glendora, California 91740

Model airplanes

American Academy of Model Aeronautics
806 15th Street, NW
Washington, DC 20005

Appendix

MIC STATIONARY TEST MIC/E-76
SOUND LEVELS OF MOTORCYCLES
(Approved December 21, 1976)

1. *Scope*

This document establishes the test procedure, environment, and instrumentation for determining sound levels of motorcycles under field conditions. This procedure is designed to be incorporated as part of a motorcycle exhaust system certification program and compliance inspections.

2. *Instrumentation*

   2.1. The following instrumentation shall be used:

      2.1.1. A sound level meter meeting all requirements for type 1, type 2, type S1A or type S2A of American National Standards Institute S1.4-1971 (ANSI S1.4-1971).

   2.2. A windscreen which does not affect microphone response more than ±1 dBA for frequencies of 63-4000 Hz and ±1½ dBA for frequencies of 4000–10,000 Hz, taking into account the orientation angle of the microphone.

   2.3. If the motorcycle under test is not provided with a tachometer, then an engine speed tachometer with a steady state accuracy of ±5 percent shall be used. The tachometer may be a pointer type or a vibrating reed type as long as the accuracy specification is met.

   2.4. A calibrator which is capable of checking the sound level meter to an accuracy of ±0.5 dBA.

3. *Test Site*

   3.1. The test site shall be a flat open space free of large sound-reflecting surfaces (other than the ground) such as parked vehicles, signboards, buildings, or hillsides, located within 5m (16 ft.) radius of the motorcycle being tested and the location of the microphone.

   3.2. The surface of the ground within the area described in Section 3.1., shall be free of loose or powdered snow, plowed soil, grass of a height greater than 15 cm (6 in.), brush, trees, or other extraneous material.

   3.3. The microphone shall be located behind, 0.5m (20 in.) ±.01m (½ in.) from, and at the same height as, the rearmost exhaust outlet and a 45-

degree angle (±10 degree) to the normal line of travel of the motorcycle. The longitudinal axis of the microphone shall be in a plane parallel to the ground plane.

3.4. No wire or other means of distance measurement shall be attached to the microphone. (This may lead to erroneous reading.)

4. *Procedure*

The rider shall sit astride the motorcycle in normal riding position with both feet on the ground and run the engine with the gearbox in neutral at a speed equal to ½ of the manufacturer's recommended maximum engine speed (red line). If no neutral is provided the motorcycle shall be operated either with the rear wheel 5–10 cm (2–4 in.) clear of the ground, or with the drive chain or belt removed. If no red line is published for the particular motorcycle then an engine speed equal to 60 percent of the engine speed at which maximum horsepower is developed shall be used. If neither red line nor maximum horsepower engine speed is published, then the test speed N shall be calculated from the following formulae:

$$N = 306{,}000 \div \text{stroke, mm} \textit{ or } N = 12{,}000 \div \text{stroke, inches}$$

5. *Measurements*

5.1. The sound level meter shall be set for the A-weighting network and should be set for "slow" response. ("Fast" may be used.)

5.2. The sound level recorded shall be that measured during steady state operation at the above-mentioned engine speed, measured on the loudest side of the motorcycle. If tests are to be made on one side of the motorcycle only then they shall be made on the exhaust outlet side. The test RPM shall also be recorded.

5.3. The ambient sound level (including wind effects) at the test site due to sources other than the motorcycle being measured shall be at least 7 dBA lower than the sound level produced by the motorcycle under test.

5.4. Wind speed at the test site during test should be less than 32 km/hr (20 mph). If this is not possible, then the motorcycle and measuring microphone shall be positioned so that the prevailing wind direction is parallel to the normal direction of travel of the motorcycle.

6. *General Comments*

6.1. While making sound level measurements, not more than one person other than the rider and the measurer shall be within 3m (10 ft.) of the motorcycle under test or the microphone, and that person shall be directly behind the measurer on a line through the microphone and the measurer.

6.2. Proper use of all test instrumentation is essential to obtaining valid measurements. Operating manuals or other literature furnished by the instrument manufacturer should be referred to for both recommended operation of the instrument and precautions to be observed.

6.3. Specific items for consideration:

6.3.1. The type of microphone, its directional response characteristics, and its orientation relative to the ground plane and the sources of sound.

6.3.2. The effects of ambient weather conditions on the performance of all instruments (e.g., temperature, humidity and barometric pressure).

6.3.3. Proper acoustical calibration procedure to include the influence of extension cables, etc. Field calibration should be made immediately before the first test of each test day, and thereafter at intervals of no less than 1 hour. Internal calibration is acceptable for field use, provided that external (acoustical) calibration is accomplished immediately before and after each test day.

6.4. The procedure is intended for use as a pass-fail test. When limits are specified to be measured by this procedure, they should be set at maxima, with no additional tolerance permitted. However, it may be appropriate to consider variations in sound level meters, tachometers, or motorcycles, which although ostensibly the same, may not warrant failure of an inspection for small measurement differences.

6.5. The use of the word "shall" in the procedure is to be understood as obligatory. The use of the word "should" is to be understood as advisory.

Indexes

Author Index

Subject Index

Page numbers for definitions are in italics. References to illustrations are in boldface type.